Introduction to
Organic Electrochemistry

TECHNIQUES AND
APPLICATIONS IN ORGANIC SYNTHESIS

EDITOR: *Robert L. Augustine*

CATALYTIC HYDROGENATION
by Robert L. Augustine

REDUCTION
Robert L. Augustine, EDITOR

OXIDATION, VOLUME 1
Robert L. Augustine, EDITOR

OXIDATION, VOLUME 2
Robert L. Augustine and D. J. Trecker, EDITORS

INTRODUCTION TO
ORGANIC ELECTROCHEMISTRY
by M. R. Rifi and Frank H. Covitz

INTRODUCTION TO
Organic Electrochemistry

by M. R. Rifi and Frank H. Covitz

Union Carbide Corporation
Research and Development Department
Bound Brook, New Jersey

MARCEL DEKKER, INC. New York 1974

MARCEL DEKKER, INC.

305 East 45th Street, New York, New York   10017

LIBRARY OF CONGRESS CATALOG CARD NUMBER: 72-97484

ISBN: 0-8247-6063-8

Current printing (last digit):

10 9 8 7 6 5 4 3 2 1

PRINTED IN THE UNITED STATES OF AMERICA

CONTENTS

EDITOR'S INTRODUCTION

The synthetic organic chemist, must, of necessity, be well versed in the applications and subtleties of a wide variety of reactions. As time goes on, and the volume of literature expands, it becomes increasingly difficult for the practicing organic chemist to be aware of all of the applications of a given reaction. It also becomes much more difficult for him to select the conditions which are most suitable for each particular application of a given reaction. It is the purpose therefore of this series on techniques and applications in organic synthesis to provide chemists with concise and critical evaluations of as many reactions of synthetic importance as possible.

While the use of electrolysis in organic chemistry is not new, its synthetic application is usually limited to the Kolbe reaction on carboxylic acids. In recent years, the apparatus required for electrolysis has become more readily available and, as a result, more synthetically useful reactions have been developed. This technique unfortunately, is but one of an increasing number of synthetically useful procedures. It is the purpose of this book to provide the practicing organic chemist with a practical guide to the synthetic utility of organic electrochemistry.

Included in this book is a discussion of the various types of apparatus used, a consideration of the important reaction parameters, and a presentation of the types of reactions which can be run electrochemically. Specific experimental procedures are given for a number of these reactions.

It is hoped that by the use of this book, the synthetic organic chemist will not only become aware of the practical utility of electrochemistry but also be relieved of the necessity of surveying the original literature in the field, and, thus, have more time to spend in the pursuit of his ultimate goal.

Robert L. Augustine

PREFACE

Organic electrochemistry is a field that represents a challenge both in its theoretical as well as in its practical aspects. The researcher who is well versed in both areas has the opportunity to make significant contributions. In the past, the practical art outstripped the ability of theory to account for the observations. The opposite appears to be the case today. Electrochemical theory has been developed to the point where most mechanistic hypotheses can be put to the text of experimental verification. Sophistication in instrumentation has also undergone a recent revolution; with respect to organic synthesis, this means that the researcher has access to most of the "eyes" he needs to "see" what is really going on when a current passes between electrodes through an electrolytic medium. Electroorganic processes are "scale-up-able," a contention we hope will not escape the eyes of the industrialist.

The ideal organic electrochemist should first and foremost be a top-notch organic chemist. He also needs to understand and appreciate the fundamentals of electrochemistry and its unique advantages with respect to synthesis. He needs the ability to sort the electrochemistry from the organic chemistry and to put them back together in a realistic way. He must be able to tap on many disciplines -- besides straight organic synthesis and electrochemistry, he needs to be knowledgeable in analytical techniques, instrumentation, design of apparatus, heterogeneous and homogeneous catalysis, surface physical chemistry, and analysis of multivariable systems. Creativity and ingenuity of course are also essential to success.

The authors themselves, both of whom are organic chemists by training, recognized the desirability of being knowledgeable in the above areas when they first became involved in organic electrochemistry. At that time, however, we were unable to find a modern text which covered the subject in an introductory, yet comprehensive manner. Having acquired our knowledge the hard way, we felt a need existed for a text to serve as a guide for organic chemists interested in initating and expanding projects in organic electrochemistry. With this thought in mind, we undertook

the task of writing this book. We sincerely hope that this
book will in fact help to fulfill that need, and that
similar books will appear in the future to update the
rapidly expanding technology in this field.

M. R. Rifi

Frank H. Covitz

Introduction to
Organic Electrochemistry

INTRODUCTION

1.1. HISTORICAL PERSPECTIVE

Organic electrochemistry is not a new field of endeavor. On the contrary, it was among the first general techniques to be employed at a time when organic chemistry itself was in its infancy. The foundation of electrochemistry, as laid down by Faraday and Nernst in the nineteenth century, was well recognized by organic electrochemists of the late nineteenth and early twentieth centuries. The importance of the electrode potential in controlling the course of an electrolytic reaction was also understood and was expounded in quite definite and succinct terms by Haber (1) in 1898. The early twentieth century saw a flurry of activity in organic synthesis using electrochemical techniques by such prolific individuals as Haber, Fichter, Thatcher, Kolbe, and many others who recognized the promise of electrolysis as an important synthetic tool for both laboratory and industrial reactions. In fact, industrial electrochemical processes to produce anthraquinone, benzidine, and other intermediates were practiced within the German dye industry

on a small scale in the early 1900s (2). And yet, from
these early promising beginnings, interest and activity in
organic electrochemistry declined in the 1930s and 1940s,
a period when organic chemistry was flourishing and expanding in many directions.

What can we point to in order to understand the reasons
for a decreasing influence during this time? For one, organic electrochemistry became, to a considerable extent, a
victim of the inadequacy of the theory of reversible electron transfer to describe the irreversible events occurring
in the electrode when large currents are passing, the typical situation required for convenient synthesis. The Nernst
equation, which works so well for highly reversible systems,
is not applicable to most organic synthetic systems because
of their general irreversibility and the complexity of the
chemical and electrochemical processes occurring. Phenomena
such as polarization, overvoltage, and even electrode potential were not well understood in their connection with organic systems. The terms were so misused during the early
days of organic electrochemistry that one cannot help but
gain the impression of "black magic art" from reading the
literature of that time. The predominant use of aqueous or
partially aqueous media, with concomitant limited solubility
for organic materials, also contributed to the disenchant-
ment with organic electrochemistry. The remarkable ionizing
ability of water for inorganic salts which is largely responsible for the considerable success of electrochemistry in
producing workhorse inorganic chemicals such as sodium,
chlorine, and many metals and in the battery field, is not
necessary and is unduly restrictive for organic systems.
In many cases the evolution of hydrogen and oxygen from
electrolysis of water limits the range of accessible potentials, or causes many reported electroorganic reactions to
resemble catalytic hydrogenation or oxygenation, and would
thus tend to dampen the enthusiasm of the potential researcher
in electrochemistry. The overworked terms "hydrogen

carrier" and "oxygen carrier," found in the early literature point out the obsession with aqueous media. The need
to unravel the complexity of reaction products, common to
many organic systems under uncontrolled conditions, coupled
with the added desire to understand the electrochemical aspects, had to remain largely unfulfilled without the powerful analytical and electrochemical tools of today.

The typical introduction to electrochemistry, to which
an aspiring organic chemist is subjected, is usually made
within a predominantly physical chemical framework. Normally, little or no connection with organic synthesis is
made. On the other hand, until recently, the only reference to electrochemistry made in organic chemistry courses
was the timeworn Kolbe reaction (3) of carboxylic acids.
Can the reader recall any other "name" reactions of electrochemistry?

The organic electrochemists of yesterday thus were
improperly equipped--both because of inadequacies in the
theory and in their instrumentation--to stimulate growth
of electroorganic research and to achieve the still largely
unexploited potential of electrochemistry in organic synthesis. (The "potential" of electrochemistry is a standard
pun used by electrochemists!) Perhaps starting with the
extensive work of Tafel (4) on the phenomenological description of irreversible electron-transfer reactions, and continuing on with the more advanced theoretical studies of
Frumkin, Temkin, Delahay, Bockris, and others, the understanding of the electrochemical aspects of irreversibility
and the effect of the so-called "double layer" and adsorption is now much deeper and on firmer ground than previously.
Although the synthetic organic chemist will probably not be
able to take direct advantage of these theoretical developments, the much sounder and more quantitative, although
admittedly complex, description of electrochemical events
occurring at the electrode/solution interface allows progress to be made from a solid theoretical base.

Perhaps more important has been the development of electrochemical techniques and instrumentation, initially developed for analytical uses, but capable of providing much deeper information concerning the electrode process itself, and which have finally given the organic electrochemist the "eyes" to examine properly the electrode reactions he is dealing with. By far the most immediately useful technique is that of polarography, with which we associate the names of Heyrovsky (5) and Ilkovic (6). This technique involves a recording of the current-potential relationship at a dropping mercury electrode and is normally employed for the investigation of reduction processes, since mercury itself is easily oxidized. Polarography is covered in more detail in Chap. 3. More recent developments, covered in detail in an excellent book by Adams (7), have allowed solid electrode voltammetry to be equally useful, although requiring considerably more experimental skill. A host of instrumental techniques permit the present-day worker in organic electrochemistry to study, in as sophisticated a manner as he requires, the details of organic electrochemical reaction mechanisms. Of course, the importance of the newer analytical techniques to identify the reaction products is not to be underestimated; a detailed knowledge of the identity and amounts of the products still provides the least ambiguous information about the course of the reaction.

The activity devoted to development of organic fuel cells and high-energy-density batteries using organic electrolytes has also contributed to the awakening of interest in electroorganic synthesis.

Finally, one should acknowledge the devoted efforts of people like S. Wawzonek (8) and S. Swann (9) to promote and encourage activity and interest in organic electrochemistry. The field is, in fact, on the upswing, and a definite enthusiasm and feeling of newness is evident among the workers in organic electrochemistry, even though the report of Michael Faraday (10) that ethane is produced by electroly-

tic oxidation of aqueous acetate solutions occurred well
over 100 years ago.

1.2. ADVANTAGES AND DISADVANTAGES

As in most specialized synthetic techniques, organic
electrochemistry has intrinsic advantages and disadvantages
which should be recognized and understood by the potential
and actual practitioner. Let us dispose of the disadvan-
tage first:

1. Of primary concern to the experimentalist is the
fact that organic electrochemical reactions are usually
relatively slow, i.e., high current densities (current per
unit electrode area) cannot routinely be used, when compared to typical inorganic electrolyses or conventional
homogeneous reactions. The inherent rate is often related
to a rate-limiting electron-transfer or adsorption/desorp-
tion step. More practically, the reputed slowness can
usually be attributed to the low surface-to-volume ratio
of most preparative electrochemical cells described in the
literature.

2. Cell designs for synthetic usage are not standard,
nor is such apparatus generally available from commercial
sources. In practice, the experimenter is faced with a
compromise between designing a cell with maximum flexibility
(electrode replacement, reference electrode accessibility,
cell divider, temperature control, and agitations) and one
which maximizes electrode area and minimizes electrode spacing. The cell divider, when needed, gives rise to experimentally awkward construction, can be a maintenance problem,
and usually results in high voltage drops. Also, the number
of available types of dividers are limited.

3. The requirement that the solvent be inert as well
as capable of ionizing a suitable electrolyte and dissolving
the organic substrate, is restrictive, although several

choices, such as DMF, dioxane, acetonitrile, ethanol, pyridine, and others have been used extensively. Choices of electrolytes in nonaqueous media are usually also limited, tetraalkylammonium salts being the most generally used in organic systems; they are quite soluble and have sufficient conductivity and inertness for a wide variety of electrochemical syntheses. Also, the electrolyte must be separated from the product in the work-up.

4. Finally, in oxidative studies, the number of stable anode materials are limited, since most metals are themselves easily oxidized. The noble metals, such as platinum and gold, carbon and graphite, and lead dioxide, are the only ones which had heavy usage in anodic reactions.

The advantages of electrochemical syntheses are even more formidable:

1. Precise control of the electrode potential, and hence of selectivity, is easily attainable when required. The equivalent of a continuum of oxidizing and reducing agents, all equally accessible, is an attractive (although admittedly not precise) analogy.

2. Electrochemical reactions do not require thermal energy to overcome activation barriers, and hence are applicable to thermally sensitive compounds. The driving force is the electrode potential.

3. Stoichiometric amounts of oxidants and reductants are not required, and their by-products are thus avoided. This has interesting implications when one considers the current pressure to avoid pollution by discarded by-products. In this connection, many conventional redox reactions have the possibility of being "catalytic" if the redox product can be easily reconverted in situ to the starting reagent by electrolysis.

4. It should be generally recognized that although the cost of most materials has increased steadily over the years, the cost of electricity has remained remarkably stable, and is thus becoming an ever more attractive reagent for large-

scale reactions. The use of breeder reactors for the industrial generation of electricity is forseeable in the
not-too-distant future (11), and should further enhance the
availability and utility of this highly desirable "reagent."

        5. Electrochemical synthesis, by its very nature,
and by ease of instrumentation, is eminently suitable for
continuous and automatic operations, another industrially
attractive feature.

        6. The ease of quantitatively monitoring the course
of the reactions by coulometry (integral of current with
time), using electronic or electrochemical coulometers, is
unsurpassed compared to most other general synthetic techniques. The current itself is the measure of the rate of
reaction. Just as conveniently, polarography or related
techniques can generally be employed to follow the disappearance of starting material and, in some cases, the
appearance of product. Of practical convenience is the
ease by which the reaction can be instantly stopped--by
turning off the switch!

        In many cases one is tempted to reflect on some analo-
gies between organic electrochemistry and photochemistry.
consider the electrode potential and the current in the
same vein as the wavelength and intensity, respectively.
The energies involved in electrochemistry are of the order
of 1 to 3 volts (V) [25 to 75 kcal/mole]; in photochemistry
energies of the order of 2 to 6 V [50 to 150 kcal/mole]
are encountered. The primary "reagent" is quantized (electron vs photon) and neither contributes materially to the
mass of the product. In both techniques, novel syntheses
with no conventional counterpart are often uncovered. The
recently investigated cases of electrochemiluminescence
(12) where the product of an electrolytic oxidation (E $\simeq 2$
V) and that of a reduction (E $\simeq -2$ V) are allowed to react
in the bulk of the solution where light is emitted, is a
rather neat demonstration of the overlap in the energy
ranges accessible. Of course, the reader will draw the

appropriate conclusion when he considers the form of energy
required to power most light sources.

A much more intimate connection exists between electrochemistry and heterogeneous catalysis. The importance of
adsorption and orientation of molecules on surfaces is
obvious in both fields. Heterogeneous catalysis has also
suffered in the past from the aura of "black magic" surrounding processes that are not well understood. In both
cases, many mysteries concerning the nature of the adsorbed
species remain to be solved, and only recently have the
tools become available to study these difficult problems
in detail.

Finally, one should consider the relationship that
organometallic chemistry has toward electroorganic chemistry. In many cases, evidence has been found for the formation of intermediate organometallics in both oxidative
and reductive electrochemistry. In fact, an impressively
wide array of organometallics has been synthesized electrolytically, mainly through the efforts of Dessy and co-
workers (13). Since metals are employed almost exclusively
as electrodes in electrolytic processes, one should always
consider the possibility of specific organometallic interactions; indeed, such interactions are difficult to disprove.

1.3. SCOPE OF THE BOOK

This book is primarily intended to be used by organic
chemists in academic and industrial environments who wish
to start a program of laboratory work in the field of
electroorganic synthesis. Although little or no previous
knowledge of electrochemistry is required, a general knowledge of organic reactions and laboratory procedures is
assumed. The basic information needed to initiate such a
study is provided. This book contains the elements of
electrochemical theory, but is specifically geared to avoid

a requirement of advanced mathematics. Extensive use is
made of analogies between the more familiar concepts in
chemistry to which the reader has undoubtedly been exposed.
It contains experimental procedures, including specific
examples of syntheses, electrode preparation, cell construction, etc., trying to foster an appreciation of the types
and varieties of equipment needed, and lists commercial
sources where such equipment may be purchased if available
at all. Emphasis is placed on reactions which have the
possibility of general application, and on the effect of
variables on the course of reactions. Literature sources
are provided and references suggested for more complete
coverage of the subject material.

The book does not include detailed information on the
more sophisticated techniques available, but rather, will
refer the reader to more advanced texts. It is not meant
to be used as a reference manual, nor is it exhaustive in
coverage. Although most of the examples presented are taken
from literature sources, no attempt will be made to review
critically the literature with respect either to experimental
procedures or conclusions. On the contrary, a conscious
effort will be made to exclude highly controversial subjects
and to include only information which is generally agreed
upon by leaders in the field of electroorganic chemistry.

Within this framework, we apologize to those who may
feel we have underemphasized certain aspects or have presented analogies which are somewhat naive. To the extent
that these analogies have been helpful in assisting electroorganic chemists understand electrochemical reactions and
predict their results, they will, in fact, be emphasized.

Organic electrochemistry has had an unusual past--the
early workers recognized the promise but were improperly
equipped to fulfill it. The latter cannot be said today.
Many unique reactions remain to be discovered; the time is
ripe--let's use it.

REFERENCES

1. F. Haber, Z. Elektrochem., 4, 506 (1898).

2. C. J. Thatcher, Trans. Am. Electrochem. Soc., 36, 337 (1919).

3. B. C. L. Wheedon, Quart. Rev. (London), 6, 300 (1952).

4. J. Tafel, Z. Phys. Chem., 50, 641 (1905).

5. J. Heyrovsky, Chem. Listy, 16, 256 (1922).

6. D. Ilkovic, Collection Czech. Chem. Commun., 6, 498 (1936).

7. R. N. Adams, Electrochemistry at Solid Electrodes, Dekker, New York, 1969.

8. S. Wawzonek, Science, 155, 39 (1967).

9. S. Swann, Jr., in Techniques of Organic Chemistry (A. Weissberger, Ed.), Vol. 2, Interscience, New York 1956, pp. 385-523.

10. M. Faraday, Pogg. Ann., 33, 438 (1834).

11. G. T. Seaborg and J. L. Bloom, Sci. Amer., 223, #5, 13 (1970).

12. T. Kuwana, Electroanalytical Chemistry, Vol. 3, A. J. Bard, Ed., Chapter on "Photoelectrochemistry and Electroluminescence," M. Dekker, Inc., New York, 1969.

13. R. E. Dessy and R. L. Pohl, J. Am. Chem. Soc., 90, 2005 (1968); and other papers in this series.

2
BASIC PRINCIPLES

2.1. FARADAY'S LAW AND OHM'S LAW

2.1.1. FARADAY'S LAW

Of extreme importance to the early development of
electrochemical synthesis was the discovery by Michael
Faraday in 1834 of the relationship between the amount of
electricity used in an electrolytic reaction and the amount
of material involved. It is comprised of two parts:

1. The amount of material transformed is proportional
to the quantity of electricity passed.

2. The weights of various materials transformed by a
given quantity of electricity are proportional to their
respective equivalent weights.

These laws, as is obvious today, are the result of the
fact that charge is quantized such that all charged species
contain simple multiples of the charge of a single electron.
Faraday's law formed the basis of the definition of the
coulomb (C) as the amount of electricity necessary to
deposit 1.11800 mg of silver, and is also taken as the
standard definition of current (rate of flow of electricity) as one ampere (A) equals one coulomb/second (C/sec).
The amount of electricity required to transform one mole
of a substance was found by Faraday to be 96,500 x nC.
where n is the number of electrons involved in the process. Faraday's law may be restated as

$$Q = \int_{o}^{t} idt \tag{2.1}$$

$$w = M(Q/96,500n) \tag{2.2}$$

where Q is amount of electricity in coulombs; i is current
in amperes; t is time in seconds; w is weight of material;
n is number of electrons transferred per molecule; and M
is molecular weight of material.

The number 96,500 is called the faraday(F) and is just
the amount of charge per mole of a singly charged species.
Since a mole of substance contains Avogadro's number of
particles, the charge on a single ion, equal to the charge
of an electron, is just $96,500/6.02 \times 10^{23} = 1.6 \times 10^{-19}$ C,
a value which is in good agreement with the results obtained
from other independent methods, e.g., Millikan's oil-drop
technique, for measuring the charge of the electron.

As a result of Faraday's laws, the electrical efficiency for formation of a particular reaction product may
be defined. It is just the ratio between the number of
equivalents of product actually formed and the total number
of faradays passed. In terms of the weight of product and
coulombs of electricity, the electrical efficiency may be
expressed as

$$\text{E.E.} = \frac{96500 \ nw}{MQ} \tag{2.3}$$

where n is number of electrons required/mole product; w is
weight of product; M is mol wt of product; and Q is total
number of coulombs passed.

The chemical efficiency or yield, on the other hand,
is of course, equal to the number of moles of product formed
divided by the equivalent number of moles of starting material consumed. The electrical efficiency is always less
than or equal to the chemical efficiency, since there may
be electrochemical side reactions (e.g., electrolysis of
solvent) which do not involve the starting material.

2.1.2. OHM'S LAW

The relationship between the voltage difference between
two points in an electronic conductor (i.e., one which con-
ducts by movement of electrons) and the current flowing between them is embodied in Ohm's law, which states that the
current is proportional to the voltage applied

$$E = iR \qquad\qquad (2.4)$$

where E is voltage in volts; i is current in amperes; and
R is resistance in ohms.

This voltage, which occurs whenever current is flowing within a material, is often simply referred to as the
iR drop to distinguish it from other types of voltages,
namely, those occurring at the interface between dissimilar
materials. It was originally thought that ionic conductors
did not obey Ohm's law, since a linear relation was not
found between the current and the voltage applied between
two electrodes immersed in an ionic solution. In fact,
ionic conductors do obey Ohm's law when the measurements
are made in such a way as to eliminate the effect of the
potentials at the electrode/solution interface. For electrolytic solutions it is often more convenient to use the term
conductivity, which is just the reciprocal of the resistance.
The unit of resistance is, of course, the ohm; the unit of
conductivity, strangely enough, is the mho. The term voltage
should be applied only to measurements made between similar
phases, and the term potential applied when the measurement
is between dissimilar phases.

The iR drop is always associated with the generation
of thermal energy, which is given at any instant by the
formula

$$P = iV \qquad\qquad (2.5)$$

where P is the thermal power developed in watts (W) [1 W
= 1 joule/sec].
The total thermal energy generated by the iR drop is simply

$$T.E. = \int_{0}^{t} P dt \qquad\qquad (2.6)$$

where T.E. is the total thermal energy in joules.

Two other elementary laws that will be useful to remember is that the current flowing in any part of a series circuit is the same and that the current flowing into a junction equals the sum of the currents flowing out of that junction, both consequences of the law of conservation of charge. The corresponding law for voltages is that the voltage drop across a circuit is equal to the sum of the individual voltage drops in the circuit.

2.2. UNITS

It is hoped that the reader is already familiar with all of the above basic laws and their usage in simple systems. The following list of units and conversion factors will be of utility in electrolytic work.

Voltage

1 volt = 1 joule/coulomb = 23.06 kcal/mole

= 1 ampere-ohm

Current

1 ampere = 1 coulomb/sec = 1.036×10^{-5} faradays/sec

= 6.237×10^{18} electrons/sec

Charge

1 coulomb = 1 ampere-sec = 1.036×10^{-5} faradays

= 6.237×10^{18} electrons

1 faraday = 96,500 coulombs = 26.806 amp-hours

Power

1 watt = 1 joule/sec = 0.8600 kcal/hour = 3.413 BTU/hr

1 kilowatt = 1000 watts = .2389 kcal/sec

Resistance

$$1 \text{ ohm} = 1 \text{ volt/coulomb} = 1 \text{ mho}^{-1}$$

2.3. ELECTRON TRANSFER

Consider the consequences of having a steady electric
current flowing between two electrodes immersed in a solution containing ions, such as sodium chloride dissolved in
water, or tetraethylammonium bromide dissolved in dimethylformamide. We know that in electronic conductors such as
metals, the electric current is carried exclusively by the
average drifting motion of free electrons in response to
an electric field. We also know that in the solution the
current must be the result of the movement of ions. Ions
cannot traverse the metal, nor can free electrons travel
through the solution. This is depicted simply in Fig. 2.1.
Given this situation, assuming that a buildup of charge
cannot occur indefinitely at the electrode/solution interface,* we can deduce that electron transfers must be occurring, at one electrode (marked -) electrons from the metal
are transferred to molecules near the interface to create
more negatively charged entities, and conversely, molecules

---

* Some current does in fact, flow initially without electron
transfer because of the capacitance associated with the
electrode/solution interface. That this cannot persist for
long is best illustrated by the following example. Assume
a current of 1 A is flowing and that the electrode has a
capacitance of 100 μF (not unrealistic values). If no
electrons were transferred, in one second the voltage drop
across this interface would reach 10,000 V! The authors
can personally assure the reader that electron transfer
occurs to relieve the "pressure" long before such potentials
are reached!

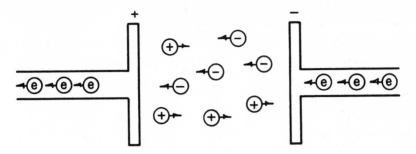

FIG. 2.1.   Motion of charges in electrolysis.

at the other electrode surface (marked +) release electrons
to the metal and are thereby transformed into more positively
charged species. By convention, the sign of the current is
positive in the direction of travel of positive charge.
Note that current flows from a region of relative positive
charge to one of relative negative charge; electrons flow
in the opposite direction--a negative flow of negative
charge is equivalent to a positive current. We can now
define the electrode toward which cations flow as the
cathode and the electrode toward which anions flow as
the anode. This is equivalent to the definition based
on the chemical changes occurring at each electrode, where
the anode is defined as the electrode at which oxidation,
or loss of electrons, is occurring, and the cathode as the
electrode at which reduction, or gain of electrons, is
occurring. Two electrodes are always involved in electrochemical processes; this duality is exactly analogous to
the situation in a redox reaction, i.e., the reducing agent
by definition reacts with an oxidizing agent. In electrolytic systems, however, the two processes are separated
spacially, so that a complete cell is comprised of separate half-cells. The chemical nature of the reactions
occurring at one half-cell can, therefore, be completely
independent of the chemical nature of reactions occurring
at the other half-cell, except that one process must be an
oxidation and the other a reduction.

The cell, in fact, can be physically separated into chambers by use of a cell divider such as a cellophane membrane or an unglazed ceramic cup. The cell divider can in principle (but not in practice) be a perfect barrier for molecules, except that it must allow the passage of at least one type of ion.

The ionic species, taken collectively, are referred to as the electrolyte. The electrolyte is required solely to insure the continuity of charge conduction across the cell--the ions making up the electrolyte are not at all necessarily the species undergoing electron transfer. The electrolyte behaves toward electrolytic current exactly as a resistor behaves toward electronic current, and a voltage drop occurs equal to the product of current and resistance, i.e., Ohm's law is obeyed.

2.4. THE ELECTRODE POTENTIAL

Let us now turn our attention to the various voltage differences which occur within an electrolysis cell. First of all, there is a distinction between the cell voltage, which pertains to the measurable difference in voltage between the two electrodes, and the electrode potential, which is a nearly discontinuous potential difference, not directly measurable, between the electrode surface and the region in the solution adjacent to the electrode. The electrode potential is mainly governed by the nature of the molecules involved in electron transfer, as is shown later. The difficulty in measuring the absolute difference in potential between two dissimilar phases is exactly analogous to the corresponding difficulty in measuring the difference in pH between aqueous and nonaqueous solutions. Of course, pH measurement is itself based on electrochemistry--think about it. In both cases the definition of a suitable reference system is required. This matter is discussed in greater detail later in this chapter.

At this point, let us accept the necessity of measuring electrode potentials by use of a reference electrode and consider what relations exist, if any, between the various potentials occurring within the cell and the current flowing between electrodes. Three possible situations are shown in Fig. 2.2, where the voltage, measured from the left-hand electrode, is plotted as a function of the distance from that electrode. In Fig. 2.2(a), typical of electrolysis, we see a nearly discontinuous potential change as we cross

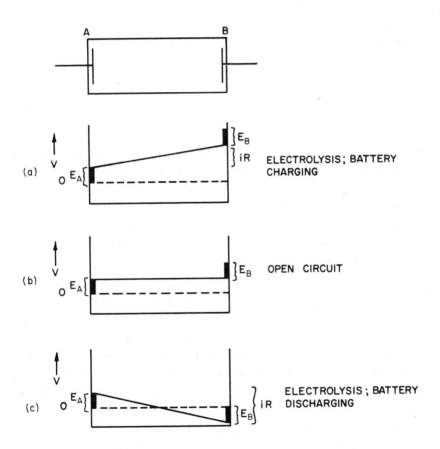

FIG. 2.2. Distribution of voltages in electrolysis cell.

the interface at electrode A, followed by a gradual increase between electrodes caused by iR drop, and another discontinuous increase at electrode B. Note that current must be passing, as evidenced by the iR drop, from right to left and, therefore, that electrode A is the cathode and B the anode. The total voltage measured between electrode B and A is simply

$$V_{cell} = (E_B - E_A) + iR$$

where E_B and E_A are measured with respect to a reference electrode placed adjacent to electrode B and A, respectively.

In Fig. 2.2(b), no current is flowing, as would necessarily be the case when the external circuit is broken, and the cell voltage is just the difference in electrode potentials

$$V_{cell} = (E_B - E_A)$$

This is just the voltage that would be measured between the terminals of an open-circuit cell with a voltmeter.

In Fig. 2.2(c), we have short-circuited the two electrodes, so that

$$V_{cell} = 0 = (E_B - E_A) - iR$$

and a current is flowing in the opposite direction to that which applied to Fig. 2.2(a), therefore the −iR term. In this case, electrode A is the anode and B the cathode, just the reverse of the situation of Fig. 2.2(a).

Note that in each case described above the individual electrode potentials have been assumed to be independent of the direction and magnitude of the current flowing. This can be the case only if the electrode reactions occurring are reversible and if no concentration gradients exist. Electrochemical batteries are specifically designed with

highly reversible electrochemical systems; so the description in Fig. 2.2 applies to the situation of battery charging, open circuit, and short circuit, respectively. Even
though reversibility of that degree is rarely the case for
organic systems, the above discussion does illustrate that
the role of the electrode as anode or cathode is not necessarily determined by electrode potential. In practice, the
electrode potential for many organic systems during electrolysis (current flow) is a strong function of the current
passing because of the irreversible nature of the electron
transfer step itself or of subsequent chemical reactions.
Of course, the region of potentials where electron transfer
can occur at all is governed primarily by the nature of the
molecules being reduced or oxidized. The cause of irreversibility in the electron-transfer reaction is exactly analogous
to the corresponding situation in chemical reactions; namely,
an activation barrier exists for the reaction. In some
cases, chemical and electrochemical irreversibility can be
separated and individual rate constants assigned to each.

In the following sections, a quantitative treatment
of electrode potential and its relationship to reversibility
is developed along lines paralleling the corresponding development of chemical equilibria and kinetics.

2.5.   THE NERNST EQUATION

Consider a general electrochemical reaction, where
the forward and reverse

$$A + ne \; \underset{\longleftarrow}{\longrightarrow} \; B \tag{2.7}$$

reactions are so rapid that equilibrium is always maintained.
For that case we may define an equilibrium constant such that

$$K = \frac{a_B}{(a_A) \cdot (a_e)^n} \tag{2.8}$$

where, in the true thermodynamic sense, activities are required rather than concentrations, and we must remember that
the activities refer to those existing at the surface of the
electrode, which may not be equal to those in the bulk of
the solution. The significance of a_e, the activity of the
electron, will hopefully become clear below. Note also,
that Eq. (2.7) is not meant to imply a mechanism for the
conversion of A to B, just as in conventional thermodynamics
the manner of achievement of equilibrium is unspecified.
The free energy change associated with this reaction can be
written as

$$-\Delta F = RT \cdot \ln K = RT \cdot \ln(a_B/a_A) - nRT \cdot \ln(a_e) \quad (2.9)$$

we now wish to change the units of the energy terms from
thermal units (kcal/mole) to electrical units (joules/coulomb
or volts), so that we divide both sides of Eq. (2.9) by nF
(the number of electrons transferred per reaction times F,
Faraday's constant, the number of coulombs per mole of
electrons) the resulting equation is

$$-\frac{\Delta F}{nF} = \frac{RT}{nF} \cdot \ln(a_B/a_A) - \frac{RT}{F} \cdot \ln(a_e) \quad (2.10)$$

We can now associate the term $\Delta F/nF$ with $-E^\circ$, the standard
potential of the half-cell measured with respect to a suitable reference electrode. By convention, the reference
system has been chosen to be the hydrogen electrode, where
the half-cell reaction is

$$H^+ + e = 1/2 \; H_2 \qquad\qquad\qquad (2.11)$$

and for this system $E^\circ = 0$, so that

$$0 = \frac{RT}{F} \cdot \ln \frac{a_{H_2}^{1/2}}{a_{H^+}} - \frac{RT}{F} \cdot \ln(a_e) \qquad (2.12)$$

Note that this also defines the reference point for electron
activity, i.e., $a_e = 1$ when $a_{H_2}^{1/2} = 1$ and $a_{H^+} = 1$. The
term $- RT/F \cdot \ln(a_e)$ is simply the electrode potential, E,
where the negative sign is included since, again by convention, we choose to measure the electrode potential from a
reference electrode to the actual electrode. This means
that as electron activity in the metal increases, the electrode potential becomes more negative. Substituting E for
the term $- RT/F \cdot \ln(a_e)$ in Eq. (2.10) and rearranging,
leads to the familiar Nernst equation

$$E = E^O - \frac{RT}{nF} \ln \frac{a_B}{a_A} \tag{2.13}$$

Several points are worth considering further. The definition
of the electrode potential as the measure of electron activity is convenient, since the activity of an entity is meant
to reflect its "escaping tendency." For example, temperature measures the escaping tendency of heat, pressure the
escaping tendency of a gas, and height the escaping tendency
of mass in a gravitational field. Thus, electrons in a
state of high activity in the metal (negative electrode potential) try to attain a state of lower activity by trans-
ferring to an electron acceptor (oxidizing agent). Conversely electrons of high activity in a molecule (reducing agent)
try to transfer into a metal of low electron activity (positive electrode potential). The activities of A and B are
normally set equal to their concentrations, so that when the
concentration of B is greater than A, the equilibrium electrode potential is more negative than E^O, and vice versa.
If, on the other hand, the electrode potential is constrained
to be at a certain value, then the concentration ratio of A
and B adjacent to the electrode is fixed, if the system is
to remain in equilibrium. The ratio of A and B in bulk of
the solution may not be equal, however, to that required
by the Nernst equation; in that case, <u>concentration grad-
ients</u> exist, and a current must flow to maintain that

gradient. To make the situation clearer, assume that we
start with a solution where the concentration of A equals
that of B, so that at equilibrium the elctrode potential E
equals the standard potential E^O, no concentration gradients
exist, and no net current is flowing. Now consider what
happens if we make the electrode potential more positive
than E^O. If the system is to remain at equilibrium, the
ratio A/B at the electrode surface must also increase. This
requires a net conversion of B to A (oxidation), electrons
must be transferred, and current flows out of the electrode.
It will continue to flow in that direction, since more B
will diffuse from the bulk to the electrode surface where
B has been depleted, until the ratio of A to B is increased
to the value required by the Nernst equation. The reverse
would occur if the electrode potential were to be constrained
to be more negative than E^O, i.e., A will be reduced to B
until the required ratio is attained. The difference between
the actual electrode potential when current is flowing and
the potential calculated using the bulk concentrations in
the Nernst equation is referred to as the concentration
overvoltage for a reversible system. Other types of overvoltage refer to irreversible reactions at the electrode.
In practice, no system can be completely at equilibrium if
a net reaction is occurring or a net current is flowing.
The degree to which a system can be said to be reversible
depends on how closely the Nernst equation is obeyed (re-
membering that surface concentrations are required rather
than bulk concentrations).

2.6. ELECTRODE KINETICS AND OVERVOLTAGE

Just as in the case of reversible systems discussed
above, we would now like to develop the simple theory of
electrode kinetics in a manner completely analogous to conventional chemical kinetics. Kinetics should be considered

whenever a net reaction is occurring, i.e., whenever a net
current is flowing. Consider the following representation
of an electron transfer reaction

$$A + ne \underset{k_b}{\overset{k_f}{\rightleftharpoons}} B \qquad\qquad (2.14)$$

where, as before, the reaction is written as a reduction so
that k_f represents the forward rate constant and k_b represents the reverse rate constant (oxidation). Since the reactions are occurring at an interface, the k's are expressed
as the formal <u>heterogeneous</u> rate constants in units of
cm/sec, and the concentrations refer to the <u>surface</u> concentrations. If we assume the electrode process is first-order
in species A and B, the rate expression can be written as
follows

$$\frac{-dN_A}{dt} = \frac{dN_B}{dt} = k_f\,(A) - k_b\,(B) \qquad\qquad (2.15)$$

where N_A and N_B are the number of moles of A and B, respectively, reacting per unit area per unit time.

The effect of the electrode potential, or electron
activity, is included in the rate constants and must be
considered separately now. To do so requires a consideration of the energetics involved in electrode processes.
Consider the energy diagram, Fig. 2.3

Here the energy coordinate is plotted in terms of volts in
a direction such that as E becomes more negative, the reduction is favored, as indicated by Eq. (2.14). The lower
diagram represents the situation when the electrode potential
of the system is at the standard value E^o. There is also a
characteristic activation energy and a transition state at
the maximum. The reaction coordinate here is used in the
normal, nebulous way, namely, it represents the progressive
extent of conversion as A is converted to B. Just as in a
conventional bimolecular reaction it is useful to consider
this coordinate as the distance between the bond-forming

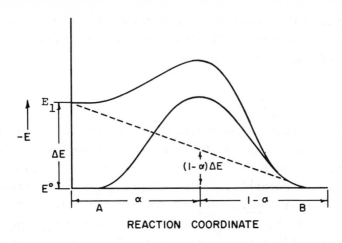

FIG. 2.3.   Energy diagram for electron transfer.

centers, it will be useful, although oversimplified, to consider it in the same way in electrochemistry. Namely, it can
be taken to represent the distance between the electron and
molecule A. Now let us increase the potential of the electrode to E_1, or a change of ΔE; the increase is indicated
by raising the potential of the left-hand side to E_1. If
we assume that the potential falls linearly with respect to
the position at B, by using the principle of superposition,
we arrive at the upper energy diagram. If the transition
state occurs at a fraction of the distance to B, the starting point has been raised in potential by ΔE and the transition state by an amount $(1-\alpha)\Delta E$. Therefore, the net change
in activation energy for the forward reaction is $\alpha\Delta E$, and
$-(1-\alpha)\Delta E$ for the reverse reaction. Restated in more direct
terms, a change in potential ΔE favors the reduction by
$\alpha\Delta E$, and retards the oxidation by $(1-\alpha)\Delta E$. From these considerations, we can now write the effect of electrode poten-
tial on the rate constants in the following manner:

$$k_f = k_s e^{\dfrac{-\alpha n F \Delta E}{RT}} \tag{2.16}$$

$$k_b = k_s e^{\dfrac{(1-\alpha)nF\Delta E}{RT}} \qquad\qquad (2.17)$$

where n is the number of electrons transferred in rate-determining step, $\Delta E = E-E^{\circ}$, and k_s is rate constant at $E = E^{\circ}$.

Note that since the potential enters as ΔE, the difference between the electrode potential and the standard potential, the results of these derivations are independent of the choice of reference electrode. Before going further, it will be helpful to reconsider the significance of α, the cathodic transfer coefficient, since it has historically been a source of confusion. Originally conceived as a phenomenological factor to describe the fact that irreversible electron-transfer rates often increased more slowly than expected, the transfer coefficient has been interpreted in several ways (1) and has been incorporated into most theoretical models. For the purposes of the organic chemist involved in electrochemistry, it will be extremely helpful to think about the transfer coefficient as exactly analogous to the familiar Bronsted coefficient for acid-base catalysis. In general acid catalysis, the rate constant does not increase directly as the acidity, Ka, increases, but rather increases as the acidity to some fractional exponent, also referred to coincidently as α, the Bronsted coefficient. This coefficient has been interpreted as representing the degree of protonation, or the extent to which the proton has been transferred in the transition state. If we substitute the words "charge transfer" for "protonation" and "electron" for "proton" in the former statement, the analogy will be seen to be complete.

It is usual in interpreting how results apply to reaction mechanism to assume that similar reactions will have the same value of α. On the other hand, differences in values of α have been taken to imply differences in reaction mechanisms.

Finally, it should be pointed out that the derivation above and continued below is only applicable to the case of a single rate-determining electron transfer, with n electrons transferred (not necessarily the same as the number of electrons transferred in the over-all reaction). Also remember that the concentrations A and B are, as before, those which exist at the electrode surface. We are now in a position to go further. The net current due to Eq. (2.14) is simply the net rate times the electrode area times the amount of charge transferred per mole. The equation is

$$i = nFSk_s \left\{ Ae^{\dfrac{-\alpha nF \Delta E}{RT}} - Be^{\dfrac{(1-\alpha)nF \Delta E}{RT}} \right\} \qquad (2.18)$$

where S is surface area.

This relationship is obtained by combining Eqs. (2.16) and (2.17) and can be seen to contain two contributions, a cathodic current due to the reduction and an anodic current due to the oxidation. This is exactly analogous to the situation in conventional kinetics, where one considers both the forward and reverse rates, characterized by k_1 and k_{-1}. At equilibrium

$$\text{where } i_c = nSFk_s \ (Ae^{-\alpha nF \Delta E/RT}) \qquad (2.19)$$

$$\text{and } i_A = -nSFk_s \ (Be^{(1-\alpha)nF \Delta E/RT}) \qquad (2.20)$$

the net current is 0, so that

$$Ae^{\dfrac{-\alpha nF(E_e - E^o)}{RT}} = Be^{\dfrac{(1-\alpha)nF(E_e - E^o)}{RT}} \qquad (2.21)$$

where E_e is potential at equilibrium or,

$$\frac{A}{B} = \frac{e^{(1-\alpha)nF(E_e - E^o)/RT}}{e^{-\alpha nF(E_e - E^o)/RT}} \qquad (2.22)$$

Simplifying we have,

$$A/B = e^{nF(Ee-E^O)/\overline{RT}}$$
(2.23)

Taking logarithms of both sides leaves

$$\ln(A/B) = \frac{nF}{RT} (E_e - E^O)$$
(2.24)

or

$$E_e = E^O - \frac{RT}{nF} \ln(B/A)$$
(2.25)

Equation (2.25) will be recognized as simply the Nernst equation, which is now seen as a special case of the more general kinetic Eq. (2.18). Just as in conventional thermodynamics, electrochemical equilibrium is seen to be a dynamic situation, and is characterized by an exchange current density, i_O/S such that at the standard potential ($\Delta E = 0$, and A = B) from Eq. (2.18),

$$i_O/S = nFk_sA$$
(2.26)

These results may perhaps be understood more easily if we examine a current-electrode potential curve as predicted by Eq. (2.18). To be consistent with the practices of polarography it is usual to plot cathodic current in the +y direction, and -E in the +x direction. Figure 2.4 is simply a plot of Eq. (2.18) and its component terms, using n = 1, S = 1 cm^2, $\alpha = 0.5$, A = B = 10^{-3}M, k_s = 10^{-3} cm/sec, and T = 298°K.

In practice, only the net current, i, can be measured experimentally and, to obtain the results of Fig. 2.4, the experiment must be conducted in such a way that no concentration overvoltage (see Sec. 2.4) occurs and so that the composition of the solution does not change during the experiment. This is normally accomplished by using a small electrode in contact with a solution containing a large excess of material and by stirring vigorously.

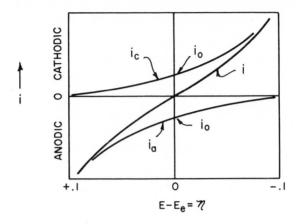

FIG. 2.4. Current-voltage relationship.

The difference between the actual electrode potential and the equilibrium electrode potential is termed the acti-vation overvoltage, η, and is a necessary consequence and function of the net current. It will be seen that η is negative for cathodic currents and positive for anodic cur-rents. It can be shown that when η exceeds $\sim 0.12/n$ V, the effect of the reverse reaction is negligible (<1% of the total current) and the system is nearly completely irre-versible. This condition occurs when, at a given current, k_s is sufficiently low, or, for a given k_s, when the current is sufficiently high. Irreversibility is then seen to be dependent on experimental conditions and on the nature of the reaction. The terms "hydrogen overvoltage" and "oxygen overvoltage" appear frequently in the early literature and are applied to the electrolysis of water at specific elec-trodes. The type of overvoltage is often not stated, but must be implied. Thus, platinum is said to be a low over-voltage electrode and mercury a high overvoltage material. These, of course, refer to hydrogen overvoltage, and to be quantitatively meaningful, the current density at which the measurement was made must be reported. In the very early

literature there are several papers in which the overvoltage
was measured at the "first noticeable sign of bubble formation." This rather qualitative observation, combined with
the fact that overvoltage is sensitive to surface impurities
of the metal, contributed to the several discrepancies in
the literature concerning the "overvoltage" of electrodes.

In practice, results of experiments designed to give
information about electrode kinetics are often plotted in
the form of log i vs η. The form of this equation is obtained from Eqs. (2.19) and (2.20), which result in the
following equations, for the case where A = B:

$$\log i_c = \log (nFSk_sA) - \frac{nF}{2.303\ RT}\ \eta \qquad\qquad (2.27)$$

$$\log (-i_a) = \log (nFSk_sA) + \frac{(1-\alpha)nF}{2.303\ RT}\ \eta \qquad\qquad (2.28)$$

and since 2.303 RT/F = 0.059 V at 25°C, both of these reduce
to the following simpler forms:

$$\log i_c = \log i_o - \frac{\alpha n}{0.059}\ \eta \qquad\qquad (2.29)$$

$$\log (-i_a) = \log i_o + \frac{(1-\alpha)n}{0.059}\ \eta \qquad\qquad (2.30)$$

where i_o is exchange current.

Thus, each gives a straight line whose intercept at
η = o gives the exchange current and whose slopes can give
α and n. These equations may be placed in the identical
form to the Tafel equation (2) which states that

$$E = a + b \log (i) \qquad\qquad (2.31)$$

and which has been used a great deal in the past.

Figure 2.5 should make the situation clearer.

The observed current is plotted at the solid lines.
Extrapolation of the linear portions of the anodic and

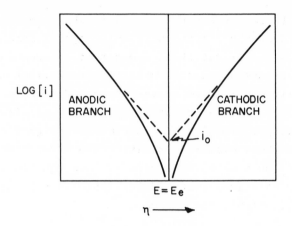

FIG. 2.5.   Tafel plot.

cathodic branches result in an intersection at η = o or
E = Ee.   The value of the current at this point is just the
exchange current.   The slopes of the linear portions give
αn and $(1 - \alpha)n$ respectively, as may be seen by inspection
of Eqs. (2.29) and (2.30).   The significance of this type
of analysis is that it can be applied to give a value for
the equilibrium potential of a system which behaves practically irreversibly at currents too low to be experimentally
measurable or meaningful.

In many organic systems, however, the degree of irreversibility, for one reason or another, is so high that only
one branch is observable (either the cathodic or the anodic
branch).   Under these conditions, the value for the equilibrium potential cannot be assigned experimentally, since the
slope of the log i or E diagram usually remains linear down
to immeasurably low currents.   The slope of the cathodic or
anodic branch, as the case may be, may still be used to determine either αn or $(1 - \alpha)n$, respectively, but not both.
(The reader should note that in this and in subsequent discussions, the transfer coefficient usually appears multiplied
by n, so that, unless other evidence exists concerning the

number of electrons transferred in the rate-determining step,
the value for α will not be determined; the uncertainty
usually arises if n can be 2.) Unfortunately, moreover,
the values of k_s, related to the exchange current, and Ee,
the equilibrium potential, are both indeterminant, since an
appropriate value of k_s can be chosen for any value of Ee
(in the region of immeasurably low current) to give the observed results. Typically, reduction and oxidation potentials for highly irreversible systems have, in fact, often
been compared to give information about the respective reaction mechanisms, but this is only strictly valid if assump-
tions concerning their equilibrium potentials are valid.

2.7. PHYSICAL CONCEPTS

2.7.1. DIFFUSION

    As has been mentioned several times above, the concen-
tration terms in the Nernst equation or in the elctrode kinetic equation always refer to the condition existing in a
region adjacent to the electrode. Thus, the electrode "sees"
only what is next to it. During electrolysis, the surface
concentrations can change with time, and therefore a concentration gradient of the various species must be present,
as is the case for all heterogeneous reactions. For the
purposes of synthesis, this gradient is usually something
to be avoided, since as starting material is consumed, the
region around the electrode becomes depleted, and the rate
of the reaction (i.e., the current) can become limited by
mass transport. Although vigorous stirring can, of course,
markedly increase the rate of mass transport, there will
always be a boundary layer, or depletion region, which must
be crossed by the process of diffusion, so that fresh material can come into contact with the electrode.

Because of the importance of diffusive mass transport
in electrolysis and in most electroanalytical techniques,
further discussion is required concerning the connection
between the passage of current and the resulting concentration gradients. First of all, we must understand Fick's
law of diffusion which states that the "flux" of material
at a given region in the solution is proportional to the
concentration gradient at that region. By flux we mean the
net rate per unit area at which material crosses the region,
in moles/cm^2/sec. In mathematical terms, this is just

$$q_A = D_A \text{ grad } (C_A) \tag{2.32}$$

where q_A is flux of material A; C_A is concentration of
A; D_A is diffusion coefficient of A; and grad is gradient.
This equation is written in terms of generalized coordinates,
where both q_A and C_A are functions of position and time.
For our purposes, it will be convenient to simplify the
situation by assuming that the electrode has a planar sur-
face, and that it is large compared to the diffusion layer.
Under these conditions of semi-infinite linear diffusion,
Eq. (2.32) may be rewritten in the following form

$$q_A(x,t) = D_A \frac{\partial C_A(x,t)}{\partial x} \tag{2.33}$$

The diffusion coefficient, D_A, is itself a function of the
medium in which A is dissolved, and hence a function of the
concentration of A. However, in many cases the concentration of electrolyte will be large and will swamp out the
effect of A on the diffusion coefficient, so that we may
consider it to to be a constant. The current, i_A, involved
in electrolysis of A is simply proportional to the rate of
consumption of A at the electrode surface, so that, convert-
ing to the appropriate units, we have the following important relationship between the current and the concentration
gradient.

$$i_A/nFS = q_A(o,t) = D_A \frac{\partial C_A(o,t)}{\partial x} \qquad (2.34)$$

or

$$i_A = nFSD_A \frac{\partial C_A(o,t)}{\partial x} \qquad (2.35)$$

This states that the current involved in the electrolysis of A is proportional to the concentration gradient of A at the electrode surface (x = o), the surface area, S, and the diffusion coefficient, D_A.

The difficulty in proceeding further with the quantitative treatment of the effect of diffusion lies in the complexity of describing hydrodynamic conditions pertinent to stirred solutions. Certain idealized systems, such as applies to a rotating disk electrode (see Sec. 3.4.3), are, in fact, subject to quantitative treatment, but these details lie beyond the scope of this book, and the reader should, if further interested, consult some of the texts and references suggested at the end of this chapter.

We can, however, progress one step further if we restrict our consideration to quiet, or unstirred, solutions. The time dependence of the concentration of A at a selected plane is just the difference between the flux at point x and the flux at x + dx. This can be shown to result in the following equation

$$\frac{dC_A(x,t)}{dt} = \frac{dq_A(x,t)}{dx} = D_A \frac{d^2C_A(x,t)}{dx^2} \qquad (2.36)$$

Equation (2.36) is known as the linear diffusion equation, and is a second-order partial differential equation, requiring the specification of "initial conditions" and "boundary conditions." An equation of this type can be written for each of the several species involved. If homogeneous chemical reactions are also involved, appropriate terms are added to the right-hand side of Eq. (2.36).

Initial conditions are usually such that the solution
is homogeneous, where the concentration of each species is
independent of position. The boundary conditions depend on
the experimental conditions, e.g., if the electrode potential
is held constant, then the Nernst equation is used as a
boundary condition for reversible reactions and the electrode
kinetic equation, relating the current (hence the flux due
to various species) to the concentration at the electrode
surface, is used for irreversible systems. Another usual
boundary condition is that sum of the fluxes of all materials is equal to zero at the electrode surface, i.e., that
no material enters or leaves the electrode. Once specified,
the initial and boundary conditions define this problem,
and solution of the equations yields the concentration dependence and in particular the dependence of the current,
the usual observable, on time.

Although the detailed solution of particular cases by
use of the above equations is beyond the scope of this book,
several cases have been solved either explicitly or through
the use of computer similation techniques, and the reader
is invited to proceed as far as his mathematical prowess
allows, through use of several of the references cited at
the end of this chapter.

It is hoped that from these simple considerations the
reader will begin to gain insight into the interrelationship
between concentration gradients, diffusion, current, and
electrode potential, and to appreciate the complexity of
their quantitative treatment. This particular field of
interest has attracted the attention of many theorists in
recent years and considerable progress has been made in
developing quantitative solutions to many complex electrochemical systems involving multistep electron transfers
with intervening chemical reactions. These studies have
often resulted in diagnostic criteria for testing possible
mechanisms by various electrochemical techniques.

2.7.2. ADSORPTION

As in all heterogeneous systems, adsorption may be
expected to play a crucial role in determining the course
of an electrochemical reaction. In a typical electrolytic
solution, either the solvent, ionic species from the electrolyte, or the substrate may be adsorbed to the electrode
surface. Therefore, it is not always safe to assume that
conditions at the electrode surface are similar to those
existing in the bulk. Various electrode materials can obviously behave quite differently from each other, and, if
specific adsorption of the substrate is involved, these
can lead to entirely different products. Adsorption can
also become a rate-limiting process and should be considered
in developing a mechanistic scheme. A simple case may be
written as

$$A_{bulk} \rightleftharpoons A_{ads} \tag{2.37}$$

$$A_{ads} + ne \longrightarrow B_{ads} \tag{2.38}$$

$$B_{ads} \rightleftharpoons B_{bulk} \tag{2.39}$$

The adsorption isotherm, giving the relationship
between the concentration of the bulk material and that of
the adsorbed species at equilibrium, needs to be considered
if further development of its effect on the kinetics is to
be made. The isotherm (constant temperature) is usually
written as a function of surface coverage, the fraction
of the surface covered by the adsorbed species. The form
of the adsorption isotherm is usually not known a priori,
nor can it usually be measured directly. The normal method
of attack is to assume various functional forms, based on
model systems, and then to determine whether the experimental results agree with the predictions of such models.

Although it is beyond the scope of this book to develop a quantitative treatment of the effect of adsorption,
several factors should be noted as to their qualitative
effects. First, a good general principle to remember is
that the least soluble material in the medium is the most
likely to be adsorbed on the electrode. For example, in
organic media, water would be a key candidate. Since only
minute quantities of material are required to achieve the
full effect of adsorption (typically monolayers are involved), there may be differences in electrolysis using
"off-the-shelf" solvents and ones which have been scrupu-
lously dried. Adsorption involving ionic or polar species
is often a function of potential in such a direction that
cations are more likely to be adsorbed at negative potentials and anions at positive potentials. In electrolysis
using tetralkylammonium salts as electrolyte, the cathode
usually has an adsorbed layer of tetralkylammonium cations;
this situation has been described as the "chicken fat"
electrode, since it would thereby have a tendency to exclude
water from the surface. Finally, one should be aware of the
possibility that adsorption may exert profound stereochemical effects, as in catalytic hydrogenation. Specific orientation may be required and, under such conditions, differences in behavior of isomeric materials can be expected.

Several diagnostic criteria may be applied to electrolysis systems to gauge whether adsorption is important in
the course of an electrolytic synthesis. First of all, the
reactions should be carried out, if possible, with different
electrode materials. If major differences in product distribution or identity are observed, under otherwise identical
conditions, then adsorption is undoubtedly a key factor.
The same is true when the product is affected by a change
in electrolyte type, e.g., organic vs inorganic salts.
Small amounts of highly surface active materials have been
used, particularly in polarography, to swamp out the effect
of other potentially adsorbable species. All of the above

changes in reaction conditions may have an effect on the
potential at which electrolysis occurs, and the experimenter will always benefit from utilizing one of the electroanalytical tools, discussed in Chap. 3, as an adjunct to
his synthesis studies. These always produce the relationship between current and voltage under specific experimental
conditions, and even if used only qualitatively give the
experimenter the "eyes" with which he may gain insight into
the reactions under study.

Although it is not possible at this time to predict
in advance the influence of adsorption on the course of the
reaction, its importance to the synthetic organic electrochemist should not be underestimated. Adsorption can make
the difference between a nonselective and a highly selective
reaction. It may also be a controlling factor in governing
the stereochemistry of the products or the nature of intermediates. Be aware of the potential influence and design
experiments to probe its effects.

2.7.3. DOUBLE LAYER

Perhaps no other topic in electrochemistry has caused
more confusion or been the source of more discouragement to
the prospective electroorganic chemist than that of the
electrical double layer. Its full understanding requires
the knowledge of many branches of science with attendant differences in terminology, and rarely is the subject discussed
in its relationship to electroorganic synthesis. Its importance, however, cannot be denied, since the double layer
is the region in which all the "action" of electron transfer
takes place. The following is a qualitative discussion of
the nature of the double layer and its possible influence
on the electroactive material.

To begin with, up to this point we have considered
the electrode potential to represent a discontinuous change
in potential in crossing the electrode-solution interface.

A closer examination of this region will show that this is
not the case when molecular dimensions are reached; in fact,
the potential develops continuously over a region of about
50 to 100 Å. Consider an electrode with a surface excess
of electrons, and hence, negatively charged. A layer of
cations will be attracted to this by electrostatic forces.
In this layer will also be specifically adsorbed material,
which may be charged or neutral species. In both cases,
the forces involved at this first layer are strong enough
so that these molecules may be considered to be essentially
unsolvated, unless, of course, it is the solvent itself
which is adsorbed. The effect of this layer, in general,
is to reduce the electric field and to produce, in turn,
a sufficiently high field to attract a second, negatively
charged layer. This second layer can be considered to be
comprised of anions which are solvated. These two layers
are usually referred to as the inner, or compact double
layer. These materials are bound so tightly to the surface that they are unaffected by agitation in the bulk of
the solution. The charged layers, taken together, constitute a capacitance which can, in fact, be measured. The
manner in which this electrode capacitance changes with
electrode potential reflects corresponding changes in the
structure of the double layer. Measurements of this sort
have been the most widely used for the study of double layer
phenomena. For example, the presence of adsorbed species
may be detected by their effect on the capacitance of the
electrode. It is within the inner double layer that the
potential determining species resides, and direct electron
transfer to and from the electrode occurs in this region.

If the electrode is deficient in electrons (more
positive electrode potential), a double layer of opposite
sense will be formed. So it can be seen that, in general,
there will exist an electrode potential at which there is
no net charge at the electrode surface. This potential,
which can be determined by capacitance measurements and is

different for different electrodes is called the <u>potential
of zero charge</u> (pzc) and has been considered for cases
where specific adsorption is absent, as a physically more
meaningful reference potential then the ones normally used.
At potentials more positive than the pzc, anion adsorption
is possible; the reverse is true for potentials more negative than the pzc. For mercury, the pzc is +0.48 V measured
with respect to a saturated calomel reference electrode.

Since the inner double layer region extends about 5 $\overset{o}{A}$
from the electrode, one can easily calculate that a one-
volt drop across this region corresponds to an electric
field of 2 x 10^7 V/cm, and polarizable molecules within this
field will have a considerably distorted electronic structure
compared to their normal field-free structure. Considerations of these effects should be made if the details of
electroorganic reaction mechanisms are to be elucidated,
particularly when systems are encountered where radically
different products are formed under different experimental
conditions, or where an electrolytic reaction produces a
product which differs from that expected from a normal redox reaction.

Beyond the inner double layer, regions of alternating
charge persist, but these now become so diffuse that no
clear boundaries can be distinguished. This region extends
roughly 50-100 $\overset{o}{A}$ into the solution and is referred to as
the outer, or diffuse, layer. At some point in this region,
the species comprising it are free to move in response to
bulk motions of the solution and the potential difference
between this <u>shear</u> layer and a region considerably further
out is referred to as the <u>zeta potential</u>, ζ. This potential is involved in various <u>electrokinetic phenomena</u> such
as electrophoresis, electroosmosis, and the stability of
colloids. Since the zeta potential occurs wholly within the
same phase, it can be measured directly without the use of
an arbitrary reference electrode.

Although the prospective electroorganic chemist normally need not delve further into these more theoretical

aspects in the course of his synthesis work, it is hoped
that the basic principles of electrochemistry are more
clearly understood, and these discussions will prompt him
to consider the more fundamental aspects of the reactions
he will be dealing with.

2.8. EFFECT OF VARIABLES

Electrolytic processes characteristically are controlled
and affected by many variables, some of which are mechanical,
electrical, and chemical, as well as combinations thereof.
The effect and relative importance of these depend on the
particular situation at hand, and it is only by understanding and considering their possible effects that electrolytic
reactions can be controlled and optimized effectively. Only
too frequently can we read in the literature examples of
electrolysis for which the effect of certain critical variables were unconsidered, and therefore, which may possibly
be far from optimum with respect to yield and/or selectivity.
The realization of the promise of electroorganic synthesis
depends heavily on the proper application and control of
those variables unique to electrochemistry.

Let us now consider what these variables are and how
they might effect the course of an electrolytic reaction.

2.8.1. CELL DESIGN (ELECTRODE GEOMETRY AND CELL TYPE)

Just as the proper equipment (usually glassware) in
conventional organic synthesis is important in bringing
about a successful synthesis, so too in electrochemistry the
electrolysis cell is the first element to be considered.
The electrolytic cell serves not just as the container for
the reactants, but is also the means whereby the electric
current is introduced. Since cells suitable for electro-

lytic reactions are rarely available from commercial sources, the experimenter will probably be directly involved in its construction. Details of actual cell designs are given in Chap. 3. Therefore, we consider here only those parameters which may have an effect on the electrolytic reaction.

The geometry of the electrode arrangement is important in insuring uniform current density over the entire surface of the electrode at which the desired reaction is to occur. The usual way of effecting this requirement is to use either a plane parallel arrangement or to utilize cylindrical sym- metry. The adverse effects of utilizing nonuniform geometry can be best visualized by referring to Fig. 2.6.

The lines between electrode A and B represent the elec- tric field between them. Although it is beyond the scope of this book to treat quantitatively the form of the elec- tric field when the electrode geometry is specified, the qualitative aspects may be grasped fairly simply. Electric field lines enter the surface of a conductor perpendicular to its surface and are parallel to the surface of an insula- ter. The electric field lines in a homogeneous resistive medium, such as an electrolytic solution, will generally be concentrated in the region where the electrodes are closest,

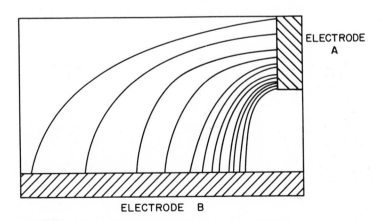

FIG. 2.6. Electric field for nonsymmetric geometry.

and further spread apart between regions where the electrodes
are distant. The current itself follows the electric field
lines and, therefore, it can be seen from the qualitative
picture in Fig. 2.6 that the current density for that arrangement of electrodes will not be uniform across the surface
of either electrode. The electrode potential will also be
nonuniform since it depends to a certain extent on the
current density. If the products of the reaction taking
place at the electrode are a function of electrode potential,
then this is an obviously undesirable effect. This cannot
be eliminated even when the electrolysis is carried out at
a controlled electrode potential (see Fig. 2.9) since the
potential being controlled will either be that at a single
point on the electrode, or the average over an extended
area.

The most common arrangement, as mentioned above, is
to have two planar electrodes with their faces parallel to
one another. To avoid possible undesirable effects on the
somewhat ill-defined (although much lower) current density
distribution on the back of the electrodes, the backs can
be insulated. Since usually only one electrode is of interest (working electrode), both sides can be utilized by
flanking it with two equidistant planar electrodes (auxiliary
electrodes) which are electrically in common (see Fig. 2.7).

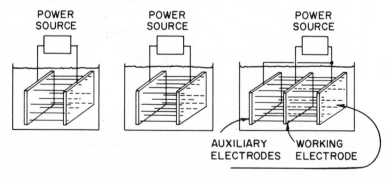

FIG. 2.7. Common arrangements for planar electrodes.

Another common arrangement which preserves uniform
current density over the surface of the working electrode
is one which has cylindrical symmetry (see Fig. 2.8). Since
the inner surface has less area than the outer, the latter
will have a lower current density than the former. If the
reactions occurring at the auxiliary electrode are unimpor-
tant, the auxiliary electrode can be a solid wire. Although
somewhat more difficult to construct, cylindrical electrodes
more efficiently utilize the space in a cylindrical container
such as a beaker.

In both cases, a more stable situation can be maintained
by the use of spacing elements made of insulating material
such as glass, ceramics, and certain plastics. The over-
all time for conducting an electrolysis at a given current
density is minimized if the surface-to-volume ratio of the
cell is maximized. Usually, this is accomplished by making
the distance between electrodes as small as feasible. This
also minimizes the iR drop, and hence, the amount of power
dissipated as heat through the electrolyte. Where available, the use of gauze electrodes are often employed to

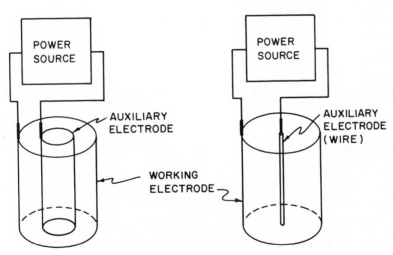

FIG. 2.8. Cells with cylindrical symmetry.

increase the available surface areas and permit better circulation, and hence, faster diffusion of material to the
electrode.

The above discussion applies to the case of the undivided cell, one in which the electrolyte, solvent, starting
materials, and products have equal contact with both electrodes. In many cases, this is undesirable, since the
reactant may undergo reaction to an undesired product at
the auxiliary electrode, or the products at the auxiliary
electrode may interact adversely with the products at the
working electrode. Consider, for example, a case where a
halide salt, such as tetraethylammonium bromide, is used as
electrolyte in the reduction of a 1,2-dibromide, such as
1,2-dibromobutane, in an undivided cell. Reduction at the
cathode will normally lead to 1-butene (see Sec. 4.3). Free
bromine will be formed at the anode. Stirring will bring
the free bromine into contact with the 1-butene and reform
the starting material, or the bromine will be reduced at
the cathode to bromide ion.

at cathode $CH_3CH_2-\underset{\underset{Br}{|}}{C}H-\underset{\underset{Br}{|}}{C}H_2 + 2e$

$$\longrightarrow CH_3CH_2CH=CH_2 + 2Br^- \qquad (2.40)$$

at anode $2\ Br^- \longrightarrow Br_2 + 2e \qquad (2.41)$

in bulk $Br_2 + CH_3CH_2CH=CH_2 \longrightarrow CH_3CH_2\underset{\underset{Br}{|}}{C}H-\underset{\underset{Br}{|}}{C}H_2 \qquad (2.42)$

possible at cathode

$$Br_2 + 2e \longrightarrow 2Br^- \qquad (2.43)$$

Reactions (2.42) and (2.43) constitute invisible reactions, or cyclic conversions, which result in a net
consumption of electricity without forming any new products.

Such reactions can be detected by their apparent (but not actual) violation of Faraday's law. Even more serious are cases where undesired side reactions result from anode and cathode reactions. For example, the reduction of nitrobenzene (see Sec. 4.2) to aniline should be carried out in a divided cell, since aniline would oxidize readily at the anode. These situations are avoided to a large extent by the use of a divided cell, one in which a semipermeable barrier is interposed between anode and cathode. The ideal divider would be chemically inert and totally impenetrable by solvent, reactants, and products, but would allow free passage of at least one ionic species. The divider separates the system into two chambers, one in which the anolyte comes into contact only with the anode and the other in which the catholyte contacts only the cathode. In practice, however, available materials for cell dividers fall far short of the ideal.

Measurable amounts of diffusion of neutral species may be expected. The membrane type, made of thin polymeric material, offers better selectivity and less diffusion of neutral species but is less chemically resistant than the ceramic type. Details of available materials and their advantages and limitations in electrolysis cells are covered in Chap. 3.

In any case, the properties of the divider material which must be considered in its use are:

1. Permeability: a measure of the rate at which specified materials can diffuse through it. Usually, lower permeability is obtained by lower porosity and greater thickness, factors which normally also increase the electrical resistance.

2. Selectivity: a measure of the ratio of diffusion rates of different species. Normally, the type of selectivity of available materials involves discrimination by molecular size, i.e., larger molecules are retarded, or discrimination by ionic type, e.g., cation exchange mem-

branes are available which have high permeability to cations.

3. Resistivity: measures the resistance of the divider in equilibrium with the electrolyte. The resistance
increases with divider thickness and with lower porosity.
At high current densities the divider resistance can give
rise to sufficient heat to destroy the divider, particularly
in the case of thin membranes . In ceramic types, excessive heat can cause the solvent within the pores to boil,
and thus the vapors can further increase the resistance.
If the divider resistance is large compared to the solution resistance, it will exert a leveling effect on the
current distribution, and can allow the auxiliary electrode
to have a nonsymmetric geometry.

4. Chemical resistance: usually high for ceramics,
low for membranes.

5. Thermal stability: usually high for ceramics, low
for membranes.

6. Rigidity: must be self-supporting (or supported)
and must withstand pressure differences across cell.

The use of divided cells is almost mandatory when conducting electrolyses for which the products are unknown.
However, the obvious additional complexity this requirement
imposes on cell construction and operation makes it worthwhile to consider the alternatives which might permit the
use of an undivided cell to carry out a known reaction.
If an undesired product is formed at the auxiliary electrode
as a result of electrolysis of the electrolyte, perhaps an
electrolyte can be found which alleviates this problem.
For example, if halides interfere by oxidation to free
halogen at the anode, the corresponding acetate might be
utilized, since the products of oxidation are mainly carbon dioxide and ethane, both gaseous and, in many cases,
chemically inert. Changes in the auxiliary electrode material, or addition of innocuous depolarizers, should also
be considered to defeat undesired reactions and to permit
the use of undivided cells.

2.8.2. ELECTRODES

The nature of the electrodes used for electrolysis should be considered a key variable. To the extent that adsorption and double layer effects may play a role in a given system, the electrode may exert a profound effect on the reaction and could, in fact, change the entire nature of the product. Although early workers in electrochemistry routinely tried different electrode materials, the more recent workers (including the authors) seem to be mesmerized by mercury and platinum. When synthesis is the goal, an effort should be made to explore the effect of electrode materials. Electron-transfer reactions may be reversible on one electrode and highly irreversible on another, although little work on this subject in organic systems has been reported. An inorganic example of this is the reduction of hydronium ion to hydrogen, which proceeds reversibly at platinum but highly irreversibly ($\eta \approx 0.8$ V) at a mercury cathode. For systems run in aqueous media, this <u>hydrogen overvoltage</u> has historically been considered to be of central importance and can serve to set the cathodic limit of the electrode potential. Reactions carried out under conditions where visible hydrogen evolution is taking place are, in a sense, being conducted under controlled potential conditions.

Several considerations should be applied to the selection and variation of electrode materials for use in organic synthesis. In most cases, the effects on a particular electro-organic synthesis will be difficult to predict unless a fair amount of information is at hand concerning the reaction, but the following should serve as a guideline for experimental studies.

2.8.2.1. Stability

Obviously, to be of much general use, the electrode ought to be stable in the medium. In aqueous media, sta-

bility is usually a simple function of the pH of the medium,
but be careful to consider the effects of possible complexing agents in the electrolyte. For example, the presence
of chloride ion will make both silver and mercury much more
susceptible to oxidation due to the formation of insoluble
chlorides. Organic complexing agents will have a similar
effect. Intermediates formed in the electron-transfer reaction may react with the electrode, such as free radicals
with mercury or lead. In some cases (see Sec. 4.6) these
reactions can be utilized for the efficient synthesis of
organometallics.

For a given electrode-electrolyte system there will
be a limited potential region of utility. The cathodic
limit is usually set by reduction of the solvent or electrolyte, since metals themselves are not susceptible to reduction. Stable electrodes for oxidation are of limited number.
The oxidation potential of the metal in question (usually
given for aqueous media) can serve as a guide for its use
in organic systems. In aqueous media, the oxidation potentials are, to some extent, dependent on the pH of the medium
and on the presence of complexing agents. The noble metals
as well as carbon and graphite have been utilized to the
greatest extent as anodes in organic media because of their
stability toward oxidation.

2.8.2.2. Surface

One should always keep in mind when considering selection of electrodes that it is the surface of the electrode
where the electron-transfer reactions are taking place, and
that the chemical nature of the surface of the electrode
may be completely different from that of the bulk material.
For cathodes, the surface can generally be considered to
be the "clean" metal, except that they are particularly
susceptible to metal ion impurities in the electrolyte.
Many metal ions will be easily plated out on the cathode
at relatively low potentials and, of course, it does not

take much to form a monolayer of the impurity metal ($\sim 10^{-8}$ moles/in^2 will do it). It will pay to bear this in mind in considering the purification of solvent and electrolyte for use in reduction.

For a liquid metal such as mercury these effects may be expected to be far less serious, since many metals are soluble in mercury, and hence, would be diluted into the bulk of the mercury. In any case, stirring of the mercury would tend to expose a fresh surface. This lack of susceptibility to impurities contributes to the widespread usage of mercury as the cathode material in many organic syntheses.

The case for anodes is considerably more complex because of the tendency of many metals to form oxide coatings. Even the noble metals such as platinum are susceptible to the formation of oxide monolayers under anodic conditions. These are difficult to avoid except under the most stringently controlled conditions, since water (the source of oxygen) is usually present in trace amounts in most organic solvents. In some cases the oxide film is remarkably inert and electrically insulating, such as applies to aluminum oxide and tantalum oxide, materials which form the basis of the electrolytic capacitor. In some cases the oxide is an electronic conductor such as lead dioxide, thallic oxide, and silver (II) oxide. The conditions under which the oxide film is formed can have an effect, e.g., the passivation (formation of a compact, inert oxide film) of iron in nitric acid. A good source of information on the suitability of anodes and the condition for their formation may be obtained in the literature of the battery industry, e.g., the use of nickel, manganese, and silver as anodes in alkaline media.

In most cases it will be advisable for electrolysis studies to subject solid slectrodes to a suitable pretreatment process (covered in more detail in Chap. 3). This can be a simple physical pretreatment such as polishing or scouring the electrode to expose a fresh surface, or it can be chemical or electrochemical, such as pre-anodization.

In most cases pretreatment procedures will generally lead
to more reproducible results.

2.8.2.3. Adsorption and Catalysis

The consideration of adsorptive and catalytic effects
of electrodes are undoubtedly very important considerations
but, unfortunately, there do not appear to be very many
general guidelines for prediction of their effects on organic electrode reactions. Certainly a great deal of work
has gone on in the field of adsorption and catalysis of
organic reactions by metals, but it is beyond the scope of
this book to offer any but the most rudimentary of considerations. The catalytic and adsorptive effects of the noble
metals toward hydrogenation are well documented, and one may
expect to find parallels in cathodic reduction with the effects found in chemical hydrogenation. Consultation of the
literature in metal-catalyzed redox reactions may provide
clues to the better understanding of electron-transfer
reactions.

If adsorptive effects are suspected, one might do well
to examine the reaction using polished vs "black" (very
finely divided surface) electrodes, such as bright and platinized platinum electrodes, respectively. If the electron-
transfer reaction proceeds more rapidly, then adsorption is
probably the rate-limiting steps; if the nature and distribution of products change, then adsorption and/or catalysis
is probably the product-determining step. The experimenter
should also keep in mind the possibility that these effects
may also be potential dependent.

2.8.2.4. Physical Form

A variety of forms of electrode material are possible,
and although these different types may not generally lead
to significant differences in the over-all reaction, certain
forms are more convenient to use under different experimental conditions or within cells of different design. Metal

sheets are generally the most readily available and useful since they may be cut and shaped into suitable forms, usually planar or cylindrical. Wires or rods of metals (and various types of carbon) are available and are useful as central elements in cylindrical cells. Their surface areas are limited. This is usually undesirable, except when used as auxiliary electrodes. Wire mesh or "expanded" electrodes can significantly increase surface area and improve the efficiency of stirring to increase diffusion rates. A limited number of porous electrodes are available from the Clevite Corp., such as steel and nickel, which are used as fuel cell elements. These may be useful in electrolysis of gaseous materials, such as low-molecular-weight olefins, where the gas may be efficiently brought into contact with the electrode/electrolyte interface by passage through the back of a suitably designed cell. The same advantage is applicable to cases where a liquid of limited solubility in the medium is to be electrolysed. In cases where the particular metal of interest is not available in the desired form, the possibility of electroplating it onto a base metal can be considered. Further information concerning commercial sources for electrode materials is found in Sec. 3.1.

2.8.2.5. Characterization Using Voltammetry

If possible, the investigation of the effect of various electrode materials should be accompanied by the utilization of a suitable type of voltammetric measurement (see Sec. 3.4) using the same materials as in the electrolysis. It will often be possible to obtain valuable information quickly concerning electrode potential dependence, side reactions, adsorption and/or catalysis, effects of pretreatment, and kinetic and/or diffusion parameters as a function of the electrode material and medium. The presence and effect of oxide films or adsorbed hydrogen can also be ascertained. In addition, further insight may be gained into the reaction mechanism by qualitative and quantitative inspection of the results of voltammetric experiments.

2.8.3. REFERENCE ELECTRODES

If the electrolytic reaction is to be studied as a function of potential, or if the electrolysis is to be carried
out at constant electrode potential (strongly advisable),
then it will be required to measure or monitor the electrode
potential by use of a <u>reference electrode</u>. The need for
such a reference has been alluded to previously. It is hoped
that the following discussion will clarify the situation.

We wish to measure the potential difference between the
working electrode and the region adjacent to it in the solution (ideally just outside the double layer). Why not use
an ordinary voltmeter to do this? In doing so we would at-
tach one probe to the metallic electrode and place the other
in the solution near the electrode. In general, a voltage
would register on the meter, but this would be comprised of
not only the working electrode potential, but also the potential between the solution and the probe itself. Thus, we
would not know, in general, whether changes in the voltage
read-out by the meter reflected changes in the working electrode potential or changes at the probe itself. To cure
this problem we could surround the probe with a solution of
known and invariant composition which formed a highly reversible electron-transfer couple with the probe metal. In
this way we would be assured that its potential would remain
invariant, and we have now satisfied the basic requirement
for a suitable reference electrode.

In principle, it would be desirable to use the hydrogen
electrode (1 N HCl, hydrogen gas at one atmosphere bubbling
over the surface of a platinized platinum electrode), since
by convention it has been assigned a potential of zero volts
at all temperatures. In practice, however, it would be awkward to use. More commonly, the <u>saturated calomel elec-
trode</u> has found the most widespread use; it consists of
mercury in contact with mercurous chloride (calomel) and
a saturated aqueous solution of potassium chloride. The
electron-transfer couple for this electrode is

$$Hg_2Cl_2 + 2e = 2Hg^O + 2Cl^-$$ \hfill (2.44)

Since the composition of the solution is maintained constant by saturation (solid KCl and Hg_2Cl_2 are present), small currents in either direction or losses of water by evaporation will not effect the potential of this highly reversible reaction. Thus, its utility as a reference electrode is enhanced. Another popular reference electrode system is the silver-silver chloride couple (details are found in Chap. 3).

The reference electrode is usually enclosed in a slender glass cylinder whose tip is a fine bore capillary, or has an asbestos fiber or fritted glass sealed to it. This prevents its electrolyte from leaking out rapidly. In use, the reference electrode is placed as close as possible to the working electrode to prevent any iR drop between the reference and working electrode from influencing the measurement.

The general considerations concerning the selection and limitations of reference electrodes with which the reader should be familiar are as follows:

2.8.3.1. Liquid Junction Potential

It is not generally possible to have the medium in the reference electrode the same as that in the electrolysis cell. This creates a liquid-junction potential of usually indeterminant magnitude at the tip of the reference electrode. Unfortunately, the effect is greatest between aqueous solutions (normally used in commercially available reference electrodes) and nonaqueous solutions, and amounts to several tens of millivolts. For precision work, it will usually be necessary to make up special reference electrodes for organic media, but it will then be generally invalid to precisely compare potentials measured in different media. If possible, the electrolyte should also be common to both reference and electrolysis solutions.

2.8.3.2. Internal Resistance

The internal resistance of most commercially available reference electrodes will be from several hundred to several thousand ohms, usually caused by the constriction at the tip. To prevent this resistance from causing a voltage drop which would influence the measurement, a high-impedence voltmeter, such as a vacuum-tube voltmeter should always be used in measuring electrode potentials.

2.8.3.3. External Resistance

When measuring an electrode potential during electrolysis, there will generally be an iR drop between the tip of the reference electrode and the working electrode. The measured potential for cathodes will be more negative, and, for anodes, more positive than the actual electrode potential. This effect is minimized if the conductivity of the medium is high and if the tip of the reference electrode is close to the working electrode.

2.8.3.4. Leakage

Since there must be ionic contact between the electrolyte of the reference electrode and the electrolyte of the electrolysis medium, there exists the possibility of physical leakage between them. This should be kept in mind, particularly when aqueous reference electrodes are used with organic electrolysis media, since water is often an undesirable contaminant.

2.8.3.5. Temperature Coefficient

The potential of the reference electrode will be a slight function of temperature, but this variation is usually negligible for the requirements of electrolytic work at moderate temperatures. The temperature dependence of the most commonly employed aqueous reference electrodes can be found in most chemical handbooks.

2.8.3.6. Shielding Effect

The presence of the reference electrode will affect
the current flow in its vicinity. The insulating sleeve
casts a "shadow," so to speak, on the working electrode so
that the local current density, and hence, electrode potential, will be reduced. Although the effect is probably
small (the authors have not seen any reports on the effect
of this phenomenon in the organic electrochemical literature),
it will be good practice to use reference electrodes with
as slender a tip as practical.

2.8.4. AGITATION

Because electrolysis is a heterogeneous reaction, mass
transport of material toward and away from the electrode is
an important consideration. In most cases it will be desirable to agitate the solution in order to speed up mass
transport. Although stirring the bulk of the solution is
the most common and convenient practice, it is not the most
effective, and a definite stationary boundary layer will
persist. The problem is further complicated in a well designed cell where the electrodes are close together, since
not much room will be left for mechanical agitators. An
alternative method would be to circulate the electrolyte
past the electrodes by means of an external pump. More
effective stirring can be obtained by moving the electrode
itself, such as with rotating or vibrating electrodes. With
porous electrodes, efficient breakup of the stationary
boundary layer may be possible by passing the electrolyte
through it or by pumping an inert gas through.

In some cases it may be desirable to prevent disruption
of the depletion region in order to take advantage of the
high-concentration gradients at the electrode surface. For
example, if a desired reaction involves a second-order process between reactants, the higher concentrations near the

electrode may result in better yield of the product. Also,
in aqueous solution the high pH gradient possible at the
electrode may be desirable, as in some cases of electrodeposition, where a soluble species is precipitated by the
change in pH near the electrode (always basic for cathodes,
acidic for anodes).

These effects should be kept in mind in interpreting
the results of electrolysis, particularly at high current
densities. The effect of agitation should also be considered in interpreting the results of voltammetry (usually
conducted in quiet solutions) in comparison with the results of electrolysis (usually in stirred solutions).

2.8.5. ELECTRODE POTENTIAL, CURRENT DENSITY, AND CELL VOLTAGE

Without doubt, the electrode potential is the most important electrolysis variable, since it essentially controls
the type of reaction and its rate. In most cases it will
be desirable prior to electrolysis to ascertain the over-
all voltammetric behavior of the starting material by an
electroanalytical technique such as those described in Sec.
3.4. Many compounds display relatively simple behavior
where only a single electron-transfer reaction is possible,
so that electrode potential is important only in governing
the rate of the reaction. In such a system, it may make no
difference whether the reaction is carried out at controlled
potential, current density, or cell voltage. However, it
may still be advantageous to study the reaction at different
electrode potentials, since factors other than electron
transfer may be affected, such as adsorption, orientation
of adsorbed species, or competition between subsequent
second- and first-order reactions.

When more than one electron-transfer reaction is possible, proper control of electrode potential is crucial and
is the basis for the high selectivity of the electrolysis

process. (Controlled potential electrolysis methodology is
discussed in detail in Sec. 2.9.) At constant electrode
potential, the current will gradually decrease as the reaction nears completion, so that calculation of the number
of coulombs passed is more difficult than in the case when
constant current operation is used. If a coulometer is un-
available, the number of coulombs should be obtained by
taking readings of the current at appropriate time intervals and integrating the resulting i-t curve, since the
total number of coulombs is given by the following equation

$$Q_{tot} = \int_{o}^{t} idt \qquad\qquad (2.45)$$

Proper analysis of the products (100% mass balance) will
then allow the calculation of the electrical efficiency
for each product. An appropriate goal for electroorganic
synthesis is to achieve 100% electrical efficiency for the
desired product by proper control of the variables. (Note
that chemical efficiency, i.e., yield, is always greater
than or equal to electrical efficiency.)

   After the electrolysis has been investigated by controlled potential conditions, it may then be possible to
conduct further experiments at constant current or constant
cell voltage, a somehwat more convenient situation. Under
these conditions it may be necessary to use more than the
calculated amount of coulombs to afford completion of the
reaction, since it will not be possible to achieve 100%
electrical efficiency. As the starting material is consumed,
the electrode potential will increase until a subsequent
reaction (hopefully innocuous) occurs and is able to support the current. At constant cell voltage this effect is
somewhat alleviated, since the current will drop somewhat
as the electrode potential increases.

Only under continuous conditions is the use of constant current or cell voltage strictly appropriate. Here,
the composition of the medium remains constant once steady
state conditions are achieved, so that the electrical operating conditions (electrode potential, current density, and
cell voltage) also remain unchanged with time, unless the
electrode or cell divider deteriorate. Continuous operation
is recommended when scale-up is being considered, or when
long-term effects of materials of construction are to be
investigated.

2.8.6.   SOLVENT AND ELECTROLYTE

The solvent chosen for the electrolysis medium, in
addition to the requirement of dissolving the starting
material (but not necessarily the product), must fulfill
several other restraints.

1.   It must not only dissolve the electrolyte, but
must also have a sufficiently high dielectric constant to
ionize it. Thus, only relatively highly polar liquids
have found general utility, such as dimethylformamide,
dimethylsulfoxide, dioxane, ethanol, acetonitrile, and,
of course, water.

2.   The solvent should itself be relatively stable
toward oxidation and reduction, at least over the region
of potential of interest. Another important point to remember is that in many cases of organic electrolysis,
relatively reactive species such as radicals, carbanions,
and carbonium ions are produced as intermediates. Their
reactivity toward the solvent should therefore be considered.

3.   In some cases the solvent may influence the course
of the reaction, such as the dimerization of radical anions
vs proton abstraction and subsequent further reduction.

4.   Techniques for removing traces of water from the
solvent should also be considered, since small amounts of

water can serve as a proton donor and as the source of oxygen in the formation of oxide films on the anode.

   The supporting electrolyte is subject to much the same
considerations with regard to stability toward electrolysis
and chemical inertness. The prime function of the elctro-
lyte, of course, is to provide the source of ions to conduct
current across the cell. In general, electrolytic media of
high conductivity are desirable since they prevent internal
heating due to i^2R power dissipation and permit relatively
error-free control of electrode potential. High concentrations of salt-like material are not sufficient criteria to
insure high conductivity, since many electrolytes have a
tendency to form ion pairs at high concentrations in organic
media. This is particularly true of alkali metal halides.
A few simple measurements with a conductivity bridge (see
Sec. 3.3.3.) should serve to locate the maximum conductivity.
The ability of certain electrolytes, such as the tetraalkylammonium salts, to adsorb on the electrode and to influence
the double layer structure can play a key role in determining
the course of the reaction. Techniques for the separation
of the electrolyte and solvent from the products should, of
course, always be considered. Extraction and distillation
have proven to be of great utility in this regard.

   In some cases the electrolyte may be chosen to serve
as a buffer when acids or bases are formed during electrolysis. However, one should always bear in mind in such in-
stances that the pH at the electrode surface may be appreciably different than that in the bulk because of concentration gradients. The electrolyte may also be chosen to serve
as a reactant, such as in anodic substitution (see Sec.
5.2.1), where the anion of the electrolyte reacts with an
electrolytically generated carbonium ion. The Kolbe
reaction of carboxylate salts (Sec. 5.1) is an example of
the dual use of material as electrolyte and starting material.

2.8.7. ADDITIVES

The early electrochemical literature is full of examples where certain additives have been utilized to increase
yield and efficiency. For the most part, these so-called
oxygen and hydrogen carriers have been used in aqueous media
where their role is fairly clear. Their action in these
cases results from a mechanism which the authors choose to
refer to as electroregeneration. These materials are actual
or potential oxidizing or reducing agents, where the following mechanism may be applicable:

$$O + S \xrightarrow{\text{fast}} P + R \qquad\qquad \text{chemical} \qquad\qquad (2.46)$$

$$R \xrightarrow{\text{fast}} O + e \qquad\qquad \text{electrochemical} \qquad (2.47)$$

where S and P are starting material and product, respectively, and O and R are oxidized and reduced form of the
additive, respectively.
(This has been written as an oxidation--an equally likely
mechanism for reduction in the presence of a suitable additive is also possible.) This scheme is reminiscent of catalysis, since the rate of product formation may be increased
without the destruction of the additive. Notable examples
of this effect are the use of cerium salts in promoting the
electrochemical oxidation of anthracene to anthraquinone
and the use of chromium salts in the oxidation of toluene
to benzoic acid.

The effects of additives in detail is difficult to
predict, since very little literature has appeared on this
subject as it applies to organic electrochemistry. However,
it may be helpful to bear in mind some further general
principles concerning their effect. Metal ions may reduce
to the free metal on the cathode surface and cause it to
behave accordingly. It may be possible to use surfactants
to displace other adsorbed materials. The potential of

zero charge (Sec. 2.7.3) will determine whether anion or cation adsorption will be favored at a particular electrode potential. Small amounts of water can be used as a proton source for electrolytically generated intermediates, as can other weakly acidic materials such as phenol. Larger amounts of water can improve conductivity in organic media. In some cases the solubility of the substrate in aqueous media may be improved by the use of large amounts of an organic electrolyte such as tetraalkylammonium sulfonates. Conventional buffers can be used to control pH and suppress pH gradients. If reactive intermediates such as radicals, carboniums, and carbonium ions are generated by electrolysis, one can conceive the use of trapping agents (and thereby control the nature of the product) such as radical chain transfer agents, electrophiles, and nucleophiles, respectively, e.g., the use of CO_2 to trap carbanions.

With so many possibilities available, it would normally be wise to gain as much experience with a particular electrolytic reaction under normal conditions as possible before the use of additives is contemplated. With ingenuity, and perhaps some luck, the experimenter may very well be able to direct the reaction more closely toward the desired end by the judicious use of additives.

2.8.8. OTHER VARIABLES

2.8.8.1. Temperature

Normally, temperature should have little effect on the electron-transfer part of an electroorganic reaction. Electrode potentials (including the reference electrode potential) can shift with temperature, but these are not major effects with moderate excursions. More important would be consideration of the effect of temperature on the course of followup chemical reactions and on the stability of intermediates and products. The solubility of the various species in the medium will, of course, also be a function of temperature.

The conductivity of most electrolytes will increase with increasing temperature by 1 to 2%/^{O}C. Therefore, the amount of power dissipated as heat, and hence wasted, will be less at higher temperature. A well designed cell should have provision for temperature control and will normally have the electrodes close together to minimize power losses due to i^2R heating. In certain cases it may be desirable to operate at the reflux temperature of the medium, so that a condenser is used to remove excess heat. Under these conditions, the use of a divided cell is difficult, since boiling is likely to occur within the cell divider and thus block the passage of current.

2.8.8.2. Concentration

Normally, it is desirable to utilize the highest concentrations of starting material and electrolyte within the constraint of solubility. Higher currents, and hence greater rate of production of product, can usually be attained. However, several other factors should also be considered in assessing the effect of concentration on a particular reaction. In organic electrochemical reactions, the first step is often the formation of a relatively highly reactive, and perhaps unstable, intermediate such as a radical or ion. As such, this intermediate can decompose, react with solvent, or condense with another species. Decomposition or reaction with solvent is expected to follow first-order kinetics. Condensation, on the other hand, is usually second order, and will be relatively more effective at higher concentrations. Thus, the distribution of products from parallel reactions of this type may be expected to depend on concentration. In the same sense, the reaction may also be sensitive to current density since, at the higher current, a greater concentration of the intermediate will exist in the vicinity of the electrode. If the following reactions are fast enough to occur within the diffusion layer condensation reactions may be favored by high current density.

In that case, the rate of stirring may also be important.

Since the results of voltammetric experiments are usually obtained under conditions of relatively low concentration (10^{-2} to 10^{-3} $\underline{M}$), one cannot always assume that the same effects will pertain to the case of electrolysis at much higher concentration. A voltammetric study of the effect of as wide a variation in concentration as possible is thus recommended as a matter of standard practice.

2.8.8.3. Time

Once the reaction conditions are set, i.e., amounts of materials and current density (or electrode potential), the time of electrolysis is not usually independently controllable if the reaction is to be carried to completion. However, there are several factors to consider such that time may be a more desirable variable to control at the expense of another parameter. For example, the desired product may itself be unstable in the electrolysis medium. In that case, the possibility of continuous removal of product should also be considered. Another case where time may be considered as an important variable is the effect of undesirable diffusion of material between compartments of a divided cell. Since the composition within the separate chambers will change during electrolysis, there will be a tendency of the various products to interdiffuse through the barrier due to osmotic pressure. Since ideal membranes (which would permit passage of only the electrolyte ions) have not yet been found, the amount of interdiffusion will depend on the diffusion coefficient of the various materials within the divider, the concentration gradient, and, of course, time. In cases where this effect is critical, a three-compartment cell might be used, where the contents of the middle chamber are periodically or continuously replaced with fresh electrolyte.

Finally, after giving due consideration to the other factors, the experimenter's time may itself be a variable.

How often, during the course of a conventional reaction, have we wished to terminate quickly in order to attend to other matters, but been bound by the nature of the reaction to await its completion? Where but in electrochemistry can this be done as effectively as by the simple flick of a switch?

2.9. CONTROLLED POTENTIAL ELECTROLYSIS

    Up to this point we have emphasized the importance of the electrode potential, but we have not as yet discussed how to control it. The simplest, but least convenient, method is simply to measure it periodically by use of a reference electrode and suitable voltmeter, and adjust either cell voltage or current to bring the potential to the desired value. In general, increased cell voltage or current will make a cathode potential more cathodic (more negative), and conversely, an anode potential more anodic (more positive). However, since we can measure it and can predict the direction of its response, we ought to be able to control it automatically.

    That this can be done very simply in principle is best understood by referring to Fig. 2.9.

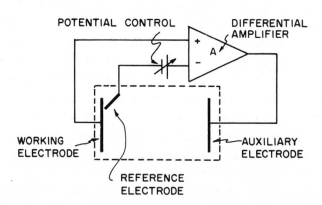

FIG. 2.9. Controlled potential electrolysis.

The differential amplifier A (commonly referred to as an operational amplifier has the property of greatly amplifying any difference in potential between its input terminals (marked + and - on the diagram) without drawing appreciable current. High-performance amplifiers having voltage gains greater than 10^5 (some as high as 10^9) and input currents less than 10^{-6} A (some less than 10^{-12} A) are readily available. Therefore, even very small potential differences, small fractions of a millivolt, will cause its output voltage and/or current to be driven to its maximum values, positive or negative depending on the sense of the input potential imbalance. To establish control, a feedback connection from its output to its negative input is required. From Fig. 2.9, it can be seen that this is accomplished through the electrolysis medium and the reference electrode in series with the potential control.

To clarify the understanding of its operation, let us consider a particular example. Suppose it is desired to control the working electrode potential at -2 V with respect to the reference. In other words, the potential of the reference electrode is to be 2 V above the potential of the working electrode. If the potential control is set to -2 V (the direction opposing the potential of the reference), the amplifier output will increase in voltage (and current) until the working electrode potential reaches -2 V. If the electrode potential changes slightly in the positive direction, the potential difference created at the input terminal will cause the amplifier output to increase until balance is restored; the converse is true if the electrode potential changes in the negative direction. Thus, the amplifier strives to maintain its input terminals in balance, a condition which is maintained only when the electrode potential is at its preset value of -2 V.

Once these principles are clearly understood, many other modes of operation using operational amplifiers can be set up. These are useful in various electroanalytical techniques

to be discussed in Sec. 3.4 (polarography, linear sweep
voltammetry, chronopotentiometry, etc.). At this point let
us consider only those factors important in controlled
potential electrolysis power supplies (called potentiostats).
First of all, the anticipated current and cell voltage required by the electrolysis must fall within the range of
compliance of the amplifier. Also, the frequency response
of the amplifier should be sufficient to follow transient
phenomena such as might be created by turbulent stirring in
the cells. The input current should be a fraction of a
microamp to avoid loading the reference electrode, and the
offset voltage between inputs should be (or be adjustable
to) a fraction of a millivolt. Other points to consider
are temperature coefficient of offset voltage, overload
protection, and recovery time. The utility of operational
amplifier techniques in electrochemistry has been reviewed
by Schwarz and Shain (3).

The actual potential under control is that which exists
between the tip of the reference electrode and the surface
of the working electrode. This potential includes both the
electrode potential and the iR drop between the reference
and working electrodes. For precise electrolytic work, this
iR drop should be made as small as possible or compensated
for. Although electronic compensation (operational amplifier style) is quite feasible, the usual practice is to
minimize it by experimental design. This can be done by
minimizing the distance between the tip of the reference
electrode and the working electrode--the use of a Luggin
capillary should be considered, which can be simply an extension of the normal tip of the reference electrode with
an insulating sleeve, filled with electrolyte, and tapering
to a fine capillary tip. Of course, the higher the conductivity of the electrolyte, the less will be the effect of
uncompensated iR drop. For uncritical cases, the iR drop
can usually be ignored, since the direction is such that
it makes cathodic potentials less cathodic and anodic poten-

tials less anodic than indicated by the set potential. Thus,
the actual potential is always "safe," since controlled
potential electrolyses are normally conducted at the highest
feasible potential to obtain rapid reaction.

     The potential at which the electrolysis is to be run
is chosen on the basis of literature values, from experience
with previous experiments, or from results of voltammetric
measurements. During electrolysis at controlled potential,
both the cell voltage and current decrease with time as the
reaction nears completion. If the reaction is 100% electrically efficient, the current will have decayed to essentially zero at the end of the reaction. For simple systems,
either diffusion-limited or first-order kinetically, the
current will follow an exponential decay [see Eq. (2.48)].

$$i = i_o e^{-kt} \tag{2.48}$$

where i_o is initial current; t is time; and k is effective
rate constant.

     For a typical case, k will depend on the diffusion
coefficient, the degree of agitation, and the kinetics of
the electrode process, as can be seen from the previous
discussions on diffusion and electrode kinetics.

     In any case, the total amount of coulombs passed can
be measured either graphically from an i vs t plot, or by
use of a coulometer (see Sec. 3.3). The electrical efficiency of the reaction can then be calculated (Sec. 2.1),
or if the reaction is 100% efficient, the over-all number
of electrons transferred, n, can be obtained.

$$n = \frac{QM}{96,500 \times w} \tag{2.49}$$

where Q is total number of coulombs; M is molecular weight
of starting material; and W is weight in grams of starting
material.

Keep in mind that n here may not necessarily be the same as the number of electrons transferred in the primary (or rate-determining) electron-transfer step.

The advantages of controlled potential electrolysis are considerable, and numerous examples of its utility will be seen in subsequent chapters. Through its proper use, undesired side reactions may be eliminated, specific functional groups may be electrolyzed in the presence of other electro-active groups, or multi-step electrolyses may be controlled to produce an intermediate product. In all of these cases it may not be possible to obtain comparable results by the use of conventional oxidants and reductants. Herein lies one of the greatest advantages of electrochemistry; controlled potential electrolysis may be considered a readily available "store" of a vast number of different oxidizing and reducing agents.

REFERENCES

1. H. H. Bauer, J. Electroanal. Chem., 16, 419 (1968).
2. J. Tafel, J. Phys. Chem., 50, 641 (1905).
3. W. M. Schwarz and I. Shain, Anal. Chem., 12, 1770 (1963).

ADVANCED TEXTS ON ELECTROCHEMICAL THEORY

a. J. O'M. Bockris and B. E. Conway, Modern Aspects of Electrochemistry, Vols. 1-6, Plenum, New York, 1954 to present.

b. P. Delahay and C. Tobias, Advances in Electrochemistry and Electrochemical Engineering, Vols. 1-6, Interscience, New York, 1961.

c. J. O'M. Bockris and A. Reddy, Modern Electrochemistry, Vols. 1 and 2, Plenum, New York, 1970.

d. K. J. Vetter, Electrochemical Kinetics, Academic, New York, 1967.

e. H. Eyring, (ed.), Physical Chemistry: An Advanced Treatise Vol. IXA, "Electrochemistry," Academic, New York, 1970.

f. A. J. Bard, (ed.), <u>Electroanalytical Chemistry</u>, Vols.
    1-3, Dekker, New York, 1969.

g. P. Delahay, <u>New Instrumental Methods in Electrochemistry</u>,
    Interscience, New York, 1954.

3

APPARATUS AND TECHNIQUES

3.1. THE ELECTROLYSIS CELL

3.1.1. CELL CONSTRUCTION

The design of an electrolysis cell, operating within
the constraints mentioned earlier, can be extremely simple
or quite complex, depending on the requirements of the
actual system under investigation. Although quite a few
cell designs can be constructed from ordinary glassware,
the experimenter must use ·quite a bit of ingenuity, since
completed cells suitable for organic electrolysis are
rarely available commercially.

Perhaps the simplest type is the undivided cell using
a glass beaker as the main element. With solid electrodes,
the anode and cathode may be simply draped over opposite
sides of the beaker as illustrated in Fig. 3.1. Electrical connections can be conveniently made with alligator
clips, using care not to allow electrolyte to touch them.
A magnetic stirrer can be used for agitation.

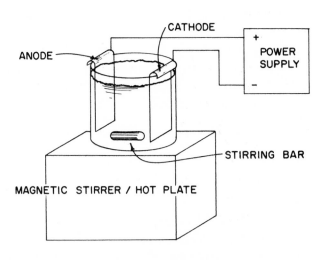

FIG. 3.1. Beaker cell.

Additional improvements could be made by utilizing a
jacketed beaker for cooling the cell (or a hot plate to
raise the temperature), a thermometer, and a reference
electrode. More efficient use of the working electrode
area could be attained by placing it in a central location,
flanked by two auxiliary electrodes with common electrical
connections (see Fig. 3.2).

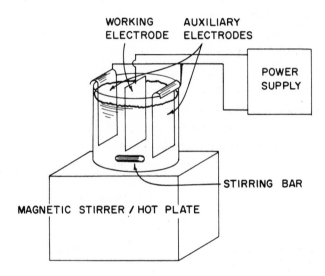

FIG. 3.2. Beaker cell with central working electrode.

A somewhat more favorable arrangement would be the use of
cylindrical electrodes.

The beaker cell can be utilized as a divided cell by
surrounding the appropriate electrode with a suitable divider material, such as a porous ceramic cup or a cellophane
membrane.

The use of a mercury pool electrode is also convenient
in a beaker cell. In this case, the magnetic stirrer will
float on the surface and provides efficient stirring of
the interface. Electrical connection to the pool may be
accomplished either by sealing a platinum wire into the

bottom of the beaker, or by inserting an <u>insulated</u> platinum wire below the surface of the pool.

Several other "glassware" electrolysis cells have been described by Allen (<u>1</u>), and of particular interest are the resin kettle cell (see Fig. 3.3) and the cell constructed of Pyrex process pipe (see Fig. 3.4). These have the advantage of being constructed from readily available materials, and have the capacity to handle moderate (100-500 g) amounts of materials. They can be disassembled readily for cleaning and have sufficient flexibility to be used for most organic electrolyses.

For microscale (<1 g) amounts, the divided polarographic H-cell is useful (see Fig. 3.5), and is readily available from commercial sources--Sargent, Beckman, etc. It usually has provision for a mercury pool electrode (platinum insert) and the electrolyte can be conveniently stirred by bubbling nitrogen through the side-arm provided for that purpose.

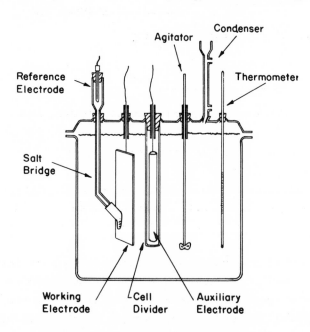

FIG. 3.3.   Resin kettle cell.

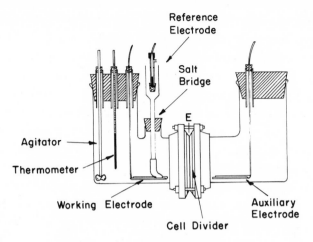

FIG. 3.4.   Process pipe cell.

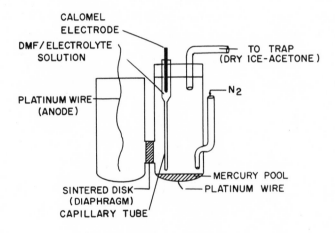

FIG. 3.5.   H-cell.

The authors have used larger H-cells, made by a commercial glass blower, with stopcocks to provide convenient draining of the electrolyte solutions. Where a relatively volatile product is formed, it may be easily recovered in a cell of this type by providing a stopper with a glass tube leading to a condenser and receiver.

For specific electrolyses, it may be convenient to purchase, or have made, beakers constructed of the electrode material. These are available in stainless steel, glassy carbon, graphite, and platinum (usually small).

Although the above cells have the advantage of being conveniently constructed of readily available parts, their disadvantages are such that they should be used only in relatively simple cases where careful control of variables is not essential. Basically, they have low surface-to-volume ratios (long electrolysis times result), their electrode geometry is difficult to maintain accurately, and their heat transfer is slow--all of which militate against their use in critical electrolyses, where potential control must be tight, or where relatively unstable materials are formed. Also, the design offers little possibility for scale-up to, let's say, kilogram amounts of materials.

A novel cell design which avoids the above disadvantages while retaining most of the conveniences has been used

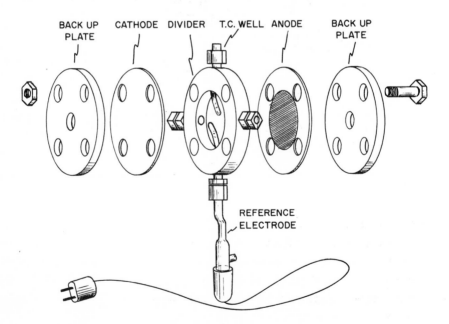

FIG. 3.6. Sketch of basic sandwich cell.

by the authors in a wide variety of electrolyses. For
reasons which become apparent below, we choose to call
this the "sandwich cell" (see Fig. 3.6).

The basic construction, shown in Fig. 3.6 is comprised
of back-up plates of steel, precut anode and cathode, and
insulating cell spacer. The back-up plates have central
threaded holes to provide a means for electrical connections.
The entire assembly, with gaskets (not shown) between electrode and spacer, is bolted together with insulating (nylon)
bolts and nuts. The central cell spacer has ports for electrolyte inlet and outlet, reference electrode, and thermometer or thermocouple well. If the spacer is ∿5/8 in. thick,
standard NPT (national pipe threads) fittings can be used,
which are available in polypropylene, nylon, and Teflon
(most labware suppliers). In addition to NPT connector
fittings, unions, tees, elbows, and step-up or step-down
adaptors are also available. Glass or hard-walled plastic
tubing (¼ in. o.d.) of Teflon or polypropylene should be
used, although inserts are available to permit the use of
flexible tubing in the fittings.

In a particular electrolysis using an aqueous electrolyte and where an undivided cell was sufficient, the system
(see Fig. 3.7) was constructed around the basic sandwich
cell.

The elements within the system shown in Fig. 3.7 were:
(1) Undivided sandwich cell with thermocouple; (2) Reference
electrode with tip bent to allow positioning at anode or cathode by rotation; (3) Circulation pump, peristaltic type;
(4) Heat exchanger made of glass condenser; (5) Gas-liquid
separator with overflow port and gas outlet; (6) Bypass filter assembly with glass wool filter; and (7) Charge tank,
feed pump, and flowmeter.

The critical dimensions were: (1) effective electrode
area, 25 cm^2; (2) effective electrode diameter, 5.7 cm;
(3) spacer thickness, 1.6 cm; and (4) total hold-up volume
of entire system, 100 ml.

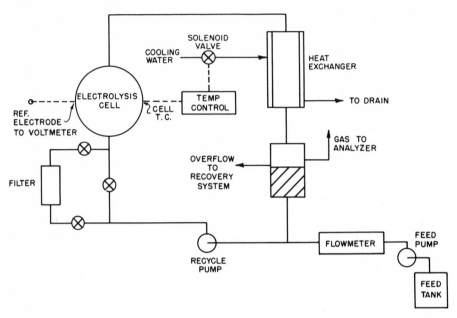

FIG. 3.7. Electrolysis assembly.

This particular system, with feed pump and overflow, was designed to operate continuously at constant current and temperature. In the continuous mode, the cell is a differential reactor (discussed in more detail in Sec. 3.6 of this chapter concerning scale-up) where conversion per pass is controlled by feed rate and total current.

The concept of the sandwich cell has been extended by the authors to encompass several other operating modes. Divided cell operation is conducted by providing two spacers between which is inserted an appropriate membrane material (gasket may be necessary). The circulation and heat exchanging system must, of course, be duplicated (two chambers). For electrolyses involving a gaseous starting material, a porous electrode material was used so that the gas could be passed from the back through the back-up plate. Internal pressures up to a few atmospheres should be attainable with reasonable care.

Although somewhat more elaborate than would usually be required by a casual user of electroorganic reactions, the sandwich cell offers the following advantages which are important to the serious researcher:

1. High surface-to-volume ratio (improves with scale-up).

2. Well-defined electrode geometry.

3. Continuous operation is feasible.

4. Can be scaled-up to much larger dimensions.

5. Adaptable to either divided or undivided operation.

6. Can easily operate in inert atmosphere.

7. Offers provision for efficient heat removal.

8. Complete flexibility for many different systems.

9. Elements can be made with standard machine shop tools--drill press, lathe, etc.

10. Fittings are available commercially in variety of types.

It is also conceivable that several elements, defining multiple chambers and manifold inlets and outlets, could be stacked together in an arrangement which might be termed the "club sandwich cell."

The reader should be aware that cell designs are practically as numerous as there are authors in electro-organic chemistry, and that consultation with original articles in the literature will lead to many practical examples. The experimenter is urged to use his own ingenuity in designing his own cells or in modifying existing designs.

3.1.2. CELL DIVIDERS

Most electroorganic reactions, for reasons previously discussed, require the use of separate chambers for anolyte and catholyte. As such, a suitable barrier material, called the cell divider, must be selected according to the following requirements:

1. It must be permeable to ions, preferably impermeable to other species.

2. It must be stable to the electrolytic medium at the temperature of the electrolysis.

3. It must be continuous (i.e., pinhole free) and mechanically strong enough to withstand any pressure differences encountered.

The most frequently used cell dividers are made of refractory materials, such as unglazed ceramics or fritted glass of various porosities. The ceramic type is commercially available in the forms of thimbles, to be used mainly with cylindrical electrodes. Also available are porous ceramic and fritted glass disks and sheets, which can be incorporated into appropriate cells with planar electrodes. These materials are convenient to use because their rigidity permits relatively easy placement within the cell. They are exceptionally inert to most electrolytic conditions and may be rigorously cleaned with oxidizing acids and by flaming. The manufacturers of porous refractory materials for use as dividers include Coors Ceramics and Corning Glass Company.

The ceramic and fritted glass types have excellent chemical resistance, but unfortunately they have little if any selectivity with respect to diffusion, i.e., solvent, starting material, products, and electrolyte can diffuse through equally well. Where this is critically to be avoided, the use of fine porosities should be considered, at the expense of higher iR drop. If the resistance is too high, passage of moderate current densities ($\sim$2-10 A/dm^2) can lead to such high local temperatures as to cause the solvent to boil within the pores of the divider. Conducting the reaction at constant current can then cause a runaway condition, and the cell voltage will rise to its maximum value.

Another peculiarity of operation with ceramic dividers is a phenomenon which the authors have observed many times but have been unable to find suitable documentation for in the literature. This occurs when high current densities (greater than $\sim$5 A/dm^2) are employed. Initially, the levels of electrolyte are made equal in the two chambers, but during the electrolysis, the level of one compartment rises at the expense of the other. A likely explanation is the unequal transport of solvent caused by the migration of the solvated ions through the cell divider. In general, the degree of solvation of anions and cations will not be equal, and the mobility of these ions will also be different. The process would continue until the hydrostatic pressure causes suffi-cient back transport of solvent to maintain the system at equilibrium. (Another possible explanation, when the catho-lyte and anolyte are different, is that the unequal trans-port is caused by osmotic pressure differences.) The effect cannot solely be due to osmotic pressure, since the phenome-non occurs even when the same concentrations of solvent and electrolyte are employed. In any case, one should avoid the use of appreciably different solvents in divided cell work, particularly if one of them interferes with the main process.

Another material which is commonly used in cell dividers is cellulose acetate or "cellophane." This material has been used extensively in dialysis and in osmometry. It has the advantages of being readily available and inexpensive enough to be discarded after each use. It can be obtained as sausage casing material in a wide variety of tubular sizes, which can be used as such (one end closed) to surround the electrode, or can be cut up to be used in sheet form. In the "sandwich-type" cell it should be supported between relatively rigid but open screens (of insulating material). When completely dry, cellophane is brittle and fragile. As supplied, it is plasticized with glycerin, which should be removed by soaking in distilled water overnight. For best results in organic media, cellophane should be "conditioned"

by successively soaking in a series of solvents, beginning
with water and then with mixed solvents, ending with the
pure solvent to be employed. Such precautions are unessen-
tial for most work, however, and the authors have successfully used cellophane directly, without conditioning, in
many cases.

Cellophane membranes offer excellent barrier properties
toward typical organic depolarizers and little resistance
to the passage of current. It is excellent for use in
aqueous media, particularly when neutral or acidic. In
basic media, and at elevated temperatures, the membrane
will deteriorate. Its suitability for use in organic media
should be ascertained experimentally.

Another type of material which has found utility as
membranes for use in electrolytic work is the ion-exchange
type. These have the unique property of allowing passage
of only one type of ion, either cations or anions, depending on the type of material. The cation exchange resins
are commonly sulfonated polystyrenes, such that the anionic
part (sulfonate anion attached to polystyrene) is immobile
and therefore only cations can readily exchange and be trans-
ported. The reverse is true for anion exchange materials,
which are usually made by attaching quaternary ammonium
groups to the polymer backbone. These materials should
find extensive use in organic electrolyses, particularly
when excessive pH gradients between compartments are to be
avoided. Thus, if acid (hydronium ion) is formed at the
anode, the use of a cation-exchange membrane will efficiently
transport the excess to the cathode chamber under the driving
force of the electric current. The anion-exchange materials
could be used in the analogous situation where hydroxide ion
is formed in the cathode chamber.

The reader should be aware that the "ideal" membrane
material has not been found as yet, and the choice of a
suitable one will depend on the particular situation, usually
involving a compromise between high conductivity and chemical

resistance on the one hand, and barrier properties on the
other. Also, one should consider the possibility of using
nondivided cells, once the reaction products and their possible side reactions are considered.

3.1.3. ELECTRODE MATERIALS AND PRETREATMENTS

The nature of the electrode used for electrolysis is
certainly a key variable. To the extent that adsorption
and double-layer effects may play a role in a given system,
the electrode may have a profound influence on the system,
and could, in fact, change the entire nature of the product.
Choice of a suitable electrode will be restricted by its
stability under electrolysis conditions. The useful potential range for a particular electrode depends on the electrode and on the medium. In aqueous media, the reduction
of water to hydrogen usually sets the cathodic limit.
However, the kinetics of the reduction of water (more exactly, hydronium ion) are very sensitive to the nature of
the electrode, and the term "hydrogen overvoltage" is used
as a measure of the difference between the reversible hydrogen potential and the potential at which hydrogen evolution
proceeds at a given rate. Conversely, the "oxygen overvoltage" of a metal refers to its ability to oxidize water to
oxygen. Table 3.1 gives an indication of the range of this
effect.

In aqueous media, the choice of cathodes can be used
to set the cathodic limit for a reaction. Thus, the use
of low overvoltage cathodes, such as nickel or platinum,
for reduction of nitro compounds (see Sec. 4.2) leads to
intermediate products, whereas high overvoltage metals,
such as lead or mercury, yield aniline. In nonaqueous media
the cathodic potential limit may be set by the electrolyte
or solvent. Although there is no a priori reason to expect
a correlation between aqueous and nonaqueous media, it would
be good practice to select both a high and low "hydrogen

TABLE 3.1

Hydrogen and Oxygen Overvoltage for Various Metals[a]

	H_2 overvoltage (1 $\underline{N}$ H_2SO_4)	O_2 overvoltage (1 $\underline{N}$ KOH)
Palladium	0.00	0.43
Gold	0.02	0.53
Iron	0.08	0.25
Smooth platinum	0.09	0.45
Silver	0.15	0.41
Nickel	0.21	0.06
Copper	0.23	--
Cadmium	0.48	0.43
Tin	0.53	--
Lead	0.64	0.31
Zinc	0.70	--
Mercury	0.78	--

[a]Adapted from M. J. Allen, Organic Electrode Processes, Rheinhold, New York, 1958.

overvoltage" metal for use as cathodes in nonaqueous media; differences in product identity or distribution are the key features to look for. Most metals will be potential candidates for use as cathodes since they have no tendency to be reduced themselves. Cathodic limits are usually set by reduction of the solvent or supporting electrolyte.

The same considerations also apply to the selection of anode materials, except that the number of choices is much more restricted. Most metals themselves are easily oxidized, and only the noble metals, such as platinum and gold, can be said to be generally useful as anodes. Carbon and graphite have also been used extensively as anodes.

Mercury

By far the most widely used cathode material has been mercury, and for good reasons. The use of a liquid as the

electrode has the advantage of uniformity and smoothness
of the surface. The electrode/solution interface can be
effectively agitated and many simple cells are suitable for
use with mercury cathodes. The metal itself has an extremely
high hydrogen overvoltage, which makes it invaluable for use
in polarography (see Sec. 3.4). A great deal of voltammetric
data for reduction of organic compounds is available in the
polarographic literature.

If other electrode configurations are desired, mercury
can be used to wet, or amalgamate, the surface of many metals,
such as lead, nickel, gold, silver, and platinum. The surface can be easily renewed after each use. In this particular case, the wetting with mercury may involve the dissolu-
tion of the base metal, so that it may be more appropriate
to consider it as an alloy electrode.

Mercury has little use in anodic work since it oxidizes
readily, particularly in the presence of halide ion or other
ions which tend to complex with mercury.

It should be pointed out that mercury is extremely toxic
and cumulative in the body, so that appropriate precautions
should be taken to avoid spillage and to clean up completely
in case spillage does occur.

Platinum

Because of its inertness to most chemical environments,
platinum has been employed extensively as anode material
for electroorganic reactions. However, it is not as "inert"
as most might think. At strongly anodic potentials or with
strong oxidizing agents, there a tendency to form a
platinum oxide film, which frequently is observable on
voltammetric curves. On the other hand, at cathodic potentials or with strong reducing agents, hydrogen will be
adsorbed on the surface. Many examples of electrolytic
hydrogenation can be found in the literature, and these
probably are similar, if not exactly analogous, to catalytic hydrogenation, with hydrogen being generated electro-

lytically from water. In neither case do these effects
prevent the use of platinum for electrolysis except that
they should be taken into account when considering mechanisms or when examining results of voltammetry.

The relatively high cost of platinum ($\sim$\$140/oz) deters
the use of the bulk metal in large-scale applications, but
this is often obviated by "platinization" (deposition of
finely divided platinum) on a base metal. When formed
properly, the platinized electrode has an extremely high
surface area and is sufficiently rugged to withstand moderate
handling. In fact, bright platinum itself is often subjected
to platinization to increase its surface area and catalytic
activity.

Directions for Platinization*

A platinum electrode is platinized, i.e., covered with
a thin layer of very finely divided metallic platinum, by
the following procedure. A platinizing solution is prepared
by dissolving 1 g of platinum in aqua regia, evaporating to
near dryness, adding about 25 ml of concentrated hydrochloric
acid and repeating the evaporation (to remove nitrate), and
finally diluting to 100 ml with water. A trace (0.1 mg) of
lead acetate added to the platinizing solution increases the
rate at which the platinum black will be deposited. The
thoroughly cleaned electrode to be platinized is now immersed
in this solution and connected to the negative terminal of
a 3-V dc power supply; a piece of scrap platinum wire attached to the positive terminal is used as the anode. The
current is allowed to flow for about 10 sec; then the electrode is removed from the solution and examined. If necessary, the electrolysis is allowed to continue until enough
platinum has been deposited to produce a dull grayish surface. Too thick a coat is harmful rather than beneficial.

---

*From L. Meites and H. C. Thomas, Advanced Analytical
Chemistry, McGraw-Hill, New York, 1958, p. 76.

The platinized electrode should then be used to elec-
trolyze a dilute solution of sulfuric acid, being made
alternately the cathode and the anode at intervals of 15
sec or so for several minutes. Again a piece of scrap
platinum wire is used as the other electrode. This treatment removes traces of chloroplatinate, chlorine, etc.,
which may have been occluded in the platinum black.

Once platinized an electrode should always be stored
in distilled water or in dilute sulfuric acid. The catalytic activity of the coating of "platinum black" disappears
rapidly if the electrode is allowed to dry out.

Pretreatment

It is to be understood that there is no "optimum" pretreatment procedure for platinum (2), and that the major
impetus for applying a pretreatment is that by duplicating
conditions each time, more reproducible results will be obtained.

The platinum electrode is stored in 10 $\underline{M}$ nitric acid.
Prior to use, it is washed thoroughly with distilled water,
then short-circuited against a calomel electrode in the
de-aerated solution for five minutes.

Carbon and Graphite

Both carbon and graphite have been used extensively
as anodes because of their relative resistance to oxidation.
Both are readily available in a variety of forms. Carbon
electrodes of high purity are sold in cylindrical shape for
use as the arc source in spectroscopy. These have found
utility in electrolysis as well as in voltammetry, where
the sides are coated with insulating wax (3). Even pencil
leads have been used (4).

In aqueous media, carbon anodes are slowly oxidized
to carbon dioxide, and it is to be expected that the electrode surface is itself in an oxidized state, probably comprising carbonyl and carboxylic acid groups (5). These

should be kept in mind when considering the adsorption of organics onto the anode surface. A possible example is in the Kolbe electrolysis, where different products are observed for carbon vs platinum anodes (see Sec 5.1).

Both carbon and graphite are porous to some extent, which may be an advantage if high surface area is desired, but porosity may also cause problems due to clogging with insoluble residues. This may be avoided, if desired, by impregnation with molten wax while in vacuo (6) followed by careful machining or sanding to expose a fresh surface.

Carbon paste electrodes have been described in detail by Adams (7). Basically, these are made by mixing graphite powder and mineral oil in about 2:1 ratio until it has the "consistency about like that of peanut butter." The electrode may be used by packing into a shallow insulating cup, followed by smoothing the top surface.

Carbon anodes have long been used in the production of chlorine by electrolysis of brine, but these are being supplanted by platinized titanium, which has a much longer useful lifetime.

Lead and Lead Dioxide

Lead is a very convenient metal to be used in electrolytic work since it is so easily cut and shaped into the desired form. It is available in sheets, bars, and as lead shot. It is very useful in cathodic work in aqueous solution because of its high hydrogen overvoltage. Lead, per se, is not particularly suited for anodic oxidation since it is itself easily oxidized (E^o = -0.126 V vs H_2 electrode). Lead dioxide, on the other hand, is extremely useful as an anode material. It is easily formed as a coating on lead or other base metals and is a good electronic conductor, better, in fact, than lead itself. It is highly stable under anodic conditions, and is, in its own right, a powerful oxidant (E^o = +1.45 V in acid media; +0.247 V in basic media). The latter fact makes it difficult to determine its role in

electrolytic oxidation, since an alternative explanation to
electron transfer from the organic molecule is direct chemical oxidation followed by electrolytic regeneration of lead
dioxide.

$$PbO_2 + organic \longrightarrow product + Pb^{II}$$

$$Pb^{II} \xrightarrow[\text{oxidation}]{\text{electrolytic}} PbO_2$$

Whatever mechanism applies, however, does little to detract
from the utility of PbO_2 in electrolytic oxidation, since
PbO_2 is itself produced electrolytically and is, in fact,
more active when generated, in situ, on the electrode surface.

Lead and lead dioxide form the basis of the most universally used storage battery, as follows:

$$PbO_2 + 2e + 4H^+ + SO_4^{2-} = PbSO_4 + 2H_2O \quad E^O = +1.4558 \text{ V}$$

$$Pb + SO_4^{2-} = PbSO_4 + 2e \quad E^O = -0.3588 \text{ V}$$

The reactions proceed in the direction indicated for discharge and in the reverse direction during charging. Thus,
initially (fully charged) the electrodes are lead and lead
dioxide, and finally (fully discharged) both electrodes are
lead sulfate. A conventional storage battery makes a convenient electrolysis cell for moderate scale work, although
not particularly suited for use as a divided cell. For
undivided cell usage, the top is cut away and the separators
(used as a mechanical barrier only to prevent shorting) may
be removed. A drain at the bottom of each cell will facili-
tate product removal and cleaning. The storage battery consists of several cells in series, so that on filling it is
necessary to avoid bridging the compartments. When used
in this way, high current (100 A) operation is possible.

Lead is also very easily alloyed with many other metals.
For this purpose, the authors have found it convenient simply
to melt together the appropriate elements in a glass crys-
tallizing dish. On cooling, the glass is broken off and a
neat circular electrode is obtained. A small amount (3-6%)
of antimony alloyed with lead makes it much stronger. Mercury amalgamates readily with lead, and in this form takes
on some of the characteristics of mercury but has the convenience of a solid. For use in anodic systems, a small
amount of silver (∿1-2%) has a dramatic effect in improving
the stability of the surface PbO_2 toward flaking away.
Alloys, in general, should not be used indiscriminately
since they can, of course, alter the electrochemical reactions.

Formation of PbO_2 on Lead

The lead is first scoured to expose fresh metal; it
is then made the anode in a cell whose electrolyte is 30%
aqueous sulfuric acid. Any convenient cathode may be used.
A current density of about 1 A/dm^2 is passed until oxygen
evolution is plainly visible and the lead has acquired the
characteristic black deposit of PbO_2. Greater surface area
can be obtained by periodic cathodization and reanodization
The electrode should be removed while current is still flowing to avoid loss by reductive processes. The electrode is
washed with distilled water and used as soon as possible.
It is preferable and convenient to rescour and reanodize
the lead electrode prior to each use.

Formation of PbO_2 on Base Metals

PbO_2 can be conveniently deposited onto nickel,
tantalum, platinum, carbon, or graphite (8). Most other
metals are unsuitable because of their inherently easy
oxidation. The plating solution contains 300 g/liter lead
nitrate, 3 g/liter copper nitrate, and 1 g/liter of a non-

ionic surfactant. The copper nitrate suppresses lead deposition on the cathode, which may be any convenient metal.
The surfactant ensures the formation of a smooth, hard deposit. The base metal is cleaned, before deposition, by
scouring with sandpaper and washing with distilled water.
A current density of 3 A/dm^2 and a temperature of $70^{\circ}C$ are
used. The amount deposited can be calculated, assuming
about 85% electrical efficiency.

The PbO_2 formed in the above two examples is presumably in the β-form. Another type, $\alpha-PbO_2$, is formed in
alkaline media. A suitable deposit of this form may be ob-
tained by anodizing with a solution of 1.5 $\underline{M}$ NH_4OH, 6.5 $\underline{M}$
ammonium oxalate, and saturated lead oxalate, at 0.3 A/dm^2
for 60 min (9).

It should be pointed out that lead is highly toxic to
the human body, and suitable care should be taken in its
use and disposal.

Nickel

Nickel is a useful cathode material, being relatively
inert in acidic media and having a moderately high hydrogen
overvoltage. It may be bright plated onto most metal substrates, and is hard and abrasion resistant. It may be useful as an anode material only in alkaline media (formation
of NiO_2). It is available in sheets, rods, and powder, in
addition to a porous form (used in fuel cells), obtainable
from the Clevite Corporation.

Nickel Plating (10)

The plating bath consists of 400 g/liter nickel sulfate,
50 g/liter nickel chloride, 35 g/liter boric acid, and 0.5
g/liter wetting agent, such as sodium lauryl sulfate. The
plating is conducted at the cathode of an electrolytic cell
operating at 7 A/dm^2 and $50^{\circ}C$. As before, the base substrate
(any convenient metal) is thoroughly cleaned (and polished,
if a mirrorlike finish is desired) and washed prior to use.

Other Materials

      In principle, the number of possible materials suitable
for use in electrolysis is enormous when alloys are also
included. The choice is dictated and restricted by the
nature of the medium and the accessible potential range,
as mentioned above. Potential cathodes are, of course,
much more numerous than anodes. Pretreatments, at least by
mechanical cleaning, are highly desirable to ensure reproducibility. The following representative list of materials
have been useful in electrolytic work, in addition to the
ones described above.

Cathodes

Aluminum	Steel
Monel metal	Silver
Copper	Chromium
Zinc	

Anodes

Manganese (MnO_2 in alkaline solution)
Silver (AgO in alkaline solution)
Iron (in alkaline solution)
Boron carbide
Gold
Tungsten hemipentoxide

      It is hoped that the potential investigator in organic
electrolysis will keep in mind the importance of considering
the electrode material as a key variable. Although the
effect is difficult to predict, the nature of the electrode
can lead to unique results in product or yield. The
tendency in recent years to utilize only mercury or platinum as electrode materials should be avoided if novel and
useful electrode reactions are to be discovered.

3.1.4.   REFERENCE ELECTRODES

Since we have emphasized the importance of measuring and controlling the electrode potential as a key variable in electrolytic systems, we will now turn our attention to an essential element in accomplishing that task, namely the reference electrode. The need for a suitable reference electrode was discussed previously (Sec. 2.8.3). To be suitable, a reference electrode should satisfy all of the following requirements:

1. Its potential should remain constant with time.

2. Small currents passed through it should have little effect on its potential, and the potential should return to its initial value on removal of polarization.

3. It should obey the Nernst equation with respect to the appropriate species.

The universally accepted standard for electrolytic potential measurement is the normal hydrogen electrode (NHE), which consists of a platinized platinum electrode in contact with an electrolyte of hydrogen ions (at unit activity) and in equilibrium with hydrogen gas at one atmosphere. A potential of zero volts <u>at all temperatures</u> has been assigned to this electrode. However, its physical form and general commercial unavailability have made it necessary for most electrolytic workers to adopt several secondary standards for use in aqueous and nonaqueous media.

3.1.4.1. Calomel Electrode

The most popular reference electrode for general aqueous work has been the saturated calomel electrode (SCE). The primary cell reaction may be written as follows:

$$Hg_2 Cl_{2(s)} + 2e = 2Hg + 2Cl^- \tag{3.1}$$

Thus, a mercury electrode in contact with solid Hg_2Cl_2 (calomel) is used, the solution being kept saturated with

excess KCl. The excess KCl maintains saturation even if a
small amount of evaporation occurs, or if the temperature
changes slightly. Thus, the composition is very well defined at all times. At 25°, the SCE has a potential of
+0.2412 V with respect to NHE. Its temperature coefficient
is approximately -0.75 millivolts/degree.

In organic media, the use of the aqueous calomel electrode, as well as most others, creates two problems of which
the reader ought to be aware. Both are related to the
creation of an ill-defined liquid junction between the organic medium of the electrolysis cell and the aqueous medium
of the reference electrode. A liquid junction potential
thereby arises, as between any two dissimilar conducting
phases, whose magnitude depends on the two phases involved,
and to some extent, the manner in which the junction is
formed. However, for most purposes, the effect of the liquid junction potential, as long as it can be held constant,
is simply to shift the potential scale by a small (but unknown) amount. Attempts to ascertain the magnitude of this
effect, both experimentally and theoretically, have led to
the conclusion (though not without controversy) that a shift
of 30-40 mV is likely (11). A shift of this magnitude is
consistent with the fact that reproducible measurements
within a few millivolts are readily attainable with this
type of reference. In any case, as long as measurements
are made within the same media, it is perfectly justifiable
to compare results and make conclusions based on differences
in potential for various species.

The second, and potentially more troublesome, effect
can arise due to leakage of some of the aqueous solution of
the reference electrode into the electrolysis compartment.
Possibly undesired water and electrolyte are thereby
introduced. By judicious arrangement of apparatus, a system
can be constructed which insures flow in the other direction
(12). In this case, however, a plug of solid KCl usually

forms within the tip of the reference and can significantly
increase its effective resistance.

A less troublesome but somewhat less convenient solution to these problems is the utilization of a nonaqueous
reference electrode, in which the solvent, and hopefully
the electrolyte, is common to both the reference and the
electrolysis cell. Since organic reference electrodes are
not available commercially, they must either be made separately or, more conveniently, by replacing the aqueous
medium. The effects of this procedure has been reported
for calomel in DMF (13), calomel in methanol (14), and for
the case where KCl is replaced by tetraethylammonium chloride (15). The latter has been used extensively and is
recommended by the authors.

3.1.4.2. Silver-Silver Chloride

Another reference electrode system which has found
considerable use is the silver-silver chloride couple:

$$AgCl + e = Ag + Cl^- \tag{3.2}$$

In aqueous solution, saturated with respect to KCl, the potential of this electrode is +0.197 V vs NHE. The silver-
silver chloride reference electrode is available commercially
in a variety of sizes (Beckman, Leeds and Northrup, etc.)

Although the reversibility of the Ag/Ag^+ couple is
quite high in non-aqueous solution, the silver-silver chloride reference is not recommended in general because of the
formation of soluble silver halide complexes, i.e.,

$$AgCl + Cl^- = AgCl_2^- \tag{3.3}$$

The equilibrium constants for the above reaction in many
different organic media have been reported by Alexander et
al. (16).

For further examples of the use and limitation of
reference electrodes in nonaqueous media, the reader is
referred to the excellent review article by Butler (17).

3.1.4.3. Experimental Technique

In use, the reference electrode, or more properly, its
tip, is placed as close as possible to the electrode whose
potential is to be measured. Under conditions where large
currents flow between the electrodes of the electrolysis
cell or where the conductivity of the electrolyte is low,
the iR drop between the tip of the reference and the working electrode may be considerable. In controlled potential
electrolysis, this is often of little significance, since,
for preparative purposes, it is usually sufficient to ensure that the potential not exceed a certain value. The
iR drop error in the measured potential is always in the
direction such that the actual potential is lower (less
cathodic for reductions, less anodic for oxidations).

The measurement itself, as noted earlier, should be
made under conditions where little or no current flows
within the reference electrode. Not only could this cause
an additional iR drop error, but it could also polarize the
reference electrode so that its potential would not be the
reversible one. In practice, this effect is easily avoided
by use of a potentiometric measuring device or by use of a
high-impedance ($>10^6 \Omega$) voltmeter, such as a vacuum-tube
voltmeter or most digital voltmeters (more details on measuring devices are found in Sec. 3.3.1).

If it is desirable that the measured potential agree
in sign with the European convention, the common (or negative) lead of the measuring device is attached to the
reference electrode and the other lead to the working electrode. In this way, more cathodic potentials will measure
in the negative direction and more anodic potentials in the
positive direction. This, however, can only be done with
impunity if the measuring device has a "floating" input,

i.e., neither lead common to the working electrode. In
cases where the measuring device is "grounded" to the working electrode, the measurement is made by attaching the
"hot" lead to the reference electrode; the sign of the
measured potential must then be reversed.

     In critical cases, where it is necessary to measure
the electrode potential accurately or where tight control
is required in controlled potential work, the tip of most
commercially available reference electrodes cannot be positioned close enough to the working electrode to ensure
negligible iR drop. For example, in an electrolysis where
10 V iR drop occurs between working and auxiliary electrode
spaced 10 cm apart with the tip of the reference just 1 mm
from the surface of the working electrode, the measured
potential will be 100 mV in error. The situation may be
considerably improved by use of a <u>Luggin capillary</u>, which
is simply a glass tube, drawn to a slender and fine tip,
filled with electrolyte, and attached with an <u>insulating</u>
sleeve to the tip of the reference electrode.

     The reference electrode can also be used to measure
iR drops within the cell. For example, the total iR drop
across the cell is easily measured by measuring the working
electrode potential in the normal way, and then measuring
the potential when the reference electrode is positioned
close to the auxiliary electrode (without changing leads).
The magnitude of the difference between these readings is
the iR drop across the cell. In a similar fashion, the
iR drop across the cell divider can also be measured--a
sure way of spotting a clogged divider (high voltage drop)
or a broken divider (low iR drop).

3.2.   SOLVENT AND ELECTROLYTE

     The solvent and electrolyte perform important functions
in electrolytic work, and should be selected only after considering several factors. The general considerations to be

applied to the selection of solvent and electrolyte may be
summarized as follows:

    1. Stability toward electrolysis conditions (range
of accessible potentials).

    2. Solubility of starting material.

    3. Conductivity.

    4. Reactivity toward products or intermediates.

    5. Ease of purification and separation.

    6. Adsorption.

    7. Toxicity and ease of handling.

In some cases the solvent or electrolyte itself can
be chosen purposely to participate in the reaction, for
example, in the case of anodic substitution (Sec. 5.2) and
in the dimerization of acrylonitrile (Sec 4.4). Proton
availability can also be an important factor, particularly
when carbanions (Sec 4.3) or radical anions (Sec 4.4) are
involved. In aqueous or partially aqueous media, the electrolyte can also serve as a buffer.

The use of nonaqueous solvents in electrochemistry has
increased dramatically over the past ten years as interest
in organic systems has grown. This subject has been reviewed admirably in a chapter by Mann (18) in the Electroanalytical Chemistry series.

The following summarizes the use of nonaqueous media
for the most widely used solvents and, together with some
additional facts, should prove to be useful to the potential electrochemist.

3.2.1. SOLVENTS

3.2.1.1. Acetonitrile (CH_3CN)

Acetonitrile has had considerable use in organic electrochemistry due to its excellent solvent properties, good
conductivity with suitable electrolytes, ease of purification, and stability toward oxidation and reduction. It is
miscible with water.

Dielectric Constant. 37.45.

Liquid Range. -45 to 82°C.

Electrolytes. Tetraalkylammonium halides, perchlorates, and tetrafluroborates; sodium perchlorate; lithium halides.

Reference Electrode. Aqueous SCE (with salt bridge, required for aqueous reference electrodes in organic media); aqueous Ag/AgCl (neither suitable for use directly in acetonitrile); Ag/0.1 $\underline{M}$ Ag$^+$.

Potential Range. +2.4 V (Pt/LiClO$_4$) to -3.5 V (LiClO$_4$) (vs SCE).

It has been reputed (19) to be stable to +4 V anodic with Pt/NaBF$_4$.

Toxicity. Recommended max conc 20 ppm (20). Lethal dose 3.8 g/kg (21).

Purification. Potential impurities include water, unsaturated nitriles, acetamide, ammonium acetate, acetic acid, aldehydes, amines, and ammonia. The recommended purification (22) procedure involves contacting, with stirring for two days over calcium hydride (10 g/liter), followed by decantation and fractional distillation over phosphorous pentoxide (5 g/liter). Further purification by refluxing for several hours over calcium hydride and very slow fractional distillation, produces polarographic grade material.

Distillation over potassium permanganate followed by reaction with a small amount of sulfuric acid [precipitates NH$_3$ as (NH$_4$)$_2$SO$_4$] then slow fractional distillation, produces material free of aromatic hydrocarbons, and is thus suitable for spectroscopy (23).

Drying with molecular sieves (Linde type 3A) and by azeotroping with methylene chloride have also been described (18).

Reactivity. Stable to sodium amalgam and lithium. Reacts with sodium to form sodium cyanide, hydrogen, and

methane (<u>24</u>). Reacts with perchlorate radical to form
succinonitrile and perchloric acid (<u>25</u>), as well as acetamide (<u>19</u>). Carbanions can abstract a proton (<u>26</u>). Can
react in anodic substitutions to give acetamides (Sec 5.2).

<u>Other Nitriles</u>. The use and properties of propioni-
trile, phenylacetonitrile, isobutyronitrile, benzonitride,
and acrylonitrile have also been described (<u>18</u>).

3.2.1.2. <u>Dimethylformamide [DMF, $HCON(CH_3)_2$]</u>

Dimenthylformamide (DMF) has been used extensively in
electroorganic systems, and its purification and properties
have recently been discussed by Visco (<u>27</u>). It has a moderately high dielectric constant, is a good solvent for most
organics, dissolves several alkali metal salts to give
moderately conductive solutions, and is stable toward reduction. DMF is miscible with water, but not with pentane.

<u>Dielectric Constant</u>. 36.7.

<u>Liquid Range</u>. -61 to $153^{\circ}C$.

<u>Electrolytes</u>. Tetraalkylammonium halides, perchlorates
and fluoroborates; sodium perchlorate; lithium chloride.

<u>Reference Electrodes</u>. Aqueous SCE; sodium amalgam/
sodium perchlorate; cadmium amalgam/cadmium chloride.

<u>Potential Range</u>. From +1.6 V anodic with Pt to -3.0 V
cathodic with tetraalkylammonium perchlorate.

<u>Toxicity</u>. Liver damage caused in animals on inhalation
of 100 ppm vapor; highly irritating to skin, eyes, and mucous
membranes.

<u>Purification</u>. DMF is unstable at its atmospheric boil-
ing point, giving dimethyl amine and carbon monoxide. This
decomposition is catalyzed by acids and bases. Commercially
available DMF contains water and dimethylamine, but is still
useful as is. A recommended purification involves distillation in vacuo (2 mm Hg) over anhydrous $CuSO_4$ [removes water
and amines (<u>27</u>)]. Use of CaH_2, KOH, and NaOH cause formation of dimethylamine. Drying has been accomplished by

the use of BaO, Al_2O_3 (28), and type 4A molecular sieves
(29).

Reactivity. DMF is quite stable toward reduction,
more so than acetonitrile. The authors have seen no con-
clusive reports that show proton abstraction or condensation reactions with carbanions. It also has a low chain
transfer coefficient (ability to donate a hydrogen atom
to a carbon radical). DMF has been reported to be less
stable than acetonitrile toward oxidation (30).

Other Amides. The use and properties of dimethyl-
acetamide, N-methylformamide (dielectric constant 182.4),
N-methylacetamide (dielectric constant 165), and formamide
(dielectric constant 110), have been summarized by Mann
(18).

3.2.1.3. Ammonia (NH_3)

The use of anhydrous ammonia has become more prevalent
in recent years, mainly due to the interest in its known
property of forming relatively stable solvated electrons,
which can be produced electrochemically, as for example in
the electrolytic Birch reaction (Sec 4.4). Although its low
boiling point requires special handling techniques, its
special properties often obviate the inconvenience. Ammonia
is relatively stable toward reduction, but not oxidation,
giving nitrogen and protons (31).

Dielectric Constant. 23.7.
Liquid Range. -77 to $-33.4^{\circ}C$.
Electrolytes. Alkali iodides, nitrates, and perchlorates; ammonium nitrate; tetraalkylammonium salts are
slightly soluble.
Reference Electrodes. Pb/Pb^{+2} (32) in ammonia; zinc
amalgam/zinc chloride in ammonia (33).
Potential Range (vs Pb/Pb^{2+}). In the absence of reducible ions, the cathodic region is limited by the disso-

lution of electrons at -2.0 V. With platinum, the anodic
region is limited by oxidation of ammonia to nitrogen (**31**).

Toxicity. Maximum tolerable vapor concentration, 100
ppm; limit of human perception 53 ppm.

Purification. Distillation into receiver containing
sodium metal, followed by redistillation, has been used to
remove water, reducible impurities, and nonvolatile components.

3.2.1.4. Hexamethylphosphoramide: HMPA, $[(CH_3)_2N]_3PO$

HMPA is a solvent which should see much greater use in
the future because of its very low proton availability and
highly polar character. It forms stable solutions of electrons (**34**) and has been used in mixtures with ethanol (**35**).

Electrolytes. Lithium chloride and perchlorate; sodium
perchlorate; tetraethyl and tetrabutylammonium perchlorate
(**18**).

Reference Electrodes. Aqueous SCE and Ag/AgCl; Ag/Ag$^+$;
Ag/AgCl not suitable in HMPA because of complex formation
(**18**).

Potential Range. The anodic limit was found to be
0.75 V vs Ag/Ag$^+$ (**36**), presumably due to oxidation of HMPA.
The cathodic limit usually depends on the cation and water
concentration.

Purification. Distillation in vacuo (bp 97-102 at
6 mm Hg), followed by treatment with molecular sieves has
been used. The purified solvent retained the characteristic
blue color for at least 12 hr with a small amount of sodium
metal (**18**).

3.2.1.5. Pyridine (C_5H_5N)

Pyridine is a useful organic solvent when a basic
medium is desired. Although it has a low dielectric constant, it dissolves many salts to give solutions of good
conductivity. It is miscible with water.

Dielectric Constant. 13.2

Liquid Range. -42 to 115°C

Electrolytes. Lithium salts; tetraalkylammonium salts; potassium thiocyanate, sodium iodide and tetraphenylborate.

Reference Electrodes. Cathodic to -2.3 V vs Ag/Ag$^+$ (37); anodic to +1.4 with graphite/perchlorate electrolytes (38). In the presence of acid, pyridine (as the pyridinium salt) can be reduced to a radical which, in turn, dimerizes. Oxidation can lead to N-pyridyl pyridinium salts.

Toxicity. Eye, skin, and respiratory irritant; large doses may produce liver and kidney damage; lethal conc.for rats 4000 ppm in air (39).

Purification. Type 4A molecular sieves to remove water; reagent grade material is satisfactory for most electrochemical usage.

Other Amines. The use of ethylene diamine and morpholine are discussed by Mann (18).

3.2.1.6. Tetrahydrofuran (THF, C_4H_8O)

THF has been used in electrolytic work because of its excellent solvent ability for most organic compounds and its ease of removal by evaporation. Its low dielectric constant, however, causes high resistance. THF is very soluble in water and forms an azeotrope (bp 60°;6%H_2O).

Dielectric Constant. 7.4.

Liquid Range. -108 to +65°C.

Electrolytes. Lithium and sodium perchlorate; tetrabutylammonium iodide.

Reference Electrodes. Aqueous SCE; Ag/Ag$^+$

Potential Range. The cathodic limit is normally reduction of the cation of the electrolyte. THF has been reported (40) to have a potential range of +1.5 to -4 V vs Ag/Ag$^+$.

Toxicity. May cause liver damage at >20 ppm.

Purification. THF is susceptible to air oxidation to form peroxides, like many ethers, and should therefore never be allowed to evaporate to dryness. Purification may be effected by distillation over sodium wire or lithium aluminum hydride with an inert atmosphere of nitrogen or argon. The distillate may contain traces of other low boiling ethers.

Reactivity. THF is inert to most reducing agents and may be used with organometallics. It will be cleaved by hot concentrated HI.

3.2.1.7. 1,4-Dioxane $(C_4H_8O_2)$

In the past dioxane has been used extensively, particularly in partially aqueous mixtures. Its very low dielectric constant makes it generally unsatisfactory for use as the neat liquid.

Dielectric Constant. 2.2.

Liquid Range. 11.8 to 101.3°C.

Electrolytes. Aqueous HCl; tetrabutylammonium iodide.

Reference Electrodes. Aqueous SCE; no satisfactory nonaqueous reference.

Potential Range. 96% dioxane/water useful to -2.3 V vs SCE; oxidation not reported.

Toxicity. Lethal dose (21) 6.0 g/kg.

Purification. Same precautions as for THF; purification by distillation over sodium.

Other Ethers. The properties and electrolytic usage of 1,2-dimethoxyethane, diethyl ether, and 1,2-epoxybutane have been reported (18).

3.2.1.8. Acetic Acid (CH_3COOH)

Acetic acid is a readily available and useful electrolytic solvent when an acidic nonaqueous medium is desired. It has been used with acetate ion electrolytes for anodic acetoxylations (Sec 5.2). Acetic acid forms a dimeric hydrogen-bonded complex and has a low dielectric constant, which makes even concentrated salt solutions fairly highly resistive. It is miscible with water.

Dielectric Constant. 6.2.

Liquid Range. 16.6 to 118°C.

Electrolytes. Sodium and ammonium acetates; hydrochloric and sulfuric acids; alkali halides and perchlorates.

Reference Electrodes. Hg/Hg_2Cl_2 in acetic acid; chloranil/Pt; aqueous SCE.

Potential Range. Cathodic to -1.7 V vs SCE due to hydrogen discharge; anodic to +2.0 V vs SCE at a platinum anode/0.5 $\underline{M}$ NaOAc.

Toxicity. Causes irritation to eyes and nose; lethal dose, 5 g/kg.

Purification. Useful, as is, in the normal reagent grade; may be purified by distillation over chromium trioxide, refluxing with triacetylborate, followed by redistillation. The purified product contains ∿10 m$\underline{M}$ water (41).

Other Acids. Acetic anhydride, formic acid, and methane sulfonic acids have been discussed by Mann (18).

3.2.1.9. Methanol (CH_3OH)

Methanol has been used widely because of its ready availability, high dielectric constant, and good stability. Aqueous mixtures have been used extensively, as well as when a highly alkaline medium is desired, because of the high solubility of alkali hydroxides and alkoxides. Methanol is miscible with water in all proportions.

Dielectric Constant. 32.6.

Liquid Range. -97.5 to 64.5°C.

Electrolytes. Sodium, lithium, and potassium hydroxides, methoxides, iodides, perchlorates; tetraalkylammonium salts, ammonium salts.

Reference Electrods. Aqueous SCE, Ag/Ag$^+$.

Potential Range. No general results reported for cathodic limitation, but presumably hydrogen discharge can occur at about -2.2 V vs SCE. Methanol has been found to display large background currents under anodic conditions, even at 0 V vs Ag/Ag$^+$(42).

Toxicity. Can cause blindness if taken internally.

Purification. Possible impurities include acetone, methylal(dimethoxymethane) methyl acetate; formaldehyde, ethanol, acetaldehyde, ether, and water. Drying may be accomplished by distilling, after refluxing over magnesium with a small amount of iodine added (20). Commercial reagent grade methanol is of satisfactory purity to be used directly in most cases.

Reactivity. Methanol is a good nucleophile, and can react with strong electrophiles, as is the case in anodic methoxylation (Sec 5.2). Methanol is a protic solvent, and cognizance should be taken of this fact when strong bases are present. Methanol reacts with strong reducing agents, such as the alkali metals, to evolve hydrogen. Strong oxidants can react to yield formaldehyde, formic acid and carbon dioxide.

3.2.1.10. Ethanol

Ethanol has much the same behavior as methanol, and has been used extensively, particularly in aqueous solutions.

Dielectric Constant. 24.3.

Liquid Range. -114.5 to 78.3°C.

Electrolytes. Same as methanol.

Reference Electrodes. Same as methanol

Potential Range. No specific data, but in general, should behave similarly to methanol.

Toxicity. Ethanol has been known to cause strange effects in humans when large quantities are ingested orally! Caution--most commercial ethanol is denatured, i.e., contains

methanol and benzene, which can cause blindness if taken
internally.

Purification. Same procedure as with methanol.

Reactivity. Similar to that of methanol.

Other Alcohols. Glycerol, ethylene glycol, 1- and 2-
propanol, and cellosolve have also been used.

3.2.1.11. Dimethylsulfoxide [DMSO, $(CH_3)_2SO$]

In recent years, DMSO has gained high utility as a
medium for electrolytic work. It is quite a versatile solvent and has a sufficiently high dielectric constant to form
highly conducting electrolytic solutions. DMSO is miscible
with water.

Dielectric Constant. 46.7.

Liquid Range. 18.6 to 189°C.

Electrolytes. Most lithium and sodium salts; potassium
perchlorates; tetraalkylammonium salts; fluorides, sulfates,
and carbonates are insoluble.

Reference Electrodes. Aqueous SCE with salt bridge;
Ag/AgCl and Hg/Hg_2Cl_2 are not suitable due to solubility of
halide complexes; thallium amalgam/thallous chloride; lithium
amalgam/lithium chloride.

Potential Range. Cathodic to -3.0 V vs SCE with tetrabutylammonium perchlorate (43); tetraalkylammonium salts;
anodic to +0.7 V vs SCE with Pt/perchlorate electrolytes
(44).

Toxicity. DMSO is in itself not toxic, however, as it
absorbs through the skin very readily, it could carry with
it substances which would ordinarily not penetrate.

Purification. Major impurities are water and dimethyl-
sulfide. Recommended purification includes shaking overnight
with heated alumina (44), and partial freezing, decantation
of the liquid, and azeotroping out water with 50 ml benzene
per liter.

Reactivity. DMSO reacts slowly with lithium and rapidly with sodium and potassium to evolve hydrogen. DMSO is potentially oxidizable to dimethylsulfone.

Other Sulfur Compounds. The use of sulfolane, dimethylsulfone, and sulfur dioxide has been described (18).

3.2.1.12. Methylene Chloride (CH_2Cl_2)

Methylene chloride has been used to some extent in electrolytic systems because of resistance to oxidation, and ease of removal. It has been recommended as the solvent of choice for oxidation of aromatic hydrocarbons (45).

Dielectric Constant. 8.9.

Liquid Range. -96.7 to $40^\circ C$.

Electrolytes. Tetrabutylammonium perchlorate, chloride, and iodide.

Reference Electrodes. Aqueous SCE with salt bridge; Ag/Ag^+ not suitable because of low solubility.

Potential Range. Reported anodic to +1.8 V vs SCE with Pt/tetrabutylammonium perchlorate. Cathodic to -1.9 V vs SCE with tetraalkylammonium salts, due to reduction of the carbon-chlorine bond.

Purification. Commercial reagent grade is satisfactory for most electrolytic work.

Reactivity. Reducing agents will reduce methylene chloride with displacement of chloride. Oxidation probably leads to chlorine. Strong bases may lead to the formation of chlorocarbons (45) (probably via chlorocarbene).

3.2.1.13. Miscellaneous

Propylene carbonate (4-methyldioxolone-2, $C_4H_6O_3$) has been used because of its high dielectric constant (69) and stability toward oxidation. A review of its use in electrochemistry has been done by Nelson and Adams (46).

Tetrahexylammonium benzoate is an interesting case because it is a liquid salt and hence serves as its own electrolyte. It is a fairly viscous material, similar to that of glycerol. It has been used for some reductions (47).

Liquid carbon dioxide has not been reported for use in an electrolytic system, presumably because of the difficulty of working with it (critical pressure, 73 atm). It should, however, be useful where carbonation of anions is desired. Liquid SO_2 is another possibility.

3.2.2. ELECTROLYTES

No specific description will be given for particular salts. The enormous number of possibilities preclude this. Some general recommendations, are, however, in order. The criteria for selection of a suitable electrolyte are as follows:

1. Must dissolve and ionize in solvent.
2. Must be inert to starting materials, intermediates, and products (unless, of course, such reactivity is desired to form specific products).
3. Must be inert over the potential range of interest.
4. Should be easily removed on product work-up.

3.2.2.1. Cations

a. Tetraalkylammonium. In general, these have been used to the greatest extent in organic electrochemistry because of their good solubility and conductivity; they are also among the most difficultly reduced cations known (most are stable to at least -2.7 V vs SCE). The tetramethyl, ethyl, and n-butyl salts have been the most widely used as halides, perchlorates, and acetates. They are somewhat reactive toward strong bases, such as carbanions--tetramethyl probably reacts to give the ylide; the others undergo β-elimination (Hoffmann elimination) to give the α-olefin and the tertiary amine.

b. <u>Alkali Metal Cations</u>. Of these, lithium salts have
been used most extensively because of their higher solubility
in organic media than the corresponding sodium or potassium
salts. In general, they reduce to the free metals (or amal-
gams) at around -2.2 V vs SCE. Lithium deposition has been
used to generate solvated electrons, in analogy with the
Birch reduction (see Sec 4.4).

c. <u>Ammonium Salts</u>. These are fairly soluble in organic systems and reduce with difficulty to nitrogen and
hydrogen. The proton availability is sufficiently high to
warrant its use for that purpose.

3.2.2.2. <u>Anions</u>

a. <u>Perchlorates</u>. These have been used quite extensively in anodic systems because of the inertness of perchlorates to oxidation and their almost complete lack of
complex formation. Perchlorate anion is one of the least
nucleophilic species known. At strongly anodic potentials,
at a platinum anode, perchlorate probably oxidizes to the
perchlorate radical, which in turn either reacts with solvent
or other organic material, eventually to yield chlorine oxides. The alkali metal perchlorates (but not transition
metal perchlorates) are not particularly unstable to shock
and decompose violently only when strongly heated. They
are, of course, potentially dangerous oxidants when concentrated, or in the presence of acid or in contact with easily
oxidizable organic material, and therefore, should be dissolved in the solvent before the starting material is added.

b. <u>Halides</u>. The halides of the alkali metals and the
tetraalkylammonium salts have been used extensively because
of their ready availability and solubility. The order of
solubility is usually iodide > bromide > chloride > fluoride.
They are all readily oxidized to their respective free halogens, the order of ease of oxidation being I > Br > Cl > F.
This factor has been used specifically in anodic halogenation, particularly in fluorination (Sec 5.3). Halides are
also potential nucleophiles themselves.

3.2.2.3. <u>Tetraphenylboride</u>   Tetraphenylboride has been
used frequently of late because of its availability (as
sodium salt), its stability toward oxidation, and good
solubility in organic media.

   a. <u>Acetates</u>. These display good solubility in organic
media, but are not in general recommended for oxidation because of the possibility of Kolbe oxidation. However, since
its oxidation products are usually carbon dioxide and ethane
or methane, acetate might be a convenient anodic depolarizer
to carry out reductions in a nondivided cell, where otherwise
the anode might adversely affect the reaction products.
Acetates are used in anodic acetoxylation (Sec. 5.2).

   b. <u>Sulfonates</u>. These are quite soluble in organic
solvents, particularly organic sulfonates such as p-toluenesulfonate. They are fairly inert to oxidation and display
little reactivity as nucleophiles. High concentrations of
sulfonates have been used to "salt in" organic compounds
which would otherwise be insoluble in water (see p. 158).

3.2.2.4. <u>Purification</u>
   Most inorganic salts are sufficiently pure, as the
commercial reagent grade, to be used directly. The tetraalkylammonium salts are usually purified by recrystallization
followed by drying in vacuo. Some, such as tetramethylammonium chloride and tetrabutylammonium chloride, are very hy-
groscopic. Direction for the purification of specific tetraalkylammonium salts may be found in the review article by
Mann (<u>18</u>).

3.3. MEASUREMENT DEVICES

   One of the inherent advantages of electrochemical operation is the ease with which the key electrical variables can
be measured and controlled with speed and accuracy. The techniques and apparatus for measuring the most commonly encountered ones are discussed in this section.

3.3.1. CURRENT AND VOLTAGE METERS

The current passing through an electrolysis cell is the measure of the rate of the electrolytic reactions occurring at the electrodes. Conventional ammeters cover a wide range of currents, from about 10 μA (microamperes) to several hundred amperes, and are usually sufficient for most electrolytic work. In practice, the ammeter is inserted in <u>series</u> with either side of the power supply, being careful to observe polarity (see Fig.3.8).

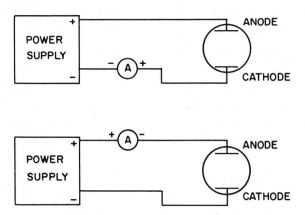

FIG. 3.8. Arrangement of ammeter in electrolysis.

The conventional pointer-type ammeters have a low resistance <u>shunt</u> across their terminals, in most cases of such a value to cause a 10 mV iR drop when the full-scale current is being passed. A voltage drop of this low magnitude is usually insignificant compared to the cell voltage. Accuracy of the order of 0.5% can be obtained.

When smaller currents are to be measured, or when the use of a shunt-type ammeter is undesirable, for example, when the current is to be displayed on an oscilloscope, alternative methods are available. One easily implemented technique is the use of an operational amplifier (<u>48</u>) in a current-to-voltage mode (see Fig. 3.9). In this case, the output voltage, E out, is equal to -iR. The accuracy is

usually determined by the accuracy of the resistor. As
shown, the working electrode is at virtual ground potential
(negative terminal of power supply). The same configuration
can be used for both anodic and cathodic currents.

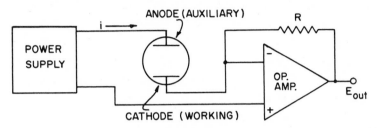

FIG. 3.9. Use of operational amplifier to measure
current.

Voltage measurement is also required for electrolytic
work. Basically, we are concerned with the voltage across
the cell and with the electrode potential, which requires
somewhat more care in measuring. The cell voltage, which
may range from a few volts to 100 volts, may be measured
with most moving-pointer meters. The voltmeter itself may
draw about 50 µA for full-scale deflection, but this is
usually insignificant compared to the cell current in most
electrolytic work. The voltmeter is connected in parallel
with the cell (observing polarity) as shown in Fig. 3.10.
When high currents are being employed, it is good practice
to connect the voltmeter, as shown in Fig. 3.10, directly
to the electrodes, rather than to the power supply terminals,
since this avoids the inclusion of the iR drop in the power
leads.

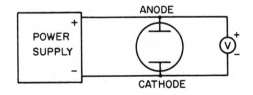

FIG. 3.10. Arrangement of voltmeter in electrolysis.

Electrode potential measurements, and voltage measurements across circuits in which small currents (milliampere
or less) are flowing, require the use of a voltmeter which
draws essentially zero current. The minimum requirement is
usually met by the VTVM (vacuum tube voltmeter) type of instrument. As commonly available, they have an internal impedance of 11 MΩ. Since electrode potentials will rarely
fall outside the range +5 to -5 V, this instrument will draw
submicroampere currents. This is usually small enough to
ensure that no interaction will occur between the electrode
potential and the measuring device. Recently, the VTVM has
been challenged by an all solid-state instrument, the FET
(field effect transistor) voltmeter, which has the advantage
of battery operation for portability and greater reliability.
Both types are available from meter manufacturers. The
measurement is made according to Fig. 3.11.

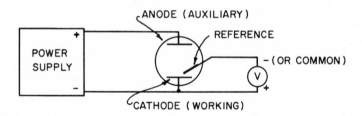

FIG. 3.11. Arrangment for measuring electrode potential.

If the common lead of the voltmeter is connected to the reference electrode, the sign of the measured voltage will be
correct. If the leads must be interchanged, as for example
if the voltmeter does not have a polarity switch, the sign
of the measured voltage must be changed.

The accuracy of both voltmeters mentioned above is at
best 0.5%, typical of moving pointer-type meters. If better
accuracy is desired, there are available (at significantly
higher costs) digital voltmeters, from which the voltage is
read directly as a 3-, 4-, 5-, or 6-digit numbers (the cost

goes up rapidly with the number of digits), whose accuracy
is usually specified as $\pm$ one count. Digital voltmeters
are also available with auto-polarity and auto-ranging
features.

All the meters mentioned above are available as multi-
ranged instruments, capable of measuring voltage, resistance,
and current over wide ranges with selector switches. At
least one such instrument should be obtained as part of
the standard equipment in electrolytic work.

Often, one is required not simply to measure the electrode potential, but also to use it for control purposes,
such as in controlled potential electrolysis. For this
purpose, the operational amplifier circuit shown in Fig.
3.12 is often employed. In this configuration, the operational amplifier is used as a voltage follower, i.e., where

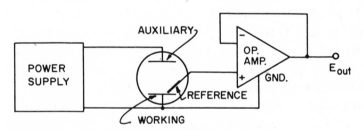

FIG. 3.12. Use of operational amplifier to monitor
electrode potential.

E_{out} = electrode potential. The advantage is that very
little current is drawn at the input, typically less than
0.1 μA, whereas appreciable current, of the order of milliamperes, may be drawn at the output to be used for control
and monitoring purposes.

3.3.2. COULOMETERS

The number of coulombs passed is central to the concept of electroorganic synthesis, since it is a measure of

the total amount of electricity consumed in an electrolytic reaction. For a reaction which is 100% electrically efficient, and where n electrons are required per mole, the number of coulombs needed for 100% conversion is

$$Q_{theo} = (nF) \cdot m \tag{3.4}$$

where m is the number of moles of starting material and F is Faraday's constant (96,500 coulombs).

Conversely, the electrical efficiency, E.E., can be calculated for reactions which are not 100% efficient, since

$$E.E. = Q_{theo}/Q_{actual} \tag{3.5}$$

Just as there exists a need for ascertaining the identity of side products for reactions which do not proceed with 100% chemical yield, so also there exists the similar need for identifying electrolytic side reactions for systems which do not give 100% electrical efficiency. An electrical balance is as important as a mass balance.

The number of coulombs passed during an electrolytic reaction run at constant current is simple to calculate, being just

$$Q = it \tag{3.6}$$

where i is the current in amperes and t is the time in seconds, so that all that is required is an ammeter and a timer.

For reactions run at constant electrode potential, the situation is somewhat more complex, since the current will, in general, be time dependent. Equation (3.6) must be replaced by Eq. (3.7):

$$Q = \int_0^t i \, dt \tag{3.7}$$

One possible way of handling this situation would be to
record the current-time response on a chart recorder, and
then obtain the area under the curve by conventional graphi-
cal means. A recorder equipped with a ball-and-disk integrator would be more convenient.

A conventional, and extremely accurate, way of
measuring coulombs is by use of a chemical coulometer, in
which an electrochemical cell with specific components is
placed in series with the electrolysis cell. Coulometers
of this type have been used for the determination of the
value of the Faraday. They usually involve evolution of
a gas, such as hydrogen and oxygen, or deposition of a
metal, such as silver or copper.

The amount of material deposited or gas evolved is,
of course, proportional to the number of coulombs passed.
For deposition the number of coulombs is calculated from:

$$Q = (\frac{w}{M.W.})\ (n)\ (96,520) \tag{3.8}$$

where w is the weight of the deposit in grams; M.W. is the
molecular weight of deposited material; and n is the number of electrons per mole.

For gases, the following equation is used:

$$Q = (\frac{V}{22,400})\ (n)\ (96,520) \tag{3.9}$$

where V is the volume of gas evolved reduced to STP.
Although chemical coulometers can be highly accurate and
precise, they have two disadvantages which prevent their
general use in electrolysis.

First, they cannot easily be used with large currents,
and second, they require a relatively large voltage drop.

A somewhat more convenient and useful type of coulometer involves the use of a precision dc motor. In this
case, the voltage developed across a shunt resistor is used

to drive a motor, which in turn drives a mechanical counter, as illustrated in Fig. 3.13.

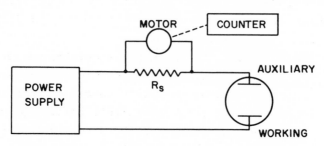

FIG. 3.13.   Electromechanical coulometer.

The system is calibrated by impressing a known voltage across the motor terminals and measuring the response in counts/sec/V. The total number of coulombs passed in the experiment in Fig. 3.13 is just

$$Q = counts/R_s K \qquad\qquad (3.10)$$

where R_s is the shunt resistance in ohms and K is the cali-bration factor in counts/sec/V.

The electromechanical coulometer is subject to the limi-tations of the motor. The linearity of speed with applied voltage is valid only over a certain dynamic range; preci-sion dc motors designed for this application use precision bearings, a very light-weight armature, and fine gold wires for brushes. A typical unit operates effectively over an applied voltage range of 50 mV to 12 V, while drawing only a few milliamperes. Sudden changes of current should be avoided, since this can cause such high accelerations to rupture the armature. The shunt resistor should be chosen to develop voltages within the range of the coulometer, preferably in the low range, and to be able to dissipate the $i^2 R$ heat. As such, this type of coulometer is useful mainly for electrolyses in which large currents are passed and where the current does not change rapidly.

It is also possible to have electronic coulometers which can cover much wider dynamic ranges and which are useful for situations where rapidly changing conditions exist. An

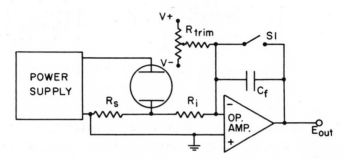

FIG. 3.14. Use of operational amplifier as coulometer.

operational amplifier type is shown in Fig. 3.14.   The configuration shown is simply a voltage integrator, where

$$V_{out} = \frac{-1}{R_i C_f} \int_0^t V_{in} \, dt$$

Switch S_1 is used to short out the feedback capacitor, i.e., to reset the coulometer. Offset currents are trimmed out with the adjustable resistor shown. The number of coulombs is given by the following equation:

$$Q = -Vout \ \frac{(R_i C_f)}{R_s} \tag{3.11}$$

Because of drift in the offset current, this type of coulometer is not useful over long periods of time; several hours would be a practical limit. Also, the capacitor used should have extremely low leakage. This coulometer is at its best when small currents are involved and where rapidly changing, or transient, conditions exist.

Another type of electronic coulometer inolves the use of digital circuitry, and is illustrated in Fig. 3.15.

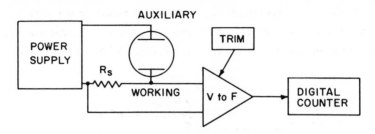

FIG. 3.15.   Digital electronic coulometers.

In this case, the voltage developed across the shunt
R_s is converted to a linearly related frequency (V to F);
the resulting frequency is counted and the total displayed
via digital readouts. This type of coulometer is usable
over long periods of time, can be extremely precise, and,
unfortunately, is also expensive.

The authors would recommend, for most electrolyses,
the use of a simple ammeter and timer, recording the current
vs time response and manually integrating the resulting
graph. For special cases, i.e., where automatic recording
is essential, or where transients are involved, the other,
more elegant types of coulometer should be used.

3.3.3.   CONDUCTIVITY

Conductivity is, in a practical sense, an important
variable in electrolyses. In general, it will always be
desirable to operate with the highest conductivity medium,
consistent with the other constraints of electrolysis. Low
conductivity media cause high operating voltages to be required, thus dissipating excessive heat in the system. Low
conductivity also makes the measurement of meaningful electrode potential difficult.

The measurement of conductivity is usually carried out
in a cell specially designed for that purpose. It consists
of two plane and parallel platinized platinum electrodes,
with fixed distance between, in a glass cell, or, more con-

veniently, in a glass cylinder which can be immersed in an
electrolyte solution. The electrode size and spacing determine the cell constant, k, which is determined by measuring
the conductivity of a known solution, usually aqueous potassium chloride. The conductivity, in mhos, is converted to
the specific conductivity in mhos/cm by multiplying by the
cell constant, i.e.,

$$\lambda = kC \tag{3.12}$$

where λ is the specific conductivity in mhos/cm; k is the
cell constant in cm^{-1}; and C is the conductivity in mhos.

The actual measurement is carried out using a conductivity bridge, or an impedance meter, which usually utilizes an ac Wheatstone bridge circuit (see Fig. 3.16).

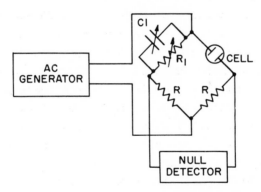

FIG. 3.16.   Circuit for measuring conductivity.

Alternating current is used to avoid polarization at
the cell electrodes, and is usually of a frequency of 1000
Hz or 60 Hz. In general, the 1000 Hz generator will give
results less subject to polarization effects. The condition of bridge null is achieved when the variable resistor
R_1 is equal to the electrolyte resistance, and the capacitor
C_1 is equal to the cell capacitance. An exact null cannot
be achieved without this compensating capacitor, although

an unsharp minimum of the bridge signal can be obtained
without it. At balance, the conductivity of the electrolyte is equal to the reciprocal of the variable resistance
arm, the value of which is usually read out directly in
mhos on a calibrated scale. Null detectors commonly used
are the "magic eye" tube (null corresponds to maximum open-
ing of "eye"), a pointer-type meter, or, in the earliest
types, a pair of headphones. Cells of various electrode
configurations are available for measuring conductivity
over a wide range.

3.3.4. RECORDING DEVICES

One of the most useful and versatile recording devices
is the strip-chart recorder. With it, signals of various
sorts can be monitored with time, for example, current,
voltage, or electrode potential during an electrolysis as
a function of time. Chart recorders are also used extensively for many electroanalytical techniques. Many instruments are available commercially, having wide ranges of
sensitivity and chart speeds. The most generally useful
is the potentiometric (i.e., input potential is balanced)
type with "floating" inputs. A ball-and-disk integrator is
a useful accessory, and most general purpose chart recorders
can be equipped with them. Pen response time should also be
considered--most general purpose recorders require $\sim\frac{1}{2}$ sec
for the pen to deflect full scale. Special purpose recorders can be obtained when writing speeds up to 100 in./sec
are required.

For certain applications, such as cyclic voltammetry,
X-Y recorders are desirable. In this case, both axes are
driven by input, so that true parametric plot can be obtained,
e.g., current vs voltage.

Recorders are often an integrated part of a specialized
instrument, such as a polarograph. Their utility is increased if the recorder can also be used as an independent
unit.

For experiments where the time scale is too short for chart recorders, an oscilloscope can be employed. Most useful will be the triggered oscilloscope, in which a single sweep can be initiated at the start of the experiment. Scopes of this type usually have as an accessory a Polaroid camera attachment, where the single trace can be recorded permanently. Also available are storage oscilloscopes, in which the single traces are temporarily stored by electronic means to retain the image of the trace, where it can be examined and measured at leisure.

Increasing use is being made of digital recording and processing equipment to monitor many variables during an experiment. This can vary in sophistication from relatively simple data logging and transcription, to computer-compatible form, such as punched tape, to complete computer control of the experiment. This type of equipment should prove to be important in monitoring and controlling continuous electrolyses, where several variables need to be controlled and long-term effects are being sought. Needless to say, this type of system and data would be valuable in the process of scale-up to industrial levels of production.

3.3.5. COMMERCIAL SOURCES (REPRESENTATIVE)

<u>Current and Voltage Meters</u> (most electrical supply houses)

Simpson Electric Company
5218 W. Kinzie Street
Chicago, Illinois 60644

Hewlett-Packard Corp.
1501 Page Mill Road
Palo Alto, Calif. 94304

Triplett Electrical Instr. Co.
286 Harmon Rd.
Bluffton, Ohio 48517

Fairchild Instrumentation
974 E. Argues
Sunnyvale, Calif. 94086

Radio Corporation of America
Electronic Products Division
Camden, New Jersey 08102

Keithley Instruments, Inc.
28775 Aurora Road
Cleveland, Ohio 44139

Coulometers (most chart recorder manufactures; ball and disk integrator)

Acromag, Inc.
 (electromechanical integrator)
30765 Wixom Road
Wixom, Michigan

Infotronics Corp.
 (digital integrator)
7800 Westglen Drive
Houston, Texas 77042

Conductivity (most chemical supply houses)

Keithley Instruments Inc.
28775 Aurora Road
Cleveland, Ohio 44139

General Radio Co.
300 Baker Avenue
West Concord, Mass. 01781

Wayne Kerr Co.
21250 10 1/2 Mile Road
Southfield, Michigan 48075

Chart Recorders (most chemical supply houses)

Beckman Instruments
2500 Harbor Blvd.
Fullerton, Calif. 92634

Leeds and Northrup Co.
Sunneytown Pike
North Wales, Pa. 19454

Hewlett-Packard Corp.
1501 Page Mill Road
Palo Alto, Calif. 94304

Heath Co.
Dept. 520-32
Benton Harbor, Mich. 49022

Clevite Corp.
Brush Instruments Div.
37th and Perkins
Cleveland, Ohio 44114

Oscilloscopes

Tektronix Inc.
Box 500
Beaverton, Oregon 97005

Heath Co.
Dept. 520-32
Benton Harbor, Mich. 49022

Fairchild Instrumentation
974 E. Aques
Sunnyvale, Calif. 90086

Hewlett-Packard Corp.
1501 Page Mill Road
Palo Alto, Calif. 94304

Dumont Oscilloscope Labs. Inc.
40 Fairfield Place
West Caldwell, New Jersey 07006

Digital Equipment

Princeton Applied Research Corp.
 (digital modules)
Box 565
Princeton, New Jersey 08540

Digital-Equipment Corp.
 (computer interface
146 Main Street
Maynard, Mass. 01754

Digital Equipment (continued)

Infotronics Corp.
 (analog/digital converters)
7800 Westglen Drive
Houston, Texas 77042

Operational Amplifiers

Philbrick/Nexus Research
 (a Teledyne Co.)
Allied Drive at Rt. 128
Dedham, Mass. 02026

Burr-Brown Research Corp.
Int'l Airport Industrial Park
Tucson, Arizona 85706

Analog Devices, Inc.
221 Fifth Street
Cambridge, Mass. 02142

Fairchild Instrumentation
974 E. Argue
Sunnyvale, Calif. 94086

Zeltex, Inc.
1000 Chalomer Rd.
Concord, Calif. 94520

3.4. VOLTAMMETRY

The researcher in electroorganic synthesis should be
continually aware of the nature of the reactions he is
conducting in terms of both the chemical and electrochemical pathways which may be possible for a given system.
Much can be done in the way of understanding these mechanisms and directing the reaction to the desired result by
careful analysis of the identity and quantity of the products formed. It would be difficult to underestimate the
importance of proper product identification. The potential
researcher in this field had best master the organic chemistry before he can hope to be successful in electroorganic
synthesis. The powerful analytical tools available today,
such as chromatrography, optical spectroscopy, NMR, mass
spectroscopy, etc., make it possible to unravel even complex mixtures of products.

However far toward understanding the course of reactions the above approach can lead, the picture cannot be
complete unless the experimenter also has at his disposal
techniques with which he can "see" the electrochemical

reactions as well. Many aspects of electroorganic reactions
cannot be inferred directly from product analysis. Phenomena
such as the formation of unstable intermediates, multistep
electron transfers, potential-dependent adsorption, compet-
ing electrolytic reactions, and so forth, can only be made
"visible" by the proper application of suitable voltammetric
experiments. In recent years, a number of powerful electro-
analytic techniques have come forward, and it is the purpose
of this section to briefly discuss some of these. The more
complete details of operating procedures and theory are best
obtained through consultation of the leading references.

Before commencing, the authors would like to make some
general comments concerning the applicability of these techniques to problems encountered in electroorganic synthesis.

1.   Although the techniques are quite powerful in giving information concerning the electrochemical mechanisms
involved, their analytical value should also be put to use.
The ability to follow quantitatively the disappearance of
starting material or appearance of product should not be
underestimated.

2.   Most of the techniques have been developed using
low concentration of electroactive material (in the milli-
molar range). In synthetic work it is usually desirable to
operate at the highest feasible concentration of starting
material. The researcher should, therefore, attempt to
conduct voltammetric experiments at or near the concentration levels of the synthesis work. It would be dangerous
to infer that the same mechanisms are operative at widely
differing concentrations, particularly when bimolecular reactions are involved.

3.   The nature of the electrode material, including
surface condition and pretreatment, should be the same for
both the synthetic and voltammetric experiments. Also, the
implications of stirring should be considered--many electroanalytical techniques are carried out in quiet solutions.

4. The time scale of the two experiments should be considered, as well as the current densities involved.

3.4.1. POLAROGRAPHY

Historically, polarography was invented by Heyrovsky in 1920 (49), and since then it has been one of the most important electroanalytical tools. In practice, it is a relatively simple technique for which many commercial instruments are available (Sargent, Beckmann, Leeds and Northrup, Radiometer, Chemtrix, etc.). The technique is capable of high reproducibility and accuracy, and although the detailed theory is quite complex, the results can be expressed in relatively simple form. Although its use is restricted to systems which use mercury as the working electrode, a polarographic apparatus should probably be acquired as one of the first electroanalytical instruments in the electroorganic chemistry laboratory. The widespread use of mercury as electrode material is sufficient justification for this recommendation, but also learning the theory and practice of polarography will be a starting point toward the use of several other more sophisticated techniques. The instrumentation involved in polarography will also be useful for other techniques. For these reasons, the subject of polarography is discussed in somewhat more detail than the other techniques.

3.4.1.1. General Principles.

In polarography, the working electrode is a small mercury droplet emerging from the tip of a fine-bore capillary tube. After the drop grows to a certain critical size, governed primarily by the diameter of the capillary and the interfacial tension between the surface of the drop and the solution, it falls and a new drop starts growing. It is this periodic, very regular, growth and fall of the mercury drop that is the essence of polarography. The expansion and renewal of the electrode surface makes polarography independent of the history of the previous droplet.

The basic features of the polarographic cell are shown in Fig. 3.17. The dropping mercury electrode (DME) is immersed in the electrolyte solution, and is connected through tubing to a mercury reservoir positioned high enough to provide sufficient pressure to form the drops at convenient intervals (2-10 sec/drop).

The counter-electrode can be formed by a mercury pool at the bottom of the cell. A reference electrode and provision for purging the cell with nitrogen (to remove dissolved oxygen) completes the assembly. Variations include the use of a divided cell, provision for blanketing the top space with nitrogen, devices which tap the capillary at regular intervals to dislodge the drop, and the use of a Luggin capillary at the reference electrode tip. Various cells are available commercially. Note that as described, this is a three-electrode system, the only type which should be considered for use in organic media.

In practice, the cell is connected to a potentiostatic instrument, and the DME potential is made to vary (between

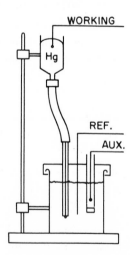

FIG. 3.17. Polarographic cell.

predetermined limits) linearly with time, over a total period
of time (5-15 min) very large compared to the drop time. The
resulting current is displayed on a strip-chart recorder.
In most commercial instruments initial potential, total scan
range, scanning speed, and recorder range are adjustable by
front-panel controls. A typical instrument set-up is shown
in Fig. 3.18

In the presence of an electroactive material, the
output of the polarograph is a "wave," shown in Fig. 3.19,
the position of which is characteristic of the material, and
the height of which is proportional to its concentration.
The oscillations of current, caused by the periodic growth
and fall of the mercury drops, are quite visible and can be
"damped out" so that only the average current is displayed.

3.4.1.2. Theory

a. The Diffusion Current. The theoretical description of the electrochemical currents flowing during the
life of an individual drop is quite complex, and resulting
equations were solved only approximately by Ilkovic in
1934 (50). He was able to derive an equation (3.13) now
bearing his name, which showed how the diffusion-limiting
current, i_D, depends on the other variables.

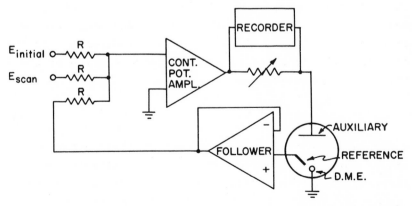

FIG. 3.18. Typical instrumental set-up for polarography.

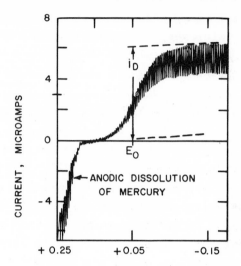

FIG. 3.19. Typical polarogram.

$$i_D = 607 \, nD^{1/2} \, Cm^{2/3} \, t^{1/6} \qquad\qquad (3.13)$$

where n is the total number of electrons transferred; D is the diffusion coefficient (cm^2/sec); C is the concentration (millimoles/liter); m is the rate of mercury flow (mg/sec); and t is the drop time (sec).

By inspection of Eq. (3.13), it can be seen that if the diffusion current, i_D, is divided by the $Cm^{2/3}t^{1/6}$, there results a parameter which is a characteristic of the particular substance involved, and is independent of the capillary characteristics and the concentration of the electroactive material, thus

$$I_D = i_D/Cm^{2/3}t^{1/6} \qquad\qquad (3.14)$$

I_D is called the diffusion current constant and its value can be compared directly with values obtained by other workers. The Ilkovic equation (3.13) can also be used for the determination of diffusion constants when the other parameters are known.

Another important feature shown by the Ilkovic equation is the fact that the diffusion-limited current is proportional to n, the <u>total</u> number of electrons transferred in the over-all process. By comparing the diffusion-limited current of a model compound where n is known, and whose size is similar (and hence diffusion constants are similar) to a compound of unknown electron-transfer mechanism, the unknown value of n can be determined.

Subsequent work has refined the original Ilkovic treatment and has included the effect of electrode curvature, the downward motion of the drop, and the shielding effect of the capillary tip. The original equation (3.13) is still accurate enough ($\sim$5%) for most work.

Aside from the purely analytical applications of polarography (due to the fact that i_D is proportional to concentration), the variation of the limiting current with the effective mercury pressure, i.e., height of the mercury column, can serve as a diagnostic criterion for several types of systems. A limiting current solely due to diffusion will vary as the square root of the height of the mercury column. If the current is limited because of a slow adsorption step, the limiting current will be directly proportional to the height of mercury. For kinetic limiting currents, i.e., where the electroactive material is produced by a slow reaction in solution, the limiting current will be essentially independent of mercury height.

b. <u>Reversible and Irreversible Waves</u>. Perhaps the most important aspect of polarographic waves which gives information concerning the nature of the electron-transfer process involved is the shape and position of the wave itself. For a reversible* electron transfer, the so-called "half-wave potential," i.e., the potential at which the current is one-half the limiting current, can be shown to be

---

*Reversibility, in polarography, means that the various species involved come to equilibrium within the life of a mercury drop.

virtually equal** to the standard potential, E^O, of the
couple. The current-voltage curve for this case is given
by

$$E = E_{1/2} - RT/nF \ln \left(\frac{i}{i_D - i}\right)$$ (3.15)

(The current is the envelope of the polarographic wave
measured at the maximum drop size, i.e., across the top;
the same results are obtained if the average currents are
used, as in the case when damping is employed.) A plot of
E vs log $(i/i_D - i)$ should therefore yield a straight line,
whose slope is $- 2.303$ RT/nF $(= \frac{0.059}{n}$ at T = 25^OC). For a
reversible system, the value of n, the number of electrons
involved in the over-all process, can be determined by this
means. N, of course, should come out to be a small (usually
1 or 2) integer.

For a totally irreversible electron transfer, similar
analysis reveals that a plot of E vs log $(i/i_D - i)$ should
also yield a straight line, but whose slope now is (0.059/
αn_a), where α is the transfer coefficient and n_a is the number of electrons transferred in the potential-determining
step. Thus, neither α nor n_a is determined with certainty
by this technique, only the product αn_a. Usually, this
ambiguity arises when n_a can be 1 or 2.

It should be noted that the above analysis cannot distinguish, for example, between a totally irreversible reaction where $\alpha n_a = 1$ and a reversible one where n = 1. The
only necessary and sufficient proof for reversibility is
the appearance of a composite wave, with diffusion-limited
currents on both the cathodic and anodic sides, when both
members of the couple are present in solution. There will
be no "kinks" as the current crosses zero, and the half-wave

**Exactly equal when the diffusion coefficients of the
oxidized and reduced forms are equal. In general, $E_{1/2}$
differs from E^O by (0.03 log D_O/D_R).

potential will be independent of the ratio of the concentrations of the two species.

The significance of the half-wave potential for totally irreversible reactions is difficult to assess, a priori, since its value depends on both the thermodynamics (i.e., E^O) and the kinetics (i.e., α and k_s) of the reaction. Nevertheless, a common assumption made in the literature is that, for a homologous series of compounds, the $E_{1/2}$'s reflect changes in the kinetics and therefore in the respective activation energies involved. Thus, the compound which reduces at the more negative potential has been considered to have a higher energy transition-state. The authors can only suggest that this line of argument be used cautiously and only when other evidence is available to confirm the conclusions.

Another commonly used assumption concerning irreversible waves is that reactions which give rise to drastically different values of α proceed via different mechanism. Conversely, reactions which have similar α's may be proceeding via a similar transition state. Again the authors will caution the reader that the simplistic concept of α (as reflecting the position of the transition state along the reaction coordinate) is not subject to rigorous proof in theory, and, in fact, much controversy still reigns concerning the significance of the transfer coefficient. For a detailed discussion of the various interpretations of the significance of α, the article by Bauer (51) is highly recommended.

c. Multiple Waves. One of the most revealing occurrences in the polarographic behavior of certain compounds is the appearance of more than one wave. This indicates that the electron-transfer reaction is comprised of separate steps, and opens the door for controlled potential electrolysis. In general, whenever the two steps are distinguishable polarographically, they also ought to be separable by controlled potential electrolysis. The electrolyses are usually carried out at the potential between the two waves

(or, when the two waves are close together at the foot of
the first wave) and the plateau region of the second wave.
Wave-shape analysis can be conducted on both waves separately, and combinations of reversible and irreversible behavior
are possible. The polarographic behavior of the system
during the electrolysis should also be investigated, since
this too can give important information. With electrolysis
carried out at the potential between the two waves, two
distinct cases are possible. (1) The first wave diminishes
and the second wave remains the same. This usually means
the intermediate is stable. (2) Both waves diminish. This
indicates that the intermediate product is unstable.

        d. Other Effects. If protons are involved in the
potential-determining step, then the resulting polarographic
wave will shift in position with pH. In general, the half-
wave potential shifts 0.059/n V/pH, where n is the number
of hydrogen ions involved per electron transferred. In such
cases, the pH of the medium ought to be well poised (either
well buffered or with high concentrations of acid or base)
in order to prevent local changes in hydrogen ion concentration at the mercury drop from distorting the polarographic
wave.

        In some cases, the electrochemical reaction may pro-
ceed via several steps where the potential of the succeed-
ing steps are not sufficiently different from that of the
first step to produce distinct waves. This may cause the
plot of E vs log $(i/i_D - i)$ to display curvature or to appear
to be composed of separate slopes. Irregularities in this
type of plot should be viewed cautiously, however, and
conclusions based on this observation should be supported
by other evidence.

        Other irregularities in polarographic behavior, such
as prewaves, maxima, changes in limiting current, nonlinear
behavior with concentration of depolarizer, etc., have been
observed and the reader should consult more advanced texts
in polarography to assess their significance.

SUGGESTED REFERENCES FOR POLAROGRAPHY

a. Kolthoff and Lingane, Polarography, 2nd ed., Inter-
    science, New York, 1952.

b. Meites, Polarographic Techniques, Interscience, New
    York, 1955.

c. J. S. Longmuir, ed., Advances in Polarography, Vols.
    1-3, Pergamon, New York, 1960.

d. C. L. Perrin, "Mechanisms of Organic Polarography," in
    Progress in Physical Organic Chemistry, (S. G. Cohen,
    A. Streitwieser, Jr., and R. W. Taft, eds.), Vol. 3,
    Interscience, New York, 1965, p. 165.

e. O. H. Miller, "Polarography," in Techniques of Organic
    Chemistry, A. Weissberger, ed., Vol. 1, Part IV, Inter-
    science, New York, 1969.

f. P. Zuman, Organic Polarographic Analysis, International
    Series of Monographs on Analytical Chemistry, (R. Belcher
    and L. Gordon, eds.), Vol. 12, Pergamon, New York, 1964.

3.4.2. LINEAR SWEEP VOLTAMMETRY

Linear sweep voltammetry differs from polarography in
two important aspects: (1) The electrode is stationary;
and (2) sweep rates are much faster.

The first difference permits one to use electrode materials other than mercury, and hence, for this reason alone,
linear sweep voltammetry (LSV) is a much more generally
applicable tool, although somewhat more complex to analyze
quantitatively, than polarography. The instrumentation required is, in principle, the same as for polarography. In
practice, however, since the sweep rates are generally much
faster, special instruments for generating the sweep and
displaying the results are usually used. A commercial
instrument which includes all the necessary electronics,
including oscilloscopic display, is available from Chemtrix,
Inc.

The cell employed in LSV is much the same as for polarography, requiring provision for reference electrode, auxiliary electrode, working electrode, and inert gas purge. In

fact, an ordinary dropping mercury assembly can be used if
the sweep rates are much faster than the drop rate; in this
case, the sweep is initiated after a specified time has
elapsed from the start of the formation of the drop.   In
general, however, solid electrodes, usually planar in shape,
are used, and in such cases, all the precautions necessary
for working with solid electrodes must be employed.

In practice, a linear potential sweep is applied to
the working electrode, and the current vs time is displayed
on an appropriate device.   In the presence of an electroactive material which undergoes reversible electron transfer
in a single step, there results a peak-shaped voltammogram
as shown in Fig. 3.20.

This characteristic shape may be qualitatively understood as follows: initially, the electrode potential is

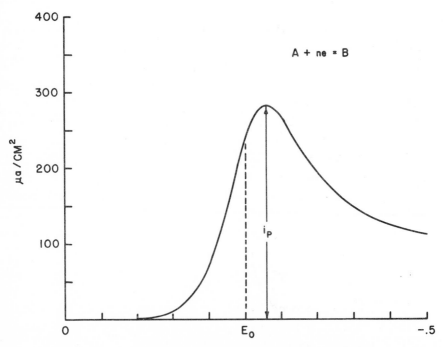

FIG. 3.20.   LSV curve for reversible reaction.

such that no current flows. As the potential becomes more
negative, the current rises exponentially. However, as
current flows, there results a progressive depletion of
material near the electrode, so that even as the potential
becomes more negative, the flux of material to the electrode
cannot increase indefinitely. When the peak current is
reached, the concentration of the reactant is essentially
zero at the electrode surface. As the depletion layer con-
tinues to expand, the current must decrease. It would continue to decrease to zero eventually, but the electrode
potential soon reaches a value where the supporting
electrolyte or solvent reduces.

The differential equations describing this system are
particularly simple to write down, involving just diffusion
and having a relatively simple boundary condition. Unfortunately, the results cannot be expressed analytically.
The simple system, a reversible electron transfer at a
planar electrode, was solved independently by Randles (52)
and Sevcik (53), using different approaches. Although the
current function can only be expressed in tabular form, the
peak current at 25°C could be shown to be equal to

$$i_p = 272 \; n^{3/2} \; AD^{1/2} \; c^o \; v^{1/2} \tag{3.16}$$

where i_p is the peak current in mA; n is the number of electrons; A is the area of electrode in cm^2; D is the diffusion coefficient in cm^2/sec; c^o is the concentration in
millimoles/liter; and v is the sweep rate in V/sec.

Proportionality to concentration opens up analytical
utility, while proportionality to square root of sweep rate
proves to be a diagnostic test for the simple electron-
transfer case.

The peak potential was shown to occur 28/n mV more
negative than the E^o of the reaction, and is independent
of concentration and sweep rate.

The corresponding case of a single irreversible electron transfer was solved by Delahay (<u>54</u>). The equation for
peak current is Eq. (3.17).

$$i_p = 301 \, n \, (\alpha n_a)^{1/2} AD^{1/2} c^o v^{1/2} \qquad\qquad (3.17)$$

where the symbols and units are the same as in Eq. (3.16),
except that n_a is the number of electrons transferred in the
potential determining step, and α is the transfer coefficient. Qualitatively, the major effect of irreversibility
is to spread out the peak-shaped profile as αn_a decreases,
and to lower the height, i.e., smaller values of i_p. The
peak potential also shifts to more cathodic potentials (for
reductions) as αn_a decreases and as the sweep rate increases.

When the electrochemical reaction occurs in discrete
steps, or when two electroactive materials are present,
multiple peaks are observed, as shown in Fig. 3.21. In
this case, i_{p2} is measured from the extension of the descend-
ing branch of the first peak, and not from the valley point,
A. This extension can be gotten experimentally by first
recording the LSV curve for the first peak, but stopping the
potential sweep just to the right of the first peak, while
continuing to record the current vs time. A second LSV
curve is generated using the full voltage span. The two
should superimpose to give a picture as in Fig. 3.21. For
the case of two independent electron-transfer reactions,
i_{p1} and i_{p2} will be separately proportional to the concentration of the respective materials. For the case of a
consecutive electron transfer, i.e., A + ne $\longrightarrow$ B, B +
me $\longrightarrow$ C, the situation is more complex, since the
current under the second peak has contributions from the
diffusion of A to the electrode and from the back diffusion
of B.

Up to this point in the discussion we can see that, at
least qualitatively, LSV gives much the same information
about electron-transfer reactions as does polarography, ex-

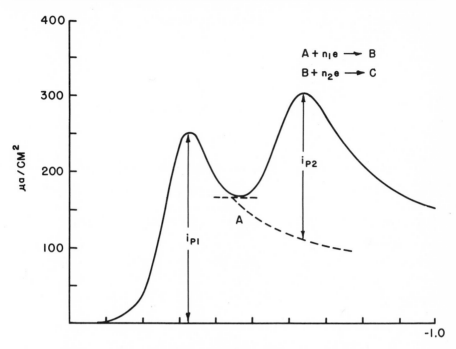

FIG. 3.21.   LSV curve for two-step reaction.

cept of course that LSV can be performed on solid electrodes.
There is, however, one further aspect of LSV which can make
it a much more informative tool.

If, after the normal linear scan has been completed,
the scan direction is reversed, products which have been
formed during the first sweep and are still in the vicinity
of the electrode may themselves undergo electron transfer.
For example, in a reversible situation the product would be
reoxidized to the starting material, resulting in a voltammogram as shown in Fig. 3.22.

In this case, the peak potential of the reverse sweep
is displaced about 57/n mV anodic of the forward peak potential (55).   The reverse peak current is equal to that of
the forward, when measured from the extension of the forward
curve.

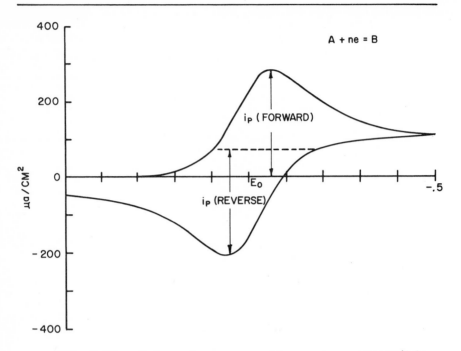

FIG. 3.22.   Triangular wave voltammogram, reversible
electron transfer.

In many organic systems, the course of the reaction can
be quite complex, with various intermediates and electron-
transfer steps possible. Many of these details may be seen
via LSV, particularly with the triangular wave technique.
Many such systems were analyzed by Nicholson, Shain, and.
Polcyn (55-60), including multistep electron transfers, both
reversible and irreversible, cases where intervening chemical
reactions are occurring, and combinations thereof. The features characteristic of each can in most cases be determined
by the way the peak currents and potentials vary with sweep
rate.

For cases which have not been analyzed in the literature,
or which may be too complex to handle by the technique used
in the above papers, the authors can recommend the digital
simulation techniques which have been documented by Covitz

(61) and Feldberg (62). In these techniques, once the
generalized computer program has been written, it is only
necessary to write down the original differential equations
and the boundary conditions; the computer then calculates
the i-V curve by numerical techniques.

In summary, the LSV technique has the following advantages:

1. Unlimited choice of electrode materials.
2. Wide range of time scale.
3. Ability to respond to intermediate products.
4. Model systems can be described analytically.

Although LSV requires somewhat more sophisticated instrumentation and more careful experimental techniques than
polarography, it is a technique which can reveal much more
about the course of reactions. It is recommended as a
technique to be adopted by the serious researcher in electroorganic synthesis.

3.4.3. ROTATING DISK VOLTAMMETRY

Up to this point, the voltammetric techniques we have
discussed were performed in "quiet," i.e., unstirred, solutions. This, of course, simplifies the mathematical formulation of the situation. In recent years, however, there
has come to the fore a technique, using stirring, which can
be handled quantitatively--namely, the technique of rotating
disk voltammetry, and the allied technique of ring-disk
voltammetry. In this technique, the electrode is a circular
disk, pressed into the end of a rod of inert material and
rotated about the axis of the rod. The hydrodynamics of
this system were first solved by Levich (63) and applied
to electrochemistry by Riddiford (64). Although the detailed
theory is far too complex to be developed here, the quantitative and qualitative results of this treatment and the applicability of this technique are discussed.

As the electrode is rotated, solution is brought up to the electrode surface and then spins out radially along the disk as shown in Fig. 3.23(a). If a current is applied to this electrode, the boundary concentration of electroactive material quickly reaches a steady state, since fresh mater-

ROTATING DISK ELECTRODE

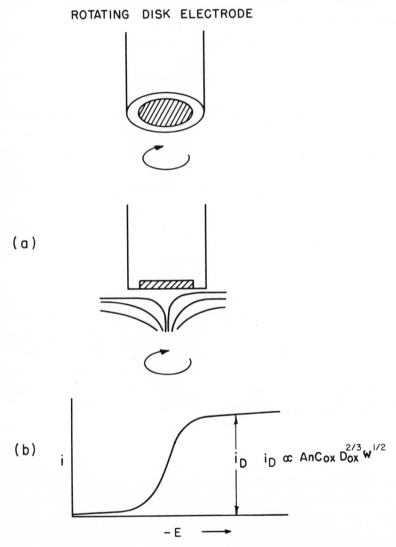

FIG. 3.23.   Rotating disk voltammetry.

ial is constantly being swept across the electrode surface.
Just as in polarography, when a slow linear potential sweep
is applied, there results a characteristic s-shaped wave,
as shown in Fig. 3.23(b). The <u>diffusion-limiting current</u>
was shown to be proportional to the square root of the rotation rate (somewhat analogous to the height of the mercury
column in polarography), and proportional to the two-thirds
power of the diffusion coefficient, to the bulk concentration
of electroactive material,to the number of electrons involved
over-all, and to the area of the electrode. Since the rotation rate can be varied over a wide range (say 10-10,000
rpm), this technique offers a stringent test for diffusion-
controlled processes, and departures from square-root dependence on rotation rate are significant in showing when other
phenomena become rate controlling.

The results of rotating disk voltammetry are quite
analogous to what can be seen by polarography, except, of
course, that one is not restricted to the use of mercury as
electrode material. Moreover, mercury itself can be used
in rotating disk voltammetry by wetting the surface of a
nickel disk with mercury (nickel does not amalgamate). Also,
some of the limitations of polarography, e.g., distortion of
waveforms at high concentrations, maxima, variations of drop
time with potential, etc., may not be present.

A recent powerful extension of this basic technique
promises to put rotating disk voltammetry on a par with
techniques such as linear sweep voltammetry in the ability
to gain insight about the intermediate products of electron-
transfer. This involves the use of a ring-shaped electrode
concentric with and in the same plane as the disk, as shown
in Fig. 3.24(a). The ring is electrically insulated from
the disk. Intermediates, or products formed at the disk
electrode, are swept past the ring electrode, where they
may be detected if they are themselves electroactive. For
unstable intermediates, the results are highly dependent on
the rotation rate.

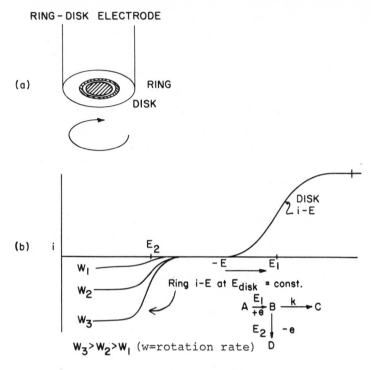

FIG. 3.24.   Ring-disk electrode.

For example, consider the case of a two-step electron transfer with an intervening chemical reaction, as illustrated in Fig. 3.24(b). The diffusion-limiting current for the oxidation of B at the ring depends on the kinetics of the intervening reaction, i.e., whether it survives the transit time from disk to ring, and on the rotation rate, i.e., the faster the rotation rate, the more of B will survive to reach the ring. Since the transit time to the ring can be of the order of 10^{-3} sec, this technique is capable of detecting quite unstable intermediates. In the above case, the ring diffusion current should increase with rotation rate.

The rotating disk and ring-disk techniques are not to be recommended for novices in the area of electroanalytical

techniques, since the apparatus and electrodes must be quite
carefully constructed before giving reproducible results.
There is no doubt, however, that this technique will continue to be developed, and may, in time, replace polarography
as the single most useful electroanalytical technique for
organic electrochemistry.

3.5. RECOMMENDED EQUIPMENT AND PROCEDURE FOR ELECTROORGANIC CHEMISTRY

As in entering any new area of endeavor, the potential
practitioner needs to start somewhere. No amount of reading
in a subject can replace the experience gained from conducting even the simplest experiment. This section suggests the
kind of equipment required in electroorganic chemistry, from
simple to more complex, and attempts to outline a general
procedure for deciding what route to follow.

3.5.1. EQUIPMENT

For carrying out electrolyses on an elementary level,
the experimenter need acquire only simple apparatus. The
standard sizes of glass beakers will serve for electrolysis
cells. Various types of sheet metal electrodes such as
lead, steel, copper, aluminum, and if possible platinum,
should be acquired. Mercury and graphite are also desirable.
Alligator clips and lengths of insulated wire are also required. A magnetic stirring hot plate will be a useful
accessory. As a start, various sizes of cellophane dialysis
membranes or ceramic (porous) cups will serve as cell dividers. Typical solvents such as DMF, acetonitrile, and
ethanol; as well as electrolytes such as lithium bromide,
tetrabutylammonium chloride, and various mineral acids
should be on hand. The power supply should be as simple as
possible, such as a Variac controlled constant voltage type,

0-50 V, and capable of delivering about 5 A. If necessary, this can be home-built and need only have a Variac, full-wave rectifier and filter capacitor ($\sim$2000 mfd). The operator should always take appropriate care in insulating and fusing the "hot" leads. For work in highly conductive media, an automobile storage battery (12 V) and rheostat (0-25Ω, 100 W) can be used. A standard volt-ohm-ammeter will be indispensable. The cell voltage and current should be monitored during the run, and the approximate number of coulombs passed should be recorded. As a start, about 10% more than the theoretical amount of coulombs should be passed. Generalized work-up apparatus for distillation and liquid-liquid extraction will be found useful.

A more advanced electrolysis laboratory will also have available various reference electrodes (calomel and Ag/AgCl) (Sec. 3.1.4) and a suitable voltmeter for measuring electrode potentials (see Sec. 3.3.1). A coulometer (Sec. 3.3.2) will also be useful. Cell designs such as are discussed in Sec. 3.1.1 will permit more flexibility and control than the beaker cells mentioned above. For the serious experimenter, a polarograph (Sec. 3.4.1) will be an indispensable starting point for understanding reactions of a novel type. The various analytical tools such as gas chromatography, NMR, and mass spectroscopy will be needed to determine the nature of all the products formed, and their relative yields. A complete mass balance will be required to permit the calculation of the electrical and chemical efficiencies. A power supply capable of automatically maintaining constant electrode potential, such as supplied by Anatrol, Hewlett-Packard, or Tacussel, will be a useful, although not essential accessory.

For the sophisticated and advanced worker in electrochemistry, cell designs of his own creation will undoubtedly be used. The "sandwich" cell construction (Sec. 3.1.1) requiring machine shop facilities, pumps, flowmeters, etc., may provide a suitable starting point for a more advanced

design. Electroanalytical instrumentation required to perform linear sweep voltammetry on solid electrodes should
also be acquired in addition to a polarograph. Fast sweep
instruments, such as supplied by Chemtrix, requiring oscilloscopic recording will be required to study the formation
of transient intermediates. Computer facilities should be
available to permit quantitative interpretation of the results of cyclic voltammetry and other more sophisticated
electroanalytical techniques. Complete identification of
all products and their yields will be essential if full
understanding and optimization of the reaction are to be
accomplished.

3.5.2. GENERALIZED PROCEDURES

Specific procedures obviously depend on the particular
reaction in question. However, the many common operations
and considerations found in electrochemical procedures permit one to sketch out a generalized plan of operation. For
convenience, such an outline is presented in Fig. 3.25.

3.6. SCALE-UP

Perhaps the greatest potential reward of work in organic electrochemistry would be its widespread use in large-
scale industrial synthesis. Realization of this goal would
undoubtedly spur on the level of research activity in this
field.

Let's begin by making it clear what we mean by scale-up
in the context of this section. For industrial purposes,
productivities in the multimillion-pound-per-year bracket
must be achieved. Because of the special nature of electrochemical operations, this almost certainly rules out any
thoughts of batch processing. Besides, as we shall soon
see, these same special features of electrochemistry are

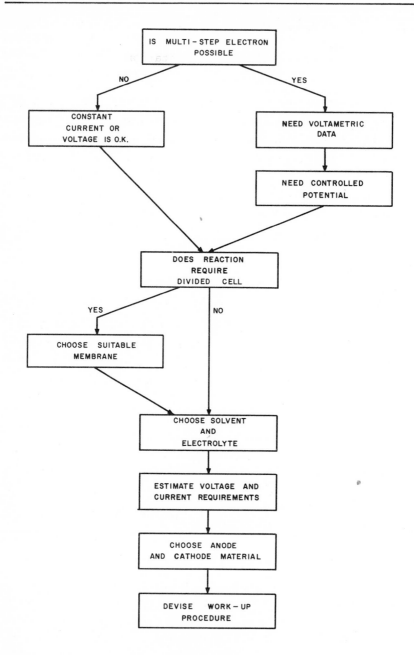

FIG. 3.25.

open invitations to those with imagination and creativity
to think in turns of the inherently more favorable continuous process. One might as well completely forget about the
commonly used cell designs for laboratory batch operations.
The standard one-liter, three-necked flask of conventional
organic chemistry scales up with almost one-to-one corres-
pondence to a 10,000-gal monster still; the industrial
electrolysis cell will be a completely different species
than a laboratory cell.

3.6.1 SPECIAL CONSIDERATIONS

In electrochemistry, one of the most easily controlled
variable is the current passing through the cell. This
is especially convenient for a continuous process, since
current (i.e., C/sec) is itself a rate of addition of a raw
material (electrons). Also, as shown in previous chapters,
and as further discussed in subsequent chapters, proper
control of electrode potential may be essential for optimal
results. Since a continuous process runs at steady-state
conditions, controlled potential operation is simply achieved
by the use of a constant current through the cells. For
noncritical cases, the production rate may be conveniently
varied over a wide range by altering the current.

1. By and large, it will probably be uneconomic and
inconvenient to operate a single continuous electrolyte cell
at near 100% conversion per pass, since efficiencies near
the end of reaction are usually low compared to what would
be obtained at the start. Since large-scale operations will
require multiple cells (can you imagine a single one-acre
electrolysis cell?), a series, or cascade, arrangement of
cells is possible. In this way, each cell operates as a
differential converter within the cell bank. Appropriate
data for such an operation can be obtained by studying the
chemical and electrical differential efficiencies as functions of conversion and current density. The approach to

steady-state conditions within the cell is exponential; at a rate depending on the total hold-up volume and the input flow rate--analogous to a continuous stirred tank reactor.

2. With cells operating in chemical cascade (in series) it is also convenient to operate with the electric current in series. In this way, the total voltage across a cell bank of n cells will be nV, where V is the individual cell voltage. The economics of voltage transformation are such that high-voltage operation, say 100 V, is desirable. Combinations of series and parallel connections are also possible.

3. Special attention, of course, must be paid to cell design. In particular, the surface-to-volume ratio must be as large as possible, and the interelectrode spacing must be minimized. For organic media, the latter is particularly important due to the inherently higher resistance of organic electrolytes. High individual cell voltage is not only costly, but may also cause excessive heat within the cell. For cells of the filter press design (multiple plane parallel electrodes sandwiched between nonconducting spacers), interelectrode spacing of $\sim\frac{1}{4}$ in. should be achievable for undivided cells. In any case, the cell should be designed such that the iR voltage drop is less than the combined electrode potentials. A possible method for increasing the active area of the electrode without increasing the geometric area involved roughening the electrode surface such that the surface irregularities are about 1/10 the interelectrode spacing. A factor of 2-3 in effective area should be achievable.

4. With such close spacing between electrodes, additional problems arise, such as suitable sized inlet and outlet ports, agitation within the cells, and possibly gas/liquid separation (i.e., foaming, if gases are produced). Entrance and exit externally through the spacing elements ($\frac{1}{4}$ in.) would result in a "plumber's nightmare" manifold. It may be possible, with suitable cutouts within the electrodes and spacers themselves, to arrange for serial or

parallel pumping of solution without external plumbing.
Also, since the throughput rate will probably be insufficient to guarantee good agitation within the cell, other
means, perhaps via a rapid recirculation loop, need to be
found. Techniques involving rotating or vibrating electrodes
are probably impractical on a large scale. In any case, the
agitation should be turbulent to avoid excessive concentration polarization within the cell.

     5. Consideration should also be given to the flow pat-
tern within the cell. Dead zones (for example, corners)
should be avoided, since they provide ideal locations for
the accumulation of particulate matter which may very likely
be present over the long term.

     6. Divided cells, of course, present a much more com-
plex design problem, are more costly, and may present additional maintenance problems. Every attempt should be made
to "trick" a reaction which apparently requires a divided
cell into a system which will be happy in a nondivided
environment. Changing the counter-electrode material,
addition of depolarizers which form easily removable products, or changing the solvent or electrolyte are worth
investigating. The nature of the counter-electrode process
must of course be thoroughly understood. If such a system
is unachievable, the potential industrial electrochemical
entrepreneur should take heart in the fact that today's
largest-scale organic electrolysis plant (the Monsanto
adiponitrile process (65)) utilizes divided cells. The problem of finding a suitable separator material (stable and
sufficiently conductive in the electrolyte) and supporting
it between electrodes is admittedly difficult, but not im-
possible.

     7. Choice of the materials of construction requires
careful consideration. The working electrode is usually
dictated by the constraints of the optimized electrosynthe-
sis as well as its freedom from corrosion under the long
term electrolysis conditions. Alloys may be of considerable

importance in attaining efficient, long-lasting, and struc-
turally sound electrodes. If this turns out to be a struc-
turally unsound material or is prohibitively expensive in
bulk, one should consider the possibility of cladding or
plating it onto a structural (and less costly) metal. If
possible, the counter-electrode should be of the same material as the working electrode, since this permits the convenient use of series electrical arrangement within the cell
bank and avoids separate electrical connections to each cell.
Flow of electrolyte between compartments of this type should
naturally be in nonconducting pipes (e.g., polyethylene,
polypropylene, Teflon, polyvinylchloride) and the electrical
path must be long enough to avoid appreciable bypass currents.
Plumbing and equipment downstream of the electrolysis cell
should also be selected with care and with an eye toward
the effect of small electrical leakage currents on corrosion.

3.6.2. WORK-UP

A continuous electrolytic system requires a continuous
product recovery and solvent/electrolyte recycling system--
the two, when "plugged" into each other, comprise the entire
plant. Most conventional recovery operations (e.g., crystallization, distillation, extraction) may "easily" be adapted
for continuous operation. The author's use of the word
"easily" in this context may be misleading; continuous re-
covery/recycle systems are "easy" in the sense that their
practice is widespread within the organic chemical industry;
handbooks exist for their design, equipment is commercially
available, and chemical engineers are not horrified at their
proposed use.

That the latter attitude is, in reality, strictly a
point of view is best illustrated by the following (true)
story. One of the authors (Covitz), in the not-too-distant

past, had occasion to consult with a group of professional
electrochemists (the inorganic variety) concerning a proposed continuous electrolytic synthesis. Being an organic
chemist by training, the author felt that professional
electrochemical advice would be helpful in properly ascertaining the feasibility of scaling up this system to indus-
trial-level production. The author discussed, in all its
gory detail, all of the electrolytic aspects, including the
interrelated effects of current density, chemical and electrical efficiencies, concentration of starting material and
electrolyte, temperature, conversion, long-term effects, and
economics. The proposed recovery/recycle system was mentioned
briefly, being dispatched with one drawing of the process,
which was comprised of two extractions and one distillation.
Within minutes of the end of the talk, the audience achieved
unaminous agreement that no particular difficulties were en-
visioned in the continuous electrolysis system, but that the
recovery/recycle system would probably create insurmountable
problems in its continuous operation. The remaining hours
of discussion were spent in ironing out the details of the
recovery system; this in turn led to a much more thorough
laboratory effort to prove its feasibility. An alternative
recovery system was the final result!

Since continuous recovery/recycle systems are usually
specific to the product being produced, the reader may get
a better feel for what is involved by considering the following examples of specific electroorganic electrolysis which
either are used, or are feasible for use in, industrial-scale
processes.

3.6.3. EXAMPLES

The following are representative examples of potential
or actual commercial electroorganic syntheses. Many other
cases may be found by consulting Swann's extensive biblio-
graphy of electroorganic patents (66).

1. Propylene oxide: A patent issued to the M. W. Kellog Company discloses the production of propylene oxide from propylene by electrolysis (67). The process involves the electrolysis of sodium chloride for example, at the anode of a divided cell to produce hypochlorous acid, which reacts with propylene to produce ethylene chlorohydrin. The anolyte is circulated into the cathode chamber where the alkaline medium (from reduction of water) converts the chloro-hydrin to propylene oxide, regenerating chloride ion. The propylene oxide is isolated by distillation, the aqueous electrolyte being returned to the cells. The reactions taking place are:

at anode

$$Cl^- + H_2O \longrightarrow HOCl + H^+ + 2e \qquad\qquad (3.18)$$

in anode chamber

$$CH_3CH=CH_2 + HOCl \longrightarrow CH_3\underset{\underset{OH}{|}}{CH}-CH_2Cl \qquad (3.19)$$

$$H^+ + OH^- \longrightarrow H_2O$$

in cathode chamber

$$CH_3\underset{\underset{OH}{|}}{CH}-CH_2Cl + OH^- \longrightarrow H_2O + Cl^- + CH_2CH\overset{O}{\overbrace{\qquad}}CH_2 \ (3.20)$$

$$2H_2O + 2e \longrightarrow H_2 + 2OH^-$$

The net reaction is therefore,

$$CH_3CH=CH_2 + H_2O \longrightarrow CH_3CH\overset{O}{\overbrace{\qquad}}CH_2 + H_2 \qquad (3.21)$$

and requires a net expenditure of electrical energy. Closer inspection reveals that this is really an example of electro-regeneration of the oxidant, hypochlorous acid. The organic chemistry, i.e., halohydrin formation and base-catalyzed epoxide formation, is well known, and in fact, forms the

basis of the conventional large-scale production of propylene oxide. In the latter, both chlorine and excess caustic
are consumed as raw materials (plus acid to neutralize the
caustic) and stoichiometric amounts of halide salts are _
formed as by-products, creating pollution problems in their
disposal. Note how neatly the electrochemical route both
avoids the use of two raw materials and also solves the pollution problems.

    2. Hydroquinone: The production of hydroquinone by
electrolytic oxidation of aqueous phenol solution is dis-
closed in a patent awarded to the Union Carbide Corporation
(68).

    The process involves oxidation of phenol to benzoqui-
none at a lead dioxide electrode in the presence of dilute
aqueous sulfuric acid. The benzoquinone, in turn, is reduced at the cathode of the undivided cell. Competitive
adsorption of phenol and benzoquinone at the anode prevents
reoxidation of the hydroquinone. The work-up is comprised
of distillation (of water and phenol) and crystallization of
the hydroquinone in the bottoms. The (simplified) reaction
scheme is as follows:

at anode

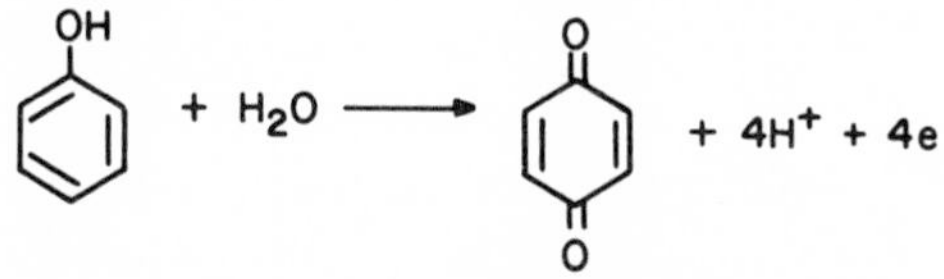

$$\text{C}_6\text{H}_5\text{OH} + H_2O \longrightarrow \text{(benzoquinone)} + 4H^+ + 4e \qquad (3.22)$$

at cathode

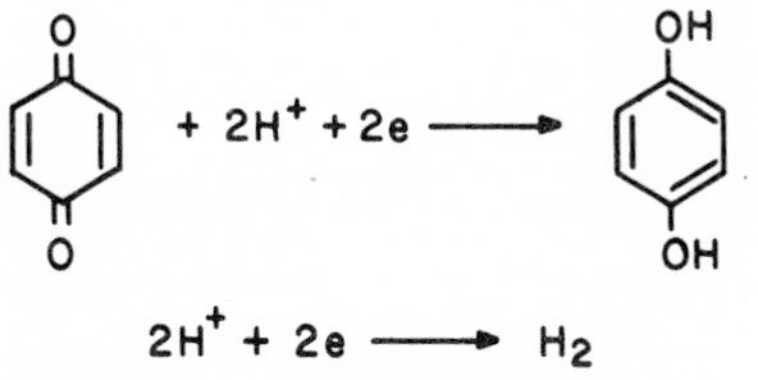

$$\text{(benzoquinone)} + 2H^+ + 2e \longrightarrow \text{(hydroquinone)} \qquad (3.23)$$

$$2H^+ + 2e \longrightarrow H_2$$

The net reaction is therefore

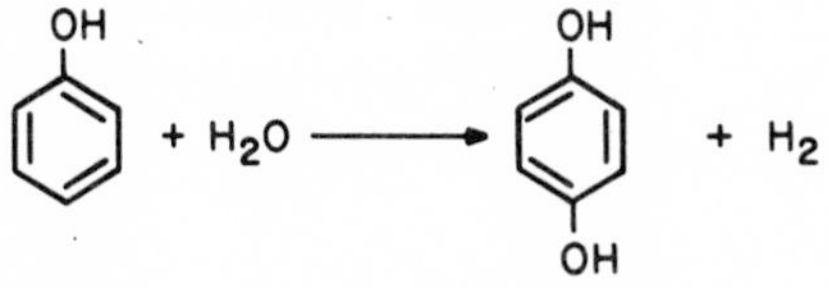

$$\text{(phenol)} + H_2O \longrightarrow \text{(hydroquinone)} + H_2 \qquad (3.24)$$

and requires the expenditure of electrical energy. The
conventional process for production of hydroquinone involves
oxidation of aniline with manganese dioxide and sulfuric
acid to produce benzoquinone. Reduction of the benzoquinone
with iron to hydroquinone completes the process. Large quantities of manganese and calcium (lime is used to neutralize
the excess acid) salts are also produced as by-products.
The electrolytic route utilizes a less expensive raw material and creates no major by-products other than hydrogen.

It may be of interest to compare the similarities of
examples 1 and 2 from a general viewpoint. Both utilize
water to introduce an oxygen atom into the starting material,
and both result in the desired product plus hydrogen gas,
with no by-product salts being formed. It is conceivable,
furthermore, that the hydrogen gas itself could be a useful
product, or that it could be "burned" chemically or electrochemically to regain some of the energy consumed in the basic
process.

3. <u>Adiponitrile</u>: The conversion of acrylonitrile to
adiponitrile, a key intermediate for the production of 6,6-
Nylon, has been accomplished by the Monsanto Company (<u>65</u>)
and is currently being carried out in large scale (in excess
of 40 MM lb/yr). The process involves reductive hydrodimerization of a partially aqueous solution of acrylonitrile with
large amounts of tetraethylammonium p-toluenesulfonate. The
quaternary ammonium salt assists in the solubilization of
the acrylonitrile and makes the medium highly conductive.
The electrolysis cells (about one square meter in cross section area) are operated at about 30 A/dm^2 and 10 V. An ion-
exchange membrane is used as the cell divider. The adiponitrile product is extracted into excess acrylonitrile and
recovered by distillation of the acrylonitrile. A second
extraction with water removes remaining acrylonitrile and
leaves crude adiponitrile. The electrolyte salt is purified
by crystallization before being recycled to the cells.

The reactions occurring are:

at cathode

$$2CH_2 = CH-CN + 2H^+ + 2e \longrightarrow NC(CH_2)_4CN \qquad (3.25)$$

through membrane

$$2H^+(anolyte) \longrightarrow 2H^+(catholyte) \qquad (3.26)$$

at anode

$$H_2O \longrightarrow 1/2\ O_2 + 2H^+ + 2e \qquad (3.27)$$

so that the net reaction is

$$2CH_2=CHCN + H_2O \xrightarrow{\text{electrical energy}} NC(CH_2)_4CN + 1/2\ O_2 \quad (3.28)$$

A discussion of how the electrolytic hydrodimerization process compares with conventional commercial processes may be found in a review article by Beck (69).

4. Tetraethyl Lead: The commercial production of tetraethyl lead (36 MM lb/yr) has been accomplished by the Nalco Chemical Company (70). The process involves electrolysis of ethylmagnesium chloride in ether solution at a lead anode to form tetraethyl lead and magnesium chloride. The magnesium is recovered at the cathode and presumably reused to form the starting Grignard reagent from ethyl chloride. A two-stage distillation serves to isolate the tetraalkyl lead and recycle solvent. The process is unusual in that the lead anode (in the form of pellets) is consumed, and in that the starting material (RMgCl) serves as its own electrolyte. The reactions occurring are:

external to cell

$$4RCl + 4Mg \longrightarrow 4RMgCl \qquad (3.29)$$

at anode

$$4RMgCl + Pb \longrightarrow 4MgCl^+ + 4e + PbR_4 \qquad (3.30)$$

at cathode

$$4MgCl^+ + 4e \longrightarrow 2Mg + 2MgCl_2 \qquad (3.31)$$

The over-all reaction is therefore

$$4RCl + Pb + 2Mg \longrightarrow PbR_4 + 2MgCl_2 \qquad (3.32)$$

It is interesting to speculate whether the system could be operated with full recovery of the magnesium. This, of course, could only be done at the expense of decreased anode efficiency, and some other oxidation would have to occur. If chlorine were evolved, the net reaction would be

$$4RCl + Pb \longrightarrow PbR_4 + 2Cl_2 \qquad (3.33)$$

3.6.4. ECONOMICS

Although detailed economics will depend heavily on the particular process under consideration, several general principles are possible to estimate the electrical costs. To do this, let us take a particular example: assume we have a system which yields a product of mol wt 100, at 90% electrical efficiency, 100% chemical efficiency, requiring 2 electrons per mole, and operating at 8 V cell(s) voltage. The electrical power to produce 1 lb of product is :

$$\frac{454 \text{ g}}{\text{lb}} \times \frac{\text{mole}}{100 \text{ g}} \times \frac{2 \times 10^5}{0.9} \frac{C}{\text{mole}} \times \frac{1 \text{ A-sec}}{1 \text{ C}} \times \frac{1 \text{ hr}}{3600 \text{ sec}} \times 8 \text{ V}$$

$$\underset{\sim}{} 2220 \text{ V-A-hr/lb or } 2.2 \text{ kWh/lb} \qquad (3.34)$$

If we assume, conservatively, a power cost of $0.01/kwh for industrial use, the cost for electricity, considered as a raw material, is $0.022 for 1 lb of product. The electrical

cost is significant, but normally low compared to the cost of
other raw materials. The above estimate is, moreover, con-
servative, since the cost of electricity can be appreciably
lower at favorable locations.

It might also be constructive to consider what size
power equipment would be required for the above example for,
let's say, a 10 MM lb/yr plant (Eq. 3.35)

$$2.2 \ \frac{kWh}{lb} \ x \ \frac{10^7 lb}{yr} \ x \ \frac{1 \ yr}{365 \ days} \ x \ \frac{1 \ day}{24 \ hr} = 2,500 \ kW \qquad (3.35)$$

Large-scale power equipment (including rectification and
filtration) operating at reasonably high voltages (>100 V)
costs in the vicinity of $20/kW (71), so that the above
would cost in the vicinity of $50,000--a minor investment
when compared to what the cell banks and recovery equipment
would probably cost.

The recovery equipment will probably be of straightforward design and can be costed accordingly. The cost of
the cells depends, of course, on their design, and the
authors have received estimates varying from $50/m^2 to
$500/m^2. If the latter figure is the more appropriate, then
it would certainly pay to operate at the highest feasible
current density, even at the expense of somewhat lower
current efficiency and higher cell voltage. At 1000 A/m^2
(10 A/dm^2), the cells for our example would cost approximately $150,000 (at $500/m^2).

The plant, assuming full automation, including current
regulation, feedback control, fail-safe shutdown, and cell
bypassing for scheduled maintenance, would probably require
no more than two operators. Presumably, one would be responsible for the cell room and the other for the recovery
operation.

Although electrochemical production of organics in the
billion-pound/year range are probably not feasible because
of the way the scale-up factors affect the cell sizes needed

(increases with cell area rather than by volume, as with conventional processes), all estimates so far point to favorable economics for the multimillion-pound/year plant.

3.6.5. FUTURE PROSPECTS

The long-term outlook for industrial electroorganic processes is extraordinarily favorable when viewed with an unbiased eye. The trend in power costs, by all estimates, is bound to decrease further with the advent of widespread nuclear power generators. This is particularly so when one considers the promise of the ultra-efficient breeder reactors, or even thermonuclear power plants (72). With the increasing concern about environmental pollution, the organic chemical industry will be put under increasing pressure to adopt processes which do not involve large quantities of chemical pollutants. Organic electrochemical techniques, in many cases, offer promising solutions to many of these problems. One can only hope that high-level management of our huge chemical corporations will recognize these significant factors.

At the present time, however, considerable difficulties and obstacles await the organic electrochemist who attempts to invade the realm of industrial-scale operation. There exist no engineering manuals for cell design, nor are any commercial cells for inorganic operations likely to be satisfactory. Pilot plant facilities generally are nonexistent, so it will be up to laboratory data to prove feasibility of many aspects which are usually ignored in laboratory batch operation.

The authors highly recommend hard work, perseverance, ingenuity, and healthy optimism (plus electricity!) as techniques to overcome some of these problems. AMEN.

REFERENCES

1. M. J. Allen, Organic Electrode Processes, Rheinhold, New York, 1950.

2. J. M. Kolthoff and N. Tanaka, Anal. Chem., 26, 632 (1954).

3. V. F. Gaylor, Paint Technology, 75, 1078 (1953).

4. S. S. Lord and L. B. Rogers, Anal. Chem., 26, 284 (1954).

5. V. Sihvonen, Trans. Faraday Soc., 34, 1062 (1938).

6. J. B. Morris and J. M. Schempf, Anal. Chem., 31, 286 (1959).

7. R. N. Adams, Electrochemistry at Solid Electrodes, Dekker, New York, 1969, p. 280.

8. C. H. Hampel, ed., The Encyclopedia of Electrochemistry, Reinhold, New York, 1964, p. 762.

9. H. B. Mark and W. C. Vosburgh, J. Electrochem. Soc., Vol. 616 (1961).

10. F. A. Lowenheim, ed., "Nickel," in Modern Electroplating, Wiley, New York, 1963.

11. J. F. Coetzee and J. J. Campion, J. Am. Chem. Soc., 89, 2517 (1967).

12. J. M. Kolthoff and J. F. Coetzee, J. Am. Chem. Soc., 79, 870 (1957).

13. C. D. Ritchie and G. H. Mergeole, J. Am. Chem. Soc., 89, 1447 (1967).

14. K. K. Barnes and C. K. Mann, Anal. Chem., 36, 2502 (1964).

15. D. H. Morman and G. A. Harlow, Anal. Chem., 39, 1869 (1967).

16. R. E. Alexander et al., J. Am. Chem. Soc., 89, 3703 (1967).

17. J. N. Butler, "Reference Electrodes in Aprotic Organic Solvents," in Advances in Electrochemistry and Electro-chemical Engineering, (P. Delahay and C. Tobias, eds.), Vol. 7, Interscience, New York, 1970.

18. C. K. Mann, "Non-Aqueous Solvents for Electrochemical Use," in Electroanalytical Chemistry, (A. J. Bard, ed.) Vol. 3, Dekker, New York, 1969, pp. 57-134.

19. J. F. O'Donell, unpublished data, 1965.

20. A. Weissberger et al., Organic Solvents, Wiley, New York, 1955.

21. P. G. Stecher, ed., The Merck Index, 7th Ed., Merck and Co., Rahway, N. J., 1960.

22. J. F. Coetzee et al., Anal. Chem., 34, 1159 (1962).

23. J. F. O'Donell, J. T. Ayres, and C. K. Mann, Anal. Chem., 37, 1161 (1965).

24. J. P. Billon, J. Electroanal. Chem., 1, 486 (1960).

25. H. Schmidt and J. Noack, Z. Anorg. Allgem. Chem., 296, 262 (1958).

26. S. Wawzonek et al., *J. Electrochem. Soc.*, 102, 235 (1955).

27. R. E. Visco, Spring Meeting, The Electrochemical Society, Dallas, Texas, May 1967.

28. A. B. Thomas and E. G. Rochow, *J. Am. Chem. Soc.*, 79, 1843 (1957).

29. S. B. Brunamer, *J. Phys. Chem.*, 42, 1636 (1965).

30. J. F. O'Donnell and C. K. Mann, *J. Electroanal. Chem.*, 13, 157 (1967).

31. H. Schmidt and H. Meinert, *Z. Anorg. Allgem. Chem.*, 295, 156 (1958).

32. V. A. Pleshov and A. M. Monossen, *Acta Physiochim*, USSR, 3, 615 (1935).

33. J. Sedlet and T. DeVries, *J. Am. Chem. Soc.*, 73, 5809 (1951).

34. G. Fraenkel, S. H. Ellis, and D. T. Dix, *J. Am. Chem. Soc.*, 87, 1406 (1965).

35. H. W. Sternberg et al., *J. Am. Chem. Soc.*, 89, 186 (1967).

36. J. E. Dubois, P. C. Lacaze, and A. M. de Fiquelmont, *Compt. Rend.*, c262, 181 (1966).

37. A. Cisak and P. J. Elving, *J. Electrochem. Soc.*, 110, 160 (1963).

38. W. R. Turner and P. J. Elving, *Anal. Chem.*, 37, 467 (1965).

39. P. G. Stecher, ed., *The Merck Index*, Merck and Co., Rahway, N. J., 1968.

40. J. Perichon and R. Buvet, *Electrochim. Acta*, 9, 567 (1964).

41. P. D. T. Coulter and R. J. Iwamoto, *J. Electroanal. Chem.*, 13, 21 (1967).

42. P. J. Smith and C. K. Manor, unpublished data, 1967.

43. R. J. Burris, Ph.D. Thesis, University of Tennessee, Knoxville, 1961.

44. I. M. Kolthoff and J. B. Redly, *J. Electrochem. Soc.*, 108, 980 (1961).

45. J. Phelps, K. S. V. Santhanom, and A. J. Bard, *J. Am. Chem. Soc.*, 89, 1752 (1967).

46. R. F. Nelson and R. N. Adams, *J. Electroanal. Chem.*, 13, 184 (1967).

47. G. G. Swain et al., *J. Am. Chem. Soc.*, 89, 2648 (1967).

48. W. M. Schwarz and I. Shain, *Anal. Chem.*, 12, 1770 (1963).

49. J. Heyrovsky, *Chem. Listy*, 16, 256 (1922).

50. D. Ilkovic, *Collection Czech. Chem. Commun.*, 6, 498 (1934).

51. H. H. Bauer, J. Electroanal. Chem., 16, 419 (1968).

52. J. E. B. Randles, Trans. Faraday Soc., 44, 327 (1948).

53. A. Sevcik, Collection Czech. Chem. Commun., 13, 349 (1948).

54. P. Delahay, J. Am. Chem. Soc., 75, 1190 (1953).

55. R. S. Nicholson and J. Shain, Anal. Chem., 36, 706 (1964).

56. R. S. Nicholson, Anal. Chem., 37, 1351 (1965).

57. D. S. Polcyn and I. Shain, Anal. Chem., 38, 370 (1966).

58. D. S. Polcyn and I. Shain, Anal. Chem., 38, 376 (1966).

59. R. S. Nicholson and I. Shain, Anal. Chem., 37, 178 (1965).

60. R. S. Nicholson, Anal. Chem., 37, 667 (1965).

61. F. H. Covitz, preprints of Symposium, The Synthetic and Mechanistic Aspects of Electroorganic Chemistry, Oct. 1968, U.S. Army Research Office, Durham, North Carolina, p. 99.

62. S. Feldberg, "Digital Simulation: A General Method for Solving Electrochemical Diffusion-Kinetic Problems," in Electroanalytical Chemistry, (A. J. Bard, ed.), Vol. 3, Dekker, New York, 1969.

63. I. G. Levich, Physicochemical Hydrodynamics, Prentice-Hall, Englewood Cliffs, N. J., 1962.

64. A. C. Riddiford, "The Rotating Disc System," in Advances in Electrochemical Engineering, (P. Delahay and C. W. Tobias, eds.), Vol. 4, Interscience, New York, 1966.

65. M. M. Baizer to Monsanto Chem. Co., U. S. Pat. 3,193,480-1, (1963), Chem. Eng. (Nov. 1965).

66. S. Swann, Electrochem. Technol., 1, 308 (1961); ibid., 4, 550 (1966); ibid., 5, 53, 101, 393, 467, 549 (1967); ibid., 6, 59 (1968).

67. J. LeDuc to M. W. Kellog Co., Belg. Pat. 637,691 (1962); Fr. Patent 1,375,298 (1964).

68. F. Covitz (to Union Carbide Corp.), U. S. Pat. 3,509,031 (1970).

69. F. Beck, Chem. Ing. Tech., 42, 153-164 (1970).

70. D. G. Braithwaite (to Nalco Chem. Co.), U. S. Pat. 3,256,161 and 3,380,899, (1961, 1964).

71. Westinghouse Electric Co., informal communication.

72. G. T. Seaborg and J. L. Bloom, Scientific American, 223, 13 (1970); "Energy and Power," in Scientific American, 224, 36-191 (1971).

4.1. CARBONYL COMPOUNDS

4.1.1. GENERAL CONSIDERATIONS

The electrochemical reduction of carbonyl groups is somewhat difficult to affect. In most aqueous systems a cathode of high hydrogen overvoltage (mercury, lead, etc.) is used in the reduction. In general, aldehydes are easier to reduce than ketones which are, in turn, easier to reduce than acids, esters, and amides. The products from the reduction of carbonyl groups vary with reaction conditions (discussed in some detail below), i.e., nature of cathode, pH of solution. Thus, an aldehyde or ketone can be reduced to the alcohol, pinacol, or hydrocarbon. Since comprehensive discussion of the reduction of all carbonyl groups could easily be the topic of a separate book, only the reduction of aldehydes and ketones are discussed here. For the reduction of other carbonyl groups, the reader should consult Ref. 1-4....9-16.

4.1.2. EFFECT OF REACTION CONDITIONS ON COURSE OF REDUCTION

Nature of Electrodes. The nature of the electrode has a pronounced effect on the product obtained from the reduction of carbonyl compounds. For example, acetone [1] with a lead cathode under acid conditions affords isopropanol [2] and pinacol [3]:

$$CH_3\text{-}\overset{\overset{\text{O}}{\|}}{C}\text{-}CH_3 \xrightarrow[\text{Pb}]{+2e} CH_3\text{-}\overset{\overset{\text{OH}}{|}}{C}H\text{-}CH_3 + CH_3\text{-}\overset{\overset{\text{OH}}{|}}{\underset{\underset{CH_3}{|}}{C}}\text{-}\overset{\overset{\text{OH}}{|}}{\underset{\underset{CH_3}{|}}{C}}\text{-}CH_3 \quad (4.1)$$

[1]   [2]   [3]

while in a reduction on a mercury cathode the main product is the alcohol with some propane (5). Leonard (26) reported that the reduction of p-aminoacetophenone in aqueous acid on a mercury cathode afforded the corresponding alcohol while reduction on a tin cathode gave the pinacol as the major product. Escherich and Moest (47) found that the reduction of Michler ketone yields the pinacol with a copper cathode, while both the alcohol and the pinacol were formed in equal amounts when nickel was used as the cathode. The reduction of 2-methylcyclohexanone [4] is another example (6), Eqs. (4.2), (4.3):

$$+ 2e \xrightarrow{Cu} \quad \text{(cis-form)} \quad (4.2)$$

$$+ 2e \xrightarrow{Pb} \quad \text{(trans-form)} \quad (4.3)$$

In some cases, the potential of the electrode as well as
its current density may have an effect on the product of
the reduction ($\underline{7}$, $\underline{8}$, $\underline{27}$). For example, Allen ($\underline{27}$) reported
that an increase in potential in the reduction of p-amino-
acetophenone increased the yield of the pinacol over the
corresponding alcohol.

pH of Medium. Reduction of aldehydes and ketones under
acidic conditions favor the formation of pinacols ($\underline{17}$, $\underline{18}$)
while, under basic conditions, the corresponding alcohols
are the main products ($\underline{19}$, $\underline{20}$):

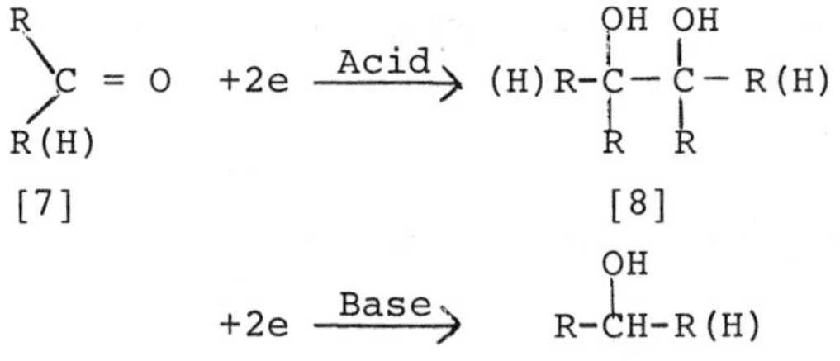

$$\begin{array}{c} R \\ \diagdown \\ C = O \\ \diagup \\ R(H) \end{array} +2e \xrightarrow{\text{Acid}} (H)R-\overset{\displaystyle OH}{\underset{\displaystyle R}{C}}-\overset{\displaystyle OH}{\underset{\displaystyle R}{C}}-R(H) \qquad (4.4)$$

[7]   [8]

$$+2e \xrightarrow{\text{Base}} R-\overset{\displaystyle OH}{C}H-R(H) \qquad (4.5)$$

[9]

There are several exceptions, however, to the above statement, which should not be considered as a rule of thumb ($\underline{21}$-
$\underline{23}$), Eq. (4.6):

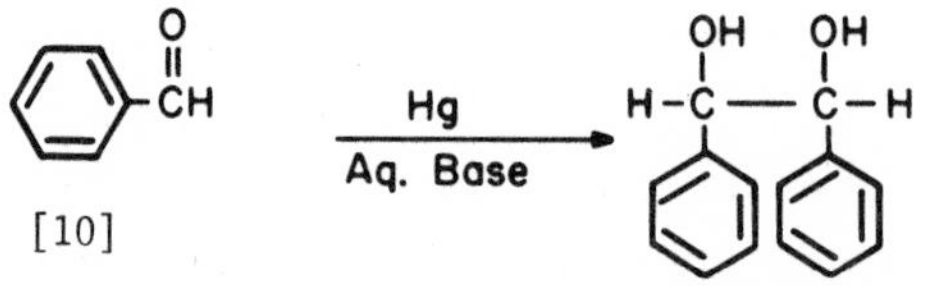

$$(4.6)$$

[10]   [11]
dl and meso-isomers

The pH of the medium may have an effect on the starting
material or its electrochemical product. For example,
under basic conditions, aliphatic aldehydes and ketones
may undergo self-condensation. Aromatic aldehydes may
also form benzoins ($\underline{29}$). On the other hand, reduction
under acidic media should be carried out at low tempera-

tures ($\sim 40^\circ$) otherwise the pinacol product may be prone
to rearrange to the corresponding pinacolone (30).

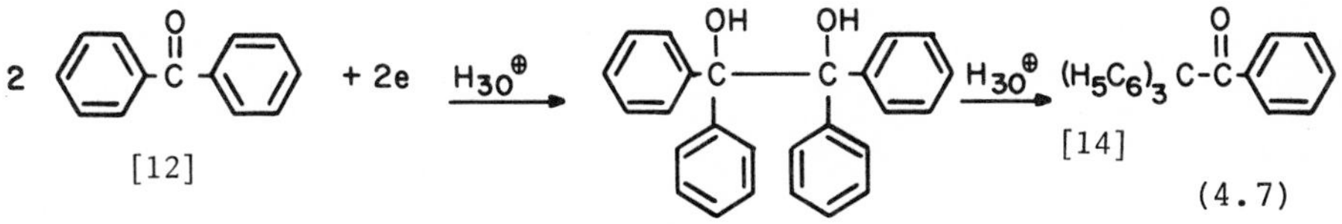

$$(4.7)$$

Other rearrangement of electrochemical reduction products
under acidic conditions have been reported (31, 32).

4.1.3. THE FORMATION OF HYDROCARBONS

Aside from pinacol and alcohols, the reductions of
aldehydes and ketones can afford the corresponding hydrocarbons (24) Eq. (4.8):

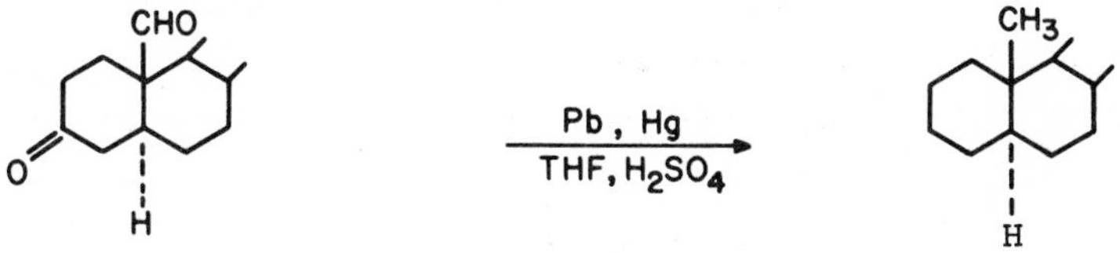

$$(4.8)$$

In some cases, a so-called "catalytic cathode" such as
cadmium has been used. Under such conditions, acetaldehyde
was reduced to ethane (25).

The formation of hydrocarbon products in appropriate
solvents allows the incorporation of deuterium in molecules
(28). This can be of great value in studying reaction
mechanisms.

4.1.4. MECHANISM OF REDUCTION

Nonaqueous Medium. The reduction of aldehydes and

ketones in aprotic solvents, e.g., DMF, exhibit two one-
electron polarographic waves (33), Eq. (4.9) and (4.10):

$$\underset{[17]}{\overset{R}{\underset{R(H)}{>}}C = O + 1e} \quad \xrightarrow{\overset{First}{Wave}} \quad \underset{[18]}{\overset{R}{\underset{R(H)}{\searrow}}C-O^{\ominus}} \qquad (4.9)$$

$$\underset{[19]}{\overset{R}{\underset{R(H)}{\searrow}}C - O^{-} + 1e} \quad \xrightarrow{\overset{Second}{Wave}} \quad \underset{[20]}{\overset{R}{\underset{R(H)}{\searrow}}C-O^{\ominus}} \qquad (4.10)$$

The appearance of a second wave, Eq. (4.10), at a
more negative potential is due to the addition of a second
electron to the negatively charged species, the ketyl radical anion. Thus the separation of the two waves will depend, to a large extent, on the stability of the ketyl radical and the availability of protons in the medium. For aliphatic aldehydes and ketones, the radical anion has little
stability and hence, undergoes protonation and further reduction to afford alcohol-type products. On the other hand,
aromatic ketyl radicals have sufficient stability to be detected by ESR spectroscopy (34). These can diffuse away
from the electrode and dimerize to afford pinacol-type products. For further discussion on the mechanism of the
electrochemical reduction of aldehydes and ketones the
reader should consult Refs. 35-41.

Aqueous Medium. The mechanism of the reduction of
aldehydes and ketones in aqueous media has been extensively discussed (42-44). Under acidic conditions (43),
two one-electron polarographic waves are observed, Eqs.
(4.11) and (4.12):

$$\underset{[21]}{\overset{R}{\underset{R}{>}}C = \overset{\oplus}{O} \quad H + 1e} \quad \xrightarrow[E_1]{\overset{First}{Wave}} \quad \underset{[22]}{\overset{R}{\underset{R}{\searrow}}C - OH} \qquad (4.11)$$

$$\underset{[23]}{\overset{R}{\underset{R}{\diagdown}}\overset{\oplus}{C}-OH} \longrightarrow \tfrac{1}{2}\ \underset{[24]}{R-\underset{\underset{R}{|}}{\overset{\overset{OH}{|}}{C}}-\underset{\underset{R}{|}}{\overset{\overset{OH}{|}}{C}}-R} \qquad (4.12)$$

$$+\ 1e\ +\ 1H^{\oplus}\ \xrightarrow[\underset{E_2}{Wave}]{Second}\ \underset{[25]}{\overset{R}{\underset{R}{\diagdown}}H-C-OH} \qquad (4.13)$$

The dimerization or further reduction of the ketyl radical, Eq. (4.12), depends on the applied potential. Thus, under macroscale conditions, with reduction at the plateau of the first polarographic wave, the main product is the pinacol, while reduction at the plateau of the second wave affords the corresponding alcohol (45).

As the pH of the medium increases, the two polarographic waves, Eqs. (4.11) and (4.13), merge and at a more negative potential a third wave appears, Eq. (4.15):

$$\underset{[26]}{\overset{R}{\underset{R}{\diagdown}}C=O}\ +\ 1e\ \xrightarrow{E_1\leftrightarrow E_2}\ \underset{[27]}{\overset{R}{\underset{R}{\diagdown}}\overset{\oplus}{C}-O^{\ominus}} \qquad (4.14)$$

$$\underset{[28]}{\overset{R}{\underset{R}{\diagdown}}\overset{\oplus}{C}-O^{\ominus}}\ +\ 1e\ \xrightarrow[\underset{E_3}{Wave}]{Third}\ \underset{[29]}{\overset{R}{\underset{R}{\diagdown}}\overset{\oplus}{C}-O^{\ominus}}\ \xrightarrow{2H^{\oplus}}\ \underset{[30]}{\overset{R}{\underset{R}{\diagdown}}CH-OH} \qquad (4.15)$$

Under these conditions, a high-energy species [29] is being produced. Consequently, it is not surprising that the polarographic waves observed under basic conditions appear at a more negative potential than the first wave observed under acidic conditions.

4.1.5.   ROLE IN ORGANIC SYNTHESIS

The electrochemical reduction of aldehydes and ketones
can be controlled to give alcohols, pinacols, or hydrocarbons.   In certain cases, electrochemical reduction affords
different products than are obtained by using conventional
reducing agents.   For example, the electrochemical reduction
of α-methyldesoxybenzoin [31] affords the erythro-1,2-
diphenylpropanol-1 [32], while reduction with sodium in alcohol yields a mixture of erythro- and threo- alcohols (48).

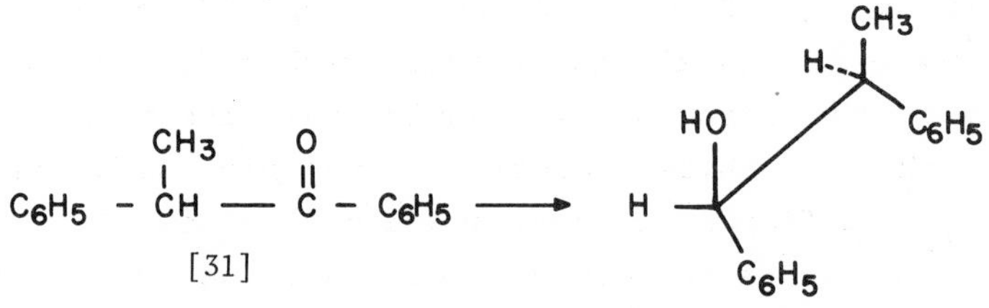

$$C_6H_5 - \overset{\overset{\displaystyle CH_3}{|}}{CH} - \overset{\overset{\displaystyle O}{||}}{C} - C_6H_5 \longrightarrow \qquad (4.16)$$

[31]   [32]

In other cases, the formation of either or both of the
erythro- and threo-isomer has been reported for electrochemical reductions.   The stereochemistry of this process
has been reviewed in Refs. 17, 56, and 57.

The electrochemical reduction of aldehydes and ketones
has not been restricted to simple molecules and the yield
of the product(s) is often above 50% (21), as shown in Eqs.
(4.17) and 4.18):

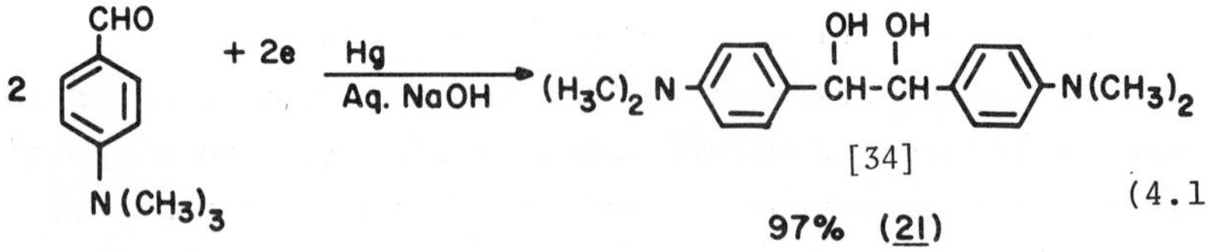

$$2 \underset{[33]}{\overset{CHO}{\underset{N(CH_3)_3}{\bigcirc}}} + 2e \xrightarrow[\text{Aq. NaOH}]{\text{Hg}} (H_3C)_2 N - \bigcirc - \overset{\overset{\displaystyle OH\ OH}{|\ \ |}}{CH-CH} - \bigcirc - N(CH_3)_2$$

[34]   (4.17)

97% (21)

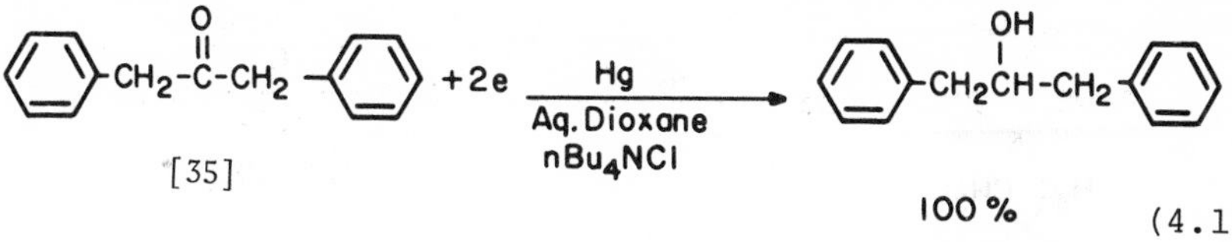

$$100\% \qquad (4.18)$$

[36]

An important application of electrochemical reduction lies in its use for the preparation of reactive compounds such as the cyclopropanediol [38] (50):

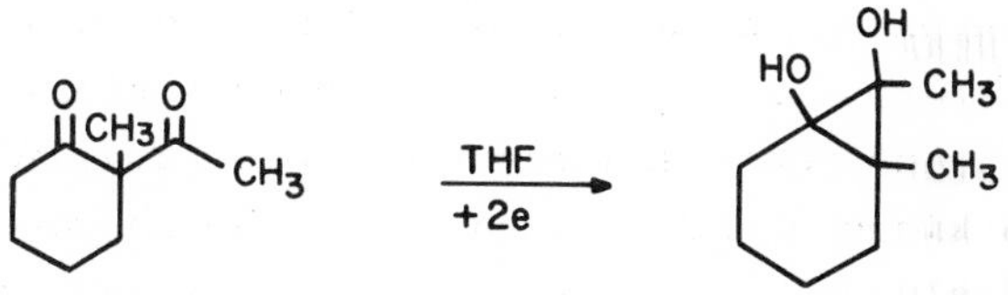

$$(4.19)$$

[37]   [38]

The marked reactivity of cyclopropanols toward acids and bases may place some stringent limitations on the use of conventional reducing agents. The diols may be transformed to other groups in situ if one places the right ingredients into the electrolysis solution. Thus, the presence of acetic anhydride during the electrolysis of [39] allowed the formation of the diacetate [40] (50).

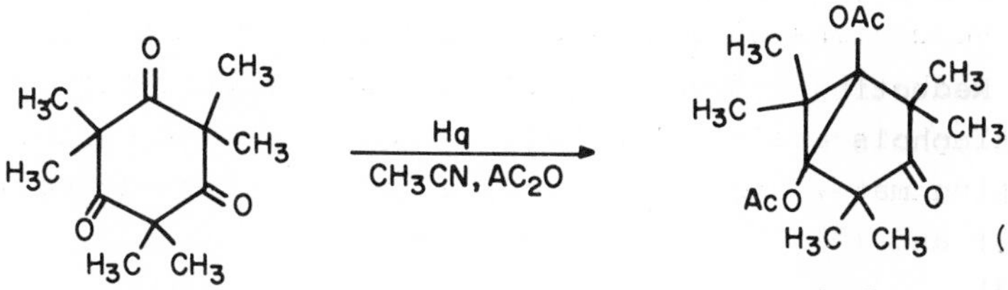

$$(4.20)$$

[39]   [40]

This same technique is being applied for a similar diol, e.g., compound [42], by one of the authors (51).

$$\underset{[41]}{\underset{\text{H}_3\text{C}\quad\text{CH}_3}{\overset{\text{H}_3\text{C}\quad\text{CH}_3}{O=\!\!\diamondsuit\!\!=O}}} + 2e \quad\xrightarrow{\text{AC}_2\text{O}}\quad \underset{[42]}{\underset{\text{H}_3\text{C}\quad\text{CH}_3}{\overset{\text{H}_3\text{C}\quad\text{CH}_3}{\text{AcO}-\!\!\diamondsuit\!\!-\text{OAc}}}} \quad (4.21)$$

4.1.6. GENERAL EXPERIMENTAL PROCEDURE

For ketones, an undivided cell may be used if the
product is known to be stable toward oxidation. The reduction of aldehydes requires the use of divided cells
(Sec. 3.1) since these compounds are themselves susceptible
to oxidation. Since carboxyl reduction in general requires
a proton source, aqueous or partially aqueous media are
usually required. Suitable organic solvents, when required
for solubility, include methanol, DMF, acetonitrile, and
DMSO (Sec. 3.2). Alkali halides or tetraalkylammonium
salts are used as electrolytes.

Reduction to Alcohols

$$\text{R}-\overset{\text{O}}{\overset{\|}{\text{C}}}-\text{R'(H)} + 2e + 2\text{ H}_2\text{O} \longrightarrow \text{R }\overset{\text{OH}}{\underset{|}{\text{CH}}}\text{ R' (H)} + 2\text{OH}^-$$

Reductions of aldehydes and ketones to the corresponding alcohols are favored by: (a) low concentration of
starting material; (b) alkaline media (KOH, $\text{Bu}_4\text{N}^+\text{OH}^-$,
etc.); and (c) high over-voltage cathodes such as Hg (Sec.
3.1.3).

Reduction to Pinacols

$$2\text{R}\overset{\text{O}}{\overset{\|}{\text{C}}}-\text{R' (H)} + 2e^- + 2\text{H}^+ \longrightarrow \text{R}\underset{\underset{\text{R' (H)}}{|}}{\overset{\overset{\text{OH}}{|}}{\text{C}}} - \underset{\underset{\text{R' (H)}}{|}}{\overset{\overset{\text{OH}}{|}}{\text{C}}} - \text{R}$$

Pinacol formation is promoted by: (a) high concentration of starting material; (b) acidic media; (c) high cur-

rent density; and (d) the use of cathodes such as tin and copper.

To obtain the pinacol, the procedure described above for the preparation of alcohols is followed, except that the catholyte is an acidic buffer solution (pH about 4), the concentration of the ketone is at least doubled, and the electrolysis is run at -1.15 V (s.c.e.).

Reduction to Hydrocarbons

$$R\text{-}\overset{\overset{\text{O}}{\|}}{C}\text{-}R'(H) + 4e \xrightarrow{2H^{\oplus}} R\text{-}CH_2\text{-}R'(CH)$$

Hydrocarbon formation from ketones and aldehydes is enhanced by the use of strongly acidic media and low hydrogen over-voltage cathodes such as cadmium or platinized platinum (Sec. 3.1.3).

4.1.7. REDUCTION OF 1,3-DIPHENYL-1,3-
PROPANEDIONE [43] (52)

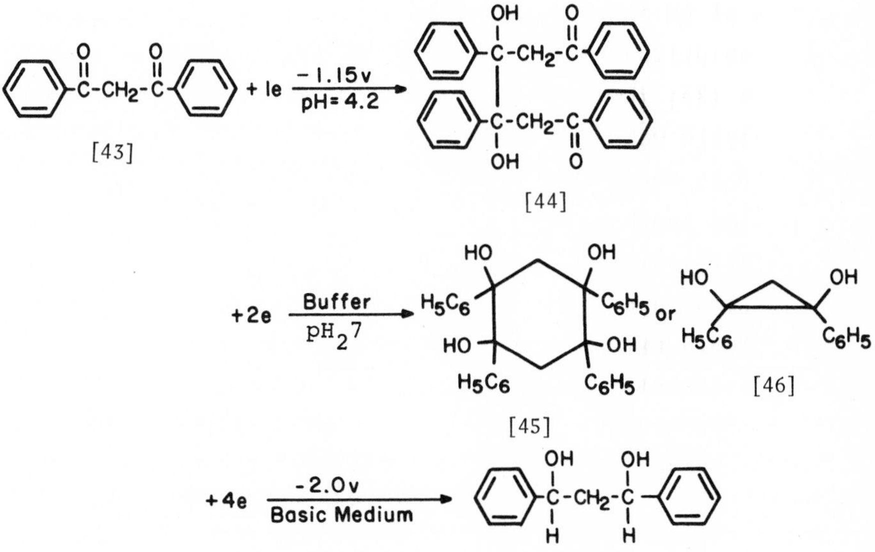

[43]   +1e $\xrightarrow[\text{pH}=4.2]{-1.15\text{v}}$ [44]

+2e $\xrightarrow[\text{pH}_2 7]{\text{Buffer}}$ [45] or [46]

+4e $\xrightarrow[\text{Basic Medium}]{-2.0\text{v}}$ [47]

The electrolysis cell used in this experiment was a
500 ml reaction kettle (Kimax, No. 33700, Owens-Illinois,
Toledo, Ohio); a mercury pool was used as the cathode.
The anode solution* consisted of 0.5 M KNO_3 in 50% ethanol/
water. The reference electrode (saturated calomel) was
placed into a bridge tube which was immersed into the
cathode solution. All electrode compartments and gas inlets
were inserted through standard taper joints in the top of
the reaction kettle. A magnetic stirrer was used for stirring. This over-all setup is similar to one shown by
Lingane (53) for chronopotentiometric measurements.

4.1.7.1. Reduction in Acid at -1.15 V

The anode compartment was charged with 500 ml of the
anolyte of pH = 4.2. The cathode compartment was charged
with a solution made up of 350 mg (0.0015 mole) of the
diketone [43] (Eastman Kodak) in 10 ml of 95% ethanol.
Electrolysis was carried out at -1.15 V (s.c.e., aqueous).
The initial current was 70 mA and fell to 8 after 2 hr, at
which time another 350 mg of the diketone [43] was added
and electrolysis continued. This procedure was repeated
until 2.1 g (0.0093 mole) was added. After 18 hr (total
electrolysis time), the current dropped to 1 mA. At this
time the catholyte and suspended product were separated from
the mercury pool and allowed to stand for 24 hr. The product was then filtered off, washed, and vacuum dried, giving
2.01 g of crude product which was identified by conventional
means as compound [44].

4.1.7.2. Reduction in Base at -2.0 V

The procedure was identical to the reduction in acid
with one exception. Tetramethylammonium hydroxide (50%
ethanol, Matheson, Coleman and Bell) 0.1 M was used as the
supporting electrolyte in the catholyte solution. The potential of the cathode was set at -2.0 V (s.c.e., aqueous).

*Placed in a suspended medium-porosity sealing tube.

Electrolysis of 2.07 g (0.009 mole) of the diketone [43] was completed in 20 hr. The catholyte was separated from the mercury pool, neutralized with hydrochloric acid and evaporated to half its original volume. Cooling allowed the separation of white crystalline material from the yellow oil. The oil was physically separated from the crystalline material which was then recrystallized from benzene-Skellysolve B. This afforded 0.4 g of the racemic alcohol [47] mp 129-130°.

The filtrate from the catholyte solution was extracted three times with ether. The extracts were combined, dried, and evaporated to leave an oil which was recrystallized to give 0.7 g of solid mp 85-90°. Recrystallization from ben-zene-Skellysolve B yielded a 0.46 g material mp 105-107°. This product was fractionally crystallized to constant melt-ing point to give a fraction mp 108-110° which was identified as the meso-form of the diol [47].

4.1.7.3.   Reduction of 1-Menthone [48](54)

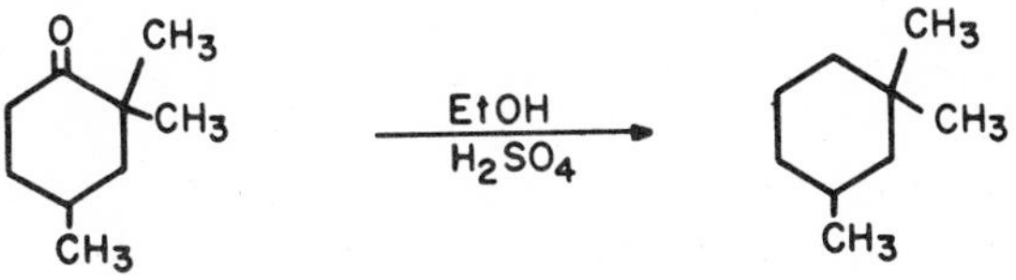

[48]   [49]

A beaker can be used for this reduction with a porous cup suspended into it and which serves as the anode chamber.

A cadmium sheet of 120 cm^2 area was used as the cathode while lead, placed in the porous cup, was used as the anode. The anolyte consisted of dilute sulfuric acid and was kept cold with a cooling coil. The catholyte consisted of a mixture of 20 g (0.14 mole) 1-menthone [48] (or dl-isomen-thone), 300 ml of ethanol, 70 ml water, and 15 ml sulfuric acid. This solution was stirred mechanically and was cooled

with an ice bath. An initial current of 3-3.3 A was passed
through the cell for 8 hr. After 2 hr from the start 10 g
menthone were gradually added to the catholyte. The temperature of the reaction remained between 10° and 15°. At the
end of the experiment (8 hr) the liquid had separated into
two layers. The isolation of the product [49] (20 g) has
been described (55) and need not be repeated in this section.

REFERENCES ON CARBONYL COMPOUNDS

1. M. J. Allen, Organic Electrode Processes, Reinhold,
    New York, 1958, p. 58.

2. C. K. Mann and K. K. Barnes, Electrochemical Reduction
    in Nonaqueous Systems, Dekker, New York, 1970, p. 70.

3. C. L. Perrin, in Progress in Physical Organic Chemistry,
    Vol. 3, Wiley, New York, London, Sidney, 1965, p. 195.

4. F. D. Popp and H. P. Schultz, Chem. Rev., 62, 29 (1962).

5. Ref. 1, p. 61.

6. P. Anziani, A. Aubry, and R. Cornubert, Compt. Rend.,
    225, 878 (1947).

7. A. Vincenz-Chodkowska and Z. R. Grabowski, Electrochim.
    Acta, 9, 789 (1964); C. A., 61 9173 (1969).

8. V. A. Smirnov, L. I. Antropov, M. G. Smirnova, and L.
    A. Demehuck, Zh. Fiz. Khim., 42, 1713 (1968); C. A.,
    69, 82877 (1968).

9. S. Ono, Hippon Kagaku Zasshi, 75, 1195 (1954); C. A.,
    51, 12704 (1957).

10. W. Oroshnik and P. E. Spoerri, J. Am. Chem. Soc., 63,
     338 (1941).

11. G. C. Whitnack, J. Reinhart, and E. St. C. Gantz, Anal.
     Chem., 27, 359 (1955).

12. R. A. Benkeser, Symposium, "Synthetic and Mechanistic
     Aspects of Electroorganic Chemistry," U. S. Army Research
     Office, Durham, N. C., Oct. 14-16, 1968.

13. R. A. Benkeser, H. Watanabe, S. J. Mels, and M. A. Sabol,
     J. Org. Chem., 35, 1210 (1970).

14. A. V. Koperina and N. I. Gavrilov, J. Gen. Chem. (USSR),
     17, 1651 (1947).

15. A. Dunet and A. Willemart, Compt. Rend., 226, 821 (1948);
     C. A., 42, 5011 (1948).

16. J. Tafel and W. Jürgens, Ber., 42, 2554 (1909).

17. J. Stocker and R. M. Jenevein, Symposium, "The Synthetic
     and Mechanistic Aspects of Electroorganic Chemistry," U.
     S. Army Office, Durham, N. C., Oct. 14-16, 1968.

18. H. Lund, Acta Chem. Scand., 11, 283 (1957).

19. T. C. Chambers and O. S. Slotterbeck, U. S. Pat.
     2,485,258 (1949); C. A., 44, 4807 (1950).

20. P. J. Elving and J. T. Leone, J. Am. Chem. Soc., 80,
     1021 (1958).

21. M. J. Allen, J. Org. Chem., 15, 435 (1950).

22. S. Swan, Jr., R. W. Benoliel, L. B. Lyons, and W. H.
     Paht, Trans. Electrochem. Soc., 79, 83 (1940).

23. E. A. Steck and W. Boehme, J. Am. Chem. Soc., 74, 4511
     (1952).

24. L. J. Throop, U. S. Pat. 3,506,549 (1970).

25. See Ref. 1, p. 59.

26. N. J. Leonard, S. Swann, Jr., and G. Fuller, J. Am.
     Chem. Soc., 75, 5127 (1953).

27. M. J. Allen and H. Corwin, J. Am. Chem. Soc., 72, 114
     (1950).

28. L. Throop and L. Töke's, J. Am. Chem. Soc., 89, 4789
     (1967).

29. W. S. Rapson and R. Robinson, J. Chem. Soc., 1935 1537.

30. S. Swann, Jr., Trans. Am. Electrochem. Soc., 64, 245
     (1933).

31. G. Sartori and C. Cattaneo, Gazz. Chim. Ital., 72, 525
     (1942)

32. L. I. Smith, I. M. Kolthoff, S. Wawzonek, and P. M.
     Ruoff, J. Am. Chem. Soc., 63, 1018 (1941).

33. S. Wawzonek and A. Gundersen, J. Electrochem. Soc.,
     107, 537 (1960).

34. N. Steinberger and G. K. Fraenkel, J. Chem. Phys., 40,
     723 (1964).

35. P. H. Given and M. E. Peover, Collection Czech. Chem.
     Commun., 25, 3195 (1960).

36. S. Wawzonek, R. Berkey, E. W. Blaha, and M. E. Runner,
     J. Electrochem. Soc., 103, 456 (1956).

37. I. M. Kolthoff and T. B. Reddy, J. Electrochem. Soc.,
     108, 980 (1961)

38. P. H. Given, M. E. Peover, and J. Schoen, J. Chem. Soc.,
     1958 2674.

39. S. Wawzonek, Talanta, 12, 1231 (1965).

40. D. E. Austen, P. H. Given, D. J. Ingram, and M. E.
     Peover, Nature, 182, 1784 (1958).

41. G. Nishi, D. Barnes, and P. Zuman, J. Chem. Soc., 1970,
     764, 771, 778.

42. I. M. Kolthoff and J. Lingane, Polarography, 2nd ed.,
     Interscience, New York, 1957.

43. S. Wawzonek and H. A. Laitinen, J. Am. Chem. Soc.,
     63, 2341 (1941).

44. C. Perrin, see Ref. 3.

45. L. Holleck and D. Marguarding, Naturwissenschaften, 49,
     468 (1962).

46. G. Gardner, Chem. Ind., 819 (1951).

47. F. Escherich and M. Moest, Z. Electrochem., 8, 849
     (1902).

48. L. Mandell, R. M. Powers, and R. A. Day, Jr., J. Am.
     Chem. Soc., 80, 5284 (1958).

49. R. M. Powers and R. A. Day, Jr., J. Org. Chem., 24,
     722 (1959).

50. T. J. Curphey, C. W. Amelotti, T. P. Layloff, R. L.
     McCartney, and J. H. Williams, J. Am. Chem. Soc., 91,
     2817 (1969).

51. M. R. Rifi, unpublished results.

52. D. H. Evans and E. C. Woodbury, J. Org. Chem., 32,
     2158 (1967).

53. J. J. Lingane, J. Electroanal. Chem., 1, 379 (1960).

54. G. H. Keats, J. Chem. Soc., 1937 2005.

55. C. Schall and W. Kirst, Z. Elektrochem., 29, 537 (1923).

56. J. H. Stocker and R. M. Jenevein, J. Org. Chem., 33,
     2145 (1968).

57. M. J. Allen, J. E. Fearnand, and H. A. Levine, J. Chem.
     Soc., 1952 2220.

4.2. NITRO GROUPS

4.2.1. GENERAL CONSIDERATION

The electrochemical reduction of nitrobenzene is perhaps the most studied reaction in electroorganic chemistry.

The reason for this is perhaps two-fold; (a) because of the complexity of the reduction, and (b) because of its importance in organic synthesis. For example, the following compounds have been obtained from the reduction of nitrobenzene: Aniline, p-aminophenol, p-chlorophenol, hydroxylamine, azoxybenzene, azobenzene, and hydrazobenzene. On the other hand, aliphatic nitro compounds have not been studied as extensively, mainly because of the instability of their reduction intermediates and limited use of their products in organic synthesis. The reduction of gem-dinitro aliphatic compounds was studied by Masui and co-workers (27). The reduction of nitro groups is extremely sensitive to a number of variables, some of which are discussed below.

4.2.2. EFFECT OF REACTION CONDITIONS ON COURSE OF REDUCTION

Nature of Electrode. In general, cathodes of high hydrogen overvoltage (in basic media) favor the formation of bimolecular products (1, 2):

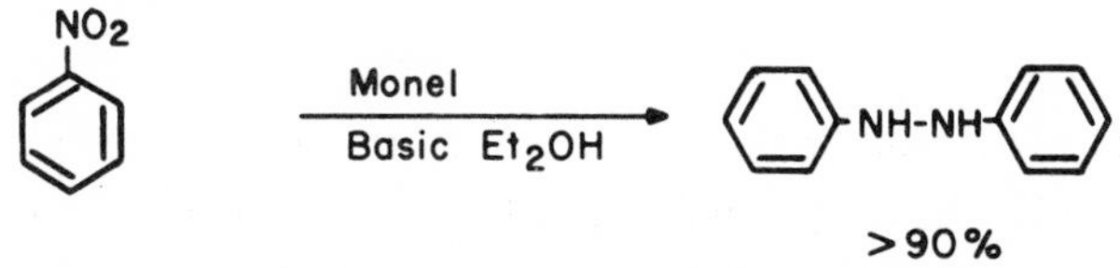

$$C_6H_5NO_2 \xrightarrow[\text{Basic Et}_2\text{OH}]{\text{Monel}} C_6H_5\text{-NH-NH-}C_6H_5$$

>90%

Aliphatic nitro groups afford the corresponding amines under acidic conditions and using a higher hydrogen overvoltage electrode (3). Under basic conditions aliphatic nitro groups afford the aci-anion, which is not susceptible to further reduction (4). The use of a low hydrogen overvoltage cathode (e.g., Pt) affords hydroxylamine, which rearranges to p-aminophenol (3, 5), Eq. (4.22):

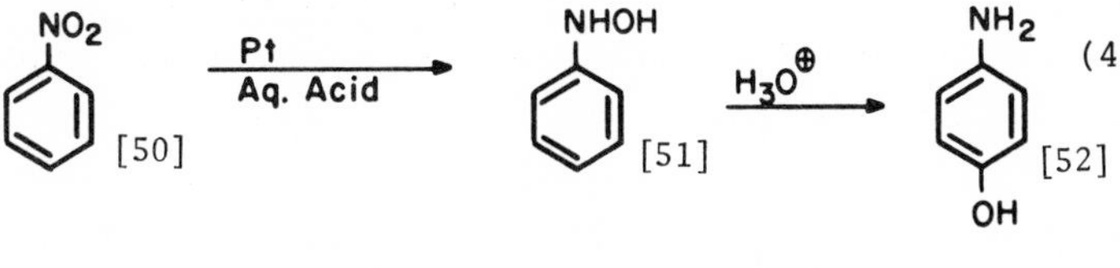

$$C_6H_5NO_2 \xrightarrow[\text{Aq. Acid}]{\text{Pt}} C_6H_5NHOH \xrightarrow{H_3O^\oplus} \text{HO-}C_6H_4\text{-NH}_2 \qquad (4.22)$$

[50]   [51]   [52]

The use of a high hydrogen overvoltage cathode does not
allow the rearrangement of hydroxylamine but reduces it
further to aniline.

Effect of Electrode Potentials. The potential of the
electrode may play an important part in determining the
nature of the product. Table 4.1 summarizes the variation
of the half-wave potential of nitrobenzene with pH of the
medium (1).

TABLE 4.1

Variation of $E_{\frac{1}{2}}$ of Nitrobenzene with pH of Reaction Medium

$-E_{\frac{1}{2}}$ (s.c.e.)	pH
0.55 V	4.5
0.65	6.5
0.75, 1.05	9.1
0.85, 0.95	14.0

It can be seen that at a pH of 9.1 two polarographic waves
are observed. At a potential of -0.7 V azobenzene is obtained, while at -1.05 V azobenzene is further reduced to
hydrazobenzene.

Effect of pH of Medium. The pH of the electrolysis
solution plays a very important role in determining the nature of the reduction product of nitro groups. Under acidic
conditions the products are generally the hydroxylamine,
amines, and products derived from their rearrangements (e.g.,
p-aminophenol from hydroxylamine (6, 7). Under basic or
neutral condition (using high overvoltage cathode) the products are generally arrived at from bimolecular reaction of
reduction intermediates (8, 9), Eq. (4.25):

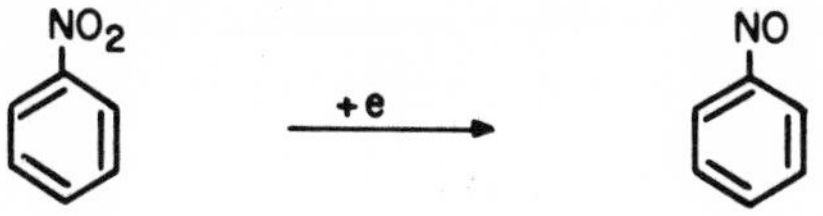

(4.23)

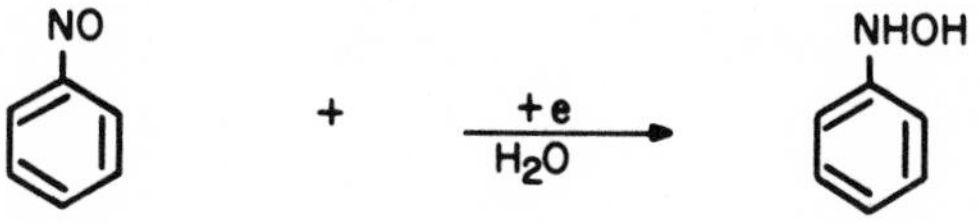

(4.24)

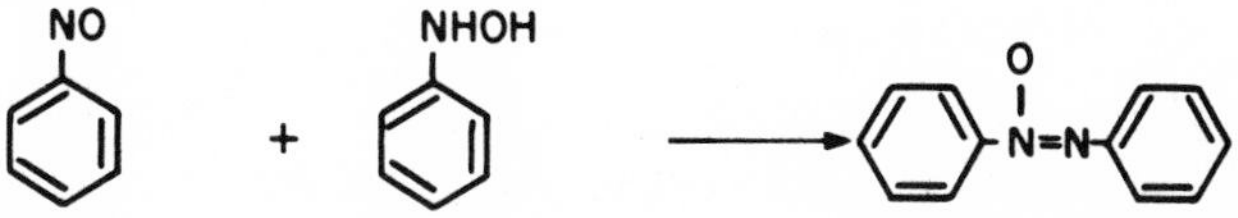

(4.25)

Aliphatic nitro groups are transformed to the aci-anion under basic conditions and do not undergo electrochemical reduction (4), Eq. (4.26):

$$RCH_2NO_2 \rightleftharpoons RCH=NO_2^{\ominus}$$
(4.26)

Nature of Medium. The nature of the reaction medium can also be of importance in determining the product formed. For example, reduction in hydrochloric acid can afford the o- and p-chloroanilines [53] (5). When ethanol is used under acidic conditions (H_2SO_4) the p-aminophenol is esterified to afford p-phenetidine (19, 20) [54]. If the reaction is carried out in fuming

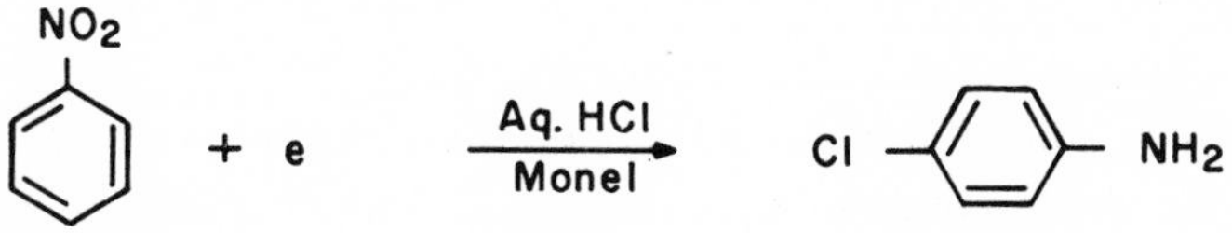

(4.27)

[53]

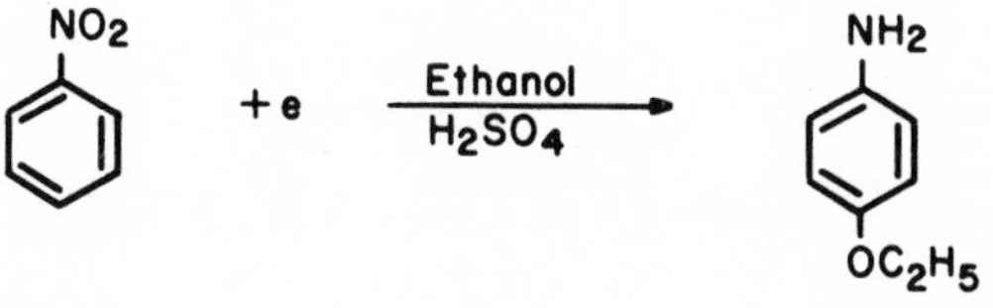

(4.28)

[54]

sulfuric acid, the p-aminophenol can be sulfonated to give
2-hydroxy-5-aminobenzenesulfonic acid [55] (15):

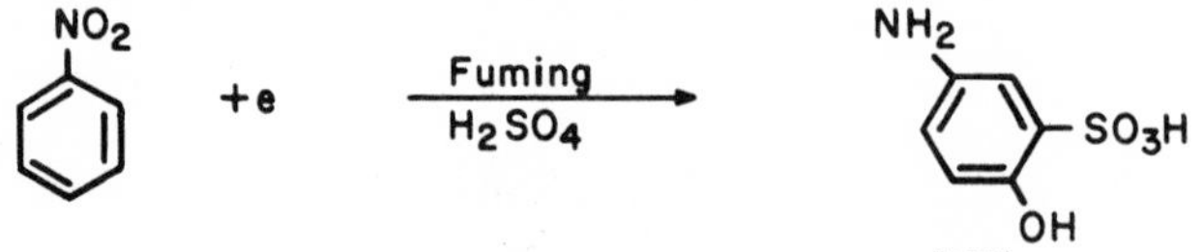

$$(4.29)$$

4.2.3. MECHANISM OF REDUCTION

Aliphatic. Nitroalkanes undergo a 4-electron reduction
in acid to form alkylhydroxylamine, Eq. (4.30):

$$R-CH_2-NO_2 + 4e \xrightarrow{H_3\overset{\oplus}{O}} R-CH_2NHOH \qquad (4.30)$$

$$[56] \hspace{5cm} [57]$$

The reduction is pH dependent and irreversible (10, 11).
Thus, nitromethane can be reduced to N-methylhydroxylamine
in about 41% yield (3). The aliphatic hydroxylamine is
not normally reduced (as is the aromatic analogue) to the
corresponding amine. There are, however, some exceptions
(12-14), Eq. (4.31):

$$C_2H_5-\overset{\overset{\displaystyle NO_2}{|}}{CH} - \overset{\overset{\displaystyle OH}{|}}{CH}-C_3H_7 \xrightarrow[10\% \ H_2SO_4]{Pb} C_2H_5-\overset{\overset{\displaystyle NH_2}{|}}{CH} - \overset{\overset{\displaystyle OH}{|}}{CH}-C_3H_7 \quad (4.31)$$

$$[58] \hspace{6cm} [59]$$
$$75\%$$

Under anhydrous conditions, aliphatic nitro groups
reduce to the unstable anion radical which undergoes cleavage to a nitrite ion and the corresponding free radical (15,
27).

$$R-NO_2 + 1e \longrightarrow R-N\overset{\overset{\displaystyle \odot}{\nearrow}O^{\ominus}}{\underset{O}{\diagdown}} \longrightarrow R^{\odot} + NO_2^{\ominus} \qquad (4.32)$$

$$2R^{\cdot} \longrightarrow R\text{-}R \tag{4.33}$$

Aromatic. A comprehensive discussion of the mechanism of the electrochemical reduction of aromatic nitro groups is beyond the scope of this book. For such a discussion the reader should consult the following excellent references, 6-8, 15-18, and 26.

Under acidic conditions nitrobenzene is reduced in a 4-electron process to hydroxylamine. Nitrosobenzene is presumably an intermediate in this reduction, but it cannot be isolated, since it reduces at a more positive potential than nitrobenzene.

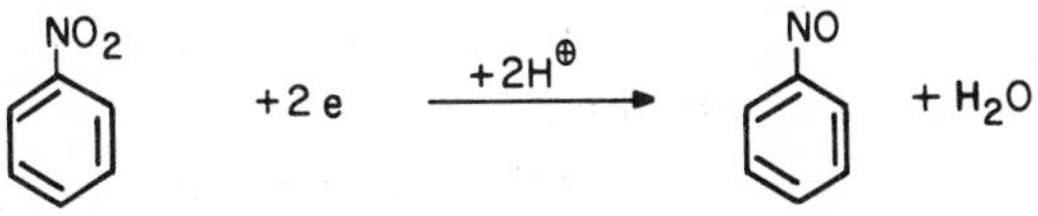

$$\text{(4.34)}$$

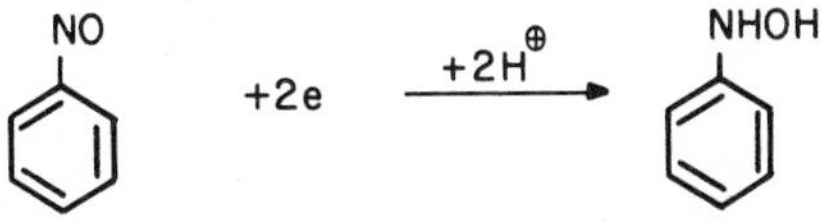

$$\text{(4.35)}$$

Phenylhydroxylamine may rearrange to p-aminophenol, or undergo further reduction, under a more negative potential, to aniline. Under basic or neutral conditions phenylhydroxylamine reacts with nitrosobenzene to form azoxybenzene, Eq. (4.36):

$$\text{(4.36)}$$

which can undergo further reduction to azo- and hydrazobenzene, Eqs. (4.37) and (4.38):

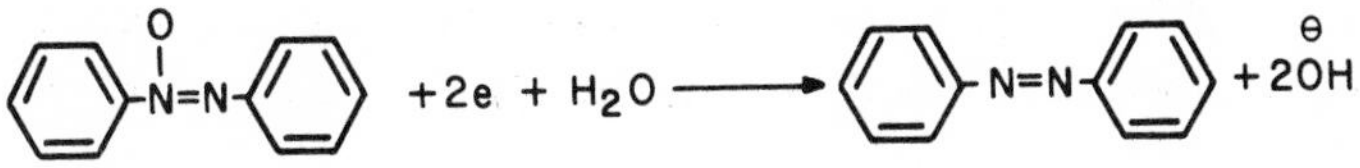

$$\text{(4.37)}$$

$$\text{Ph-N=N-Ph} + 2e + 2H_2O \longrightarrow \text{Ph-NH-NH-Ph} + 2\overset{\ominus}{O}H \quad (4.38)$$

4.2.4. ROLE IN ORGANIC SYNTHESIS

Nitrobenzene is an inexpensive compound (<10¢/lb), yet its reduction can lead to a variety of products. For example, p-aminophenol is used, as a monomer, in a number of polymerization reactions. Furthermore, by adjusting the nature of the medium, see Eq. (4.32), substituted anilines, e.g., p-chloro and p-alkoxyanilines, can be obtained. These products are generally difficult to obtain by conventional synthetic methods.

The ease of reduction of nitrobenzene to hydrazobenzene (1) allows the preparation of benzidine, an important chemical in the dye industry, in two simple steps, Eq. (4.39):

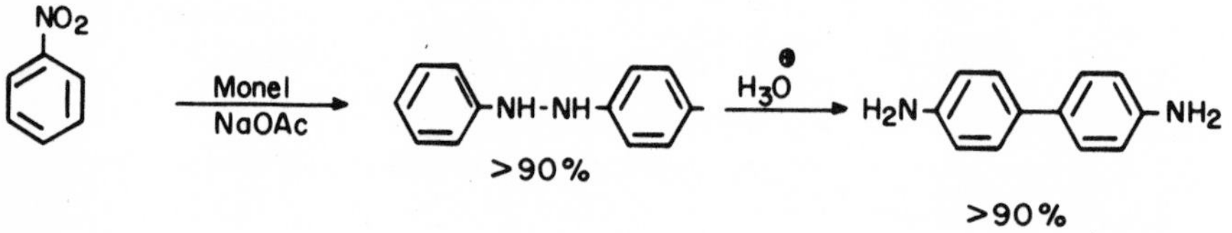

$$\overset{NO_2}{Ph} \xrightarrow[NaOAc]{Monel} \underset{>90\%}{Ph-NH-NH-Ph} \xrightarrow{H_3\overset{\oplus}{O}} \underset{>90\%}{H_2N-Ph-Ph-NH_2}$$

$$(4.39)$$

In summary, the reduction of nitrobenzene may be depicted as shown at the top of page 189.

4.2.5. EXPERIMENTAL PROCEDURE

The reduction of nitro compounds in general must be carried out in divided cells (Sec. 3.1) since most of the possible products are sensitive to oxidation. Also protic media are usually required. Partially aqueous solvents such as water-methanol, -ethanol, -DMF, and the like, have

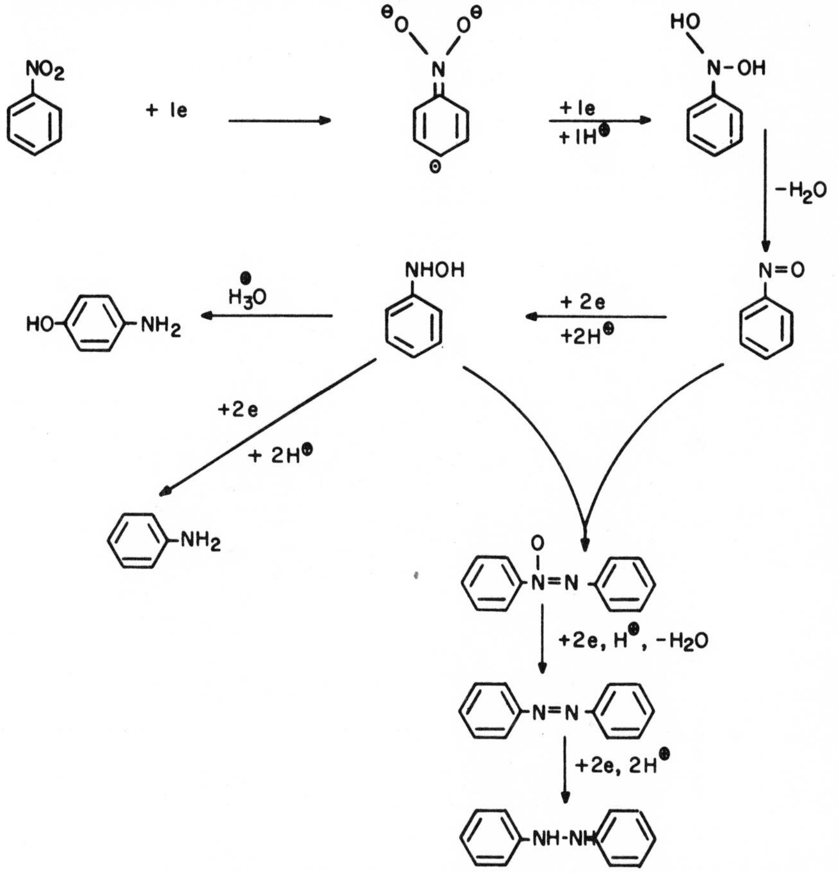

NO_2 benzene $+ 1e \longrightarrow$ $\overset{\ominus}{O}\diagdown N \diagup \overset{\ominus}{O}$ (nitrobenzene radical dianion) $\overset{+1e}{\underset{+1H^{\oplus}}{\longrightarrow}}$ $\underset{HO}{N-OH}$ phenyl

$\downarrow -H_2O$

$N=O$ nitrosobenzene $\overset{+2e}{\underset{+2H^{\oplus}}{\longleftarrow}}$ NHOH phenyl $\overset{H_3O^{\oplus}}{\longleftarrow}$ $HO-\langle\rangle-NH_2$

$+2e$
$+ 2H^{\oplus}$

$\langle\rangle-NH_2$

$\overset{O}{\langle\rangle-N=N-\langle\rangle}$ (azoxybenzene)

$\downarrow +2e, H^{\oplus}, -H_2O$

$\langle\rangle-N=N-\langle\rangle$ (azobenzene)

$\downarrow +2e, 2H^{\oplus}$

$\langle\rangle-NH-NH-\langle\rangle$ (hydrazobenzene)

been used. Controlled potential electrolysis (Sec. 2.9) may be required to optimize yields. Electrolyte restrictions are minimal, thus most alkali metal and ammonium salts have been used.

In view of the many possible products that can be obtained from the reduction of nitrobenzene, the reader should refer to Table 4.2, which may serve as a guideline for obtaining the desired product from a specific reaction.

A 1500-ml beaker was used as the electrolysis cell. A porous cup was suspended in the beaker and was used as the anode compartment. A lead sheet was placed in the

TABLE 4.2

Reduction of Nitrobenzene Under Different Reaction
Conditions

Electron per mole of Nitrobenzene	Reaction medium	Cathode	Product	Yield	Reference
6	10% Hcl	Pt	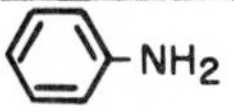 (benzene ring)–NH$_2$	91	19
4	Aq. acid	Zn	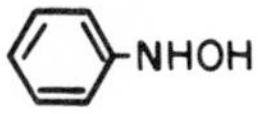 (benzene ring)–NHOH		21
4	Aq. acid	Monel	 HO–(benzene ring)–NH$_2$	72	22
3	Aq. ethanol NaOAc, reflux	Ni	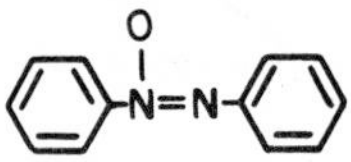 (benzene ring)–N=N–(benzene ring) with O	95	23
4	Sat. Na-salts of org. acids	Phosphor Bronze	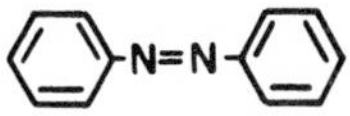 (benzene ring)–N=N–(benzene ring)	95	24
5	Basic, EtOH	Zn, Sn Monel	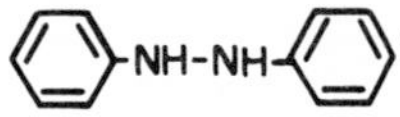 (benzene ring)–NH-NH–(benzene ring)	90	25

Reduction of 3-Nitro-4-Heptanol [60] (12)*

$$C_2H_5-\underset{[60]}{\overset{NO_2\quad OH}{CH-CH-C_3H_7}} \xrightarrow{Pb} C_2H_5-\underset{[61]}{\overset{NH_2\quad OH}{CH-CH-C_3H_7}}$$

porous cup. The cathode, a sheet of 15 x 14 cm lead, was
coated with spongy lead by placing it in a hot acidified
suspension of lead chloride with a lead anode and passing
a current until the cathode was covered with a gray coat
of the spongy lead. The catholyte and anolyte consisted

*It may be of interest to the reader to compare the
reduction of this compound by electrolysis and hydrogenation which is described in Ref. 12.

of 10% sulfuric acid. About 1000 ml of this solution was
placed in the cathode chamber together with 71 g (0.44 mole)
of 3-nitro-4-heptanol. This formed a suspension which was
stirred and electrolyzed at 8 V (over-all) and 15-17 A.
The cell was cooled by means of a water bath to maintain
the temperature of the catholyte at ⋏35°. During electrolysis the nitroalcohol went into solution. After several
hours, 47 g (0.291 mole) more nitroalcohol was added. At
the end of the reaction* only a small amount of oily material was left. The catholyte solution was filtered and extracted with toluene. It was then made strongly alkaline
with solid NaOH. This caused the separation of an oil,
which was collected. The aqueous basic solution was extracted with ether and the ether extracts were combined
with the oil. Distillation of the combined materials afforded the aminoalcohol [61] as a colorless liquid in about
75% yield.

Reduction of Nitrobenzene to Hydrazobenzene (1)

~90% [62]

The apparatus used for this reduction is shown in Fig.
4.1.
A solution of 50 g (0.48 mole) nitrobenzene in 1300 ml of
50:50 ethanol water and 10% sodium acetate was introduced
into the cathode chamber. A lead cathode, which surrounded
the inner portion of the cell, was used. The anode compartment was a porous porcelain thimble which was suspended into
the beaker and which contained a platinum anode and a
solution of aqueous sodium acetate. The level of the anolyte solution was equal to that in the catholyte solution.

---

*The time of the reaction is not given in this reference;
however, an amount of electricity which is slightly above
(5-10%) the stoichiometric amount is added.

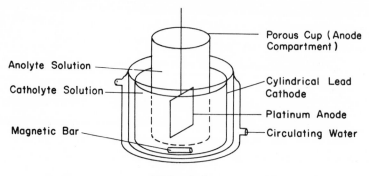

FIG. 4.1.

The cathode solution was stirred by means of a magnetic bar.
Electrolysis was carried out at 22 V (over-all) and 20 A.
Circulating water should be used to prevent overheating the
solution in the cell 50-60°. The current density was 10
A/dm^2. After 4 hr, the reaction was terminated. At this
time, white crystalline material was formed in the cathode
chamber. These were filtered, dried, and compared with an
authentic sample of hydrazobenzene. The yield from repetitive runs averaged about 90%. The electrical efficiency
was 70-75%.

Reduction of o-Chloronitrobenzene (28) [3] to the Correspond-
ding p-Aminophenol [65].

This reaction is described in detail in Organic Synthesis (28).

REFERENCES ON NITRO GROUPS

1. M. R. Rifi, unpublished results.
2. K. Sugino and T. Sekine, J. Electrochem. Soc., 104, 497 (1957).
3. M. W. Leeds and G. B. Smith, J. Electrochem. Soc., 98, 129 (1951).
4. T. DeVries and R. W. Ivett, Ind. Eng. Chem. Anal. Ed., 13, 339 (1941).
5. M. J. Allen, Organic Electrode Process, Reinhold, New York, 1958, p. 52.
6. J. Pearson, Trans. Faraday Soc., 44, 683 (1948).
7. I. Bergman and J. C. James, Trans. Faraday Soc., 50, 60 (1954).
8. S. Wawzonek and J. O. Fredrickson, J. Am. Chem. Soc., 72, 3985 (1955).
9. E. Bamberger and E. Renould, Ber., 30, 2278 (1897); 33, 271 (1960).
10. W. Radin and T. DeVries, Anal. Chem., 24, 971 (1952).
11. F. Petru, Collection Czech. Chem. Commun., 12, 620 (1947).
12. W. Gakenheimer and W. Hartung, J. Org. Chem., 9, 85 (1944).
13. G. Gever, C. O'Keefe, G. Drake, F. Ebetino, J. Michels, and K. Hayes, J. Am. Chem. Soc., 77, 2277 (1955).
14. E. Remy and G. Kümmell, Z. Elektrochem., 15, 254 (1909).
15. H. Sayo, Y. Tsukitani, and M. Masui, Tetrahedron, 24, 1717 (1968).
16. C. Perrin, Progress in Physical Organic Chemistry, Vol. 3, Wiley, New York, London, Sydney, 1965, p. 250.
17. H. L. Piette, P. Ludwig, and R. N. Adams, Anal. Chem., 34, 916 (1962).
18. M. Field, C. Valle, and M. Kane, J. Am. Chem. Soc., 71, 421 (1949).
19. L. Gatterman, Ber., 26, 1847 (1893).
20. M. Mizuguchi and S. Matsumota, Yakugaku Zasshi, 78, 129 (1958); C. A., 52, 8794 (1958).

21. F. M. Frederiksen, J. Phys. Chem., 19, 696 (1915).

22. C. L. Wilson and H. V. Udupa, Trans. Electrochem. Soc., 99, 289 (1952).

23. R. C. Snowdon, J. Phys. Chem., 15, 797 (1911); C. A., 6, 617 (1912).

24. R. H. McKee and C. J. Brockman, Trans. Electrochem. Soc., 62, 203 (1932).

25. K. Sugino and T. Sekine, J. Electrochem. Soc., 104, 497 (1957).

26. R. W. Lewis and O. W. Brown, Trans. Electrochem. Soc., 86, 135 (1943).

27. H. Sayo and M. Masui, Tetrahedron, 24, 5075 (1968).

28. R. E. Harman, Org. Syn., 35, 22 (1955).

4.3. CARBON-HALOGEN BONDS

4.3.1. GENERAL CONSIDERATIONS

The electrochemical reduction of carbon-halogen bonds has been discussed in several review articles (1-3). Unfortunately however, the potential of this reaction in organic synthesis has been, until recently, neglected. As in the case of the reduction of nitrobenzene, cleavage of carbon-halogen bonds can, under the appropriate reaction conditions, lead to a variety of products, including some very reactive intermediates such as carbenes, benzynes (4, 5), and polymeric products (7, 8). As can be seen from the discussion below, it is the cleavage of carbon-halogen bonds in aprotic solutions that has great potential in organic synthesis. The cleavage of such bonds under protic conditions often leads to the replacement of the halogen with a hydrogen atom. There are, however, some exceptions (6). Another great utility in organic synthesis is the reduction of polyhalides under controlled potential electrolysis. This technique (see discussion below) appears to have an advantage over reduction with conventional reducing agents such as zinc, sodium, and potassium.

In general, the electrochemical reduction of organic halides is independent of the nature of the cathode. Furthermore, this reduction is not affected by the pH of the medium, although some isolated examples have been mentioned (9, 10).

The products derived from the electrochemical reduction of organic halides depend on the nature of the molecule in question, i.e., whether it is a monohalide, polyhalide, etc. Consequently, each type of halide will be discussed separately.

4.3.2. MONOHALIDES

Aliphatics. The reduction of aliphatic halides in protic and aprotic solvents generally leads to the formation of the corresponding hydrocarbons. However, there are some exceptions (11-13), Eqs. (4.40)-(4.42):

$$O_2N-\langle\bigcirc\rangle-CH_2Cl \quad \xrightarrow[MeOH]{Hg} \quad O_2N-\langle\bigcirc\rangle-CH_2-CH_2-\langle\bigcirc\rangle-NO_2$$

$$(4.40)$$

$$\underset{Br-CH_2}{\overset{CH_2}{\diagup}}\overset{\diagdown}{\underset{CH_2-\overset{\oplus}{N}(CH_2CH_3)_3}{}} \overset{\ominus}{Br} \quad \xrightarrow[DMF]{Hg} \quad \triangle \quad + (CH_3CH_2)_3N$$

$$(4.41)$$

$$I-CH_2-CH_2CN \quad \xrightarrow{Hg} \quad (NC-CH_2CH_2)_2 \; Hg$$

$$(4.42)$$

In the presence of carbon dioxide, the cleavage of carbon-halogen bond gives the corresponding carboxylic acid (14).

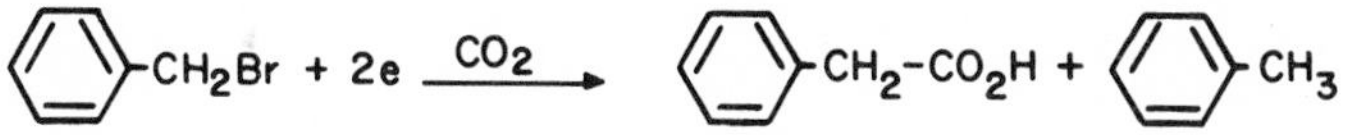

$$\langle\bigcirc\rangle-CH_2Br + 2e \xrightarrow{CO_2} \langle\bigcirc\rangle-CH_2-CO_2H + \langle\bigcirc\rangle-CH_3 \quad (4.43)$$

Mechanism of Reduction. The reduction of organic halides is irreversible, follows the Ilkovic equation (polarography), and is independent of the pH of the medium. The
ease of reduction follows the order: RI > RBr > RCl > RF.
The reduction of these bonds has been compared (15) with
SN_2 [or SN_1 depending on steric effects (16)] displacement,
where electrons may act as the nucleophile. Other investigators (17) have argued against such an analogy and claimed
that the reduction of the bicyclic bromides [66] and [67]
from the back side of the carbon would be quite difficult.

Br
[66]   Br
                                  [67]
$E_{1/2}$ (Ag/Ag⁺) − 2.17V   − 1.79V

The addition of electrons, whether to the halide or to
the back side of the carbon, may lead to the formation of
radical [68] or anionic [69] intermediates (18):

$$R-X + e \xrightarrow{slow} \left[\overset{\delta\ominus}{R} \cdots \overset{\delta\ominus}{X} \right] \qquad (4.44)$$

Transition State

$$\left[\overset{\delta\ominus}{R} \cdots \cdots \overset{\delta\ominus}{X} \right] \longrightarrow \overset{\ominus}{R} + \overset{\ominus}{X} \qquad (4.45)$$ 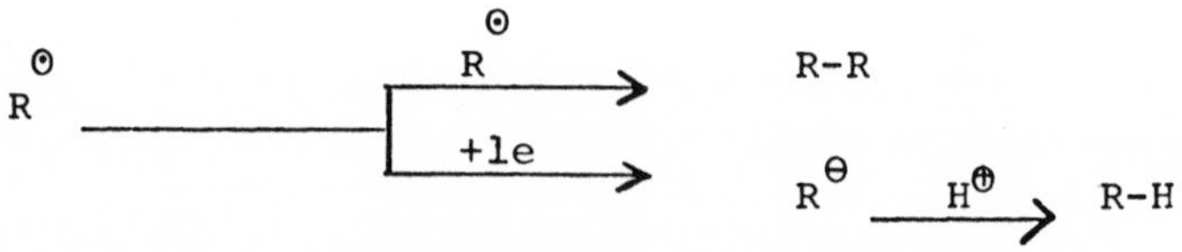
$$[68]$$

$$\overset{\ominus}{R} \xrightarrow{\quad} \overset{\ominus}{R} \longrightarrow R-R$$
$$\xrightarrow{+1e} R^{\ominus} \xrightarrow{H^{\oplus}} R-H \qquad (4.46)$$

$$(4.47)$$

To argue for or against any of these intermediates would require a discussion of a considerable amount of data. Such
a discussion is beyond the scope of this chapter. For the
interested at heart, the following key references are offered:
(11, 19-23). It would suffice to say that based on polarographic as well as product analysis, there is no compelling
evidence for the intervention of radicals as detectible
intermediates. The observed coupling products may be formed
from the attack of anionic intermediates on the alkyl halide.
Thus, the reduction of alkyl halides may be described as
follows [Eqs. (4.48)-(4.50)

$$R-X + 1e \xrightarrow{\text{slow}} \left[R \overset{\delta\ominus}{\cdots} \cdots X^{\delta\ominus} \right] \longrightarrow R^{\ominus} + X^{\ominus} \quad (4.48)$$

$$R^{\ominus} + 1e \xrightarrow{\text{very fast}} R^{\ominus} \longrightarrow \text{products} \quad (4.49)$$

$$\text{or} \quad R-X + 2e \xrightarrow{\text{slow}} \left[R \overset{\delta\ominus}{\cdots} \cdots X^{\delta\ominus} \right] \longrightarrow R^{\ominus} + X^{\ominus} \quad (4.50)$$

Role in Organic Synthesis. The electrochemical reduction of alkyl monohalides may be compared to the well-known
Grignard reaction. Its advantages, however, have not been
fully explored. For example, it was recently discovered
(24) that tetraethyl lead [70] can be formed quantitatively
from the reduction of ethylbromide on a lead electrode:

$$4CH_3-CH_2-Br \xrightarrow{\text{Pb}} (CH_3CH_2)_4Pb$$

[70]

100%

4.3.2.1. Vinyl and Aromatic Halides

The reduction of vinyl and aromatic halides is similar
to that of aliphatic halides in that similar products are

obtained.   In certain cases where the double bond is conjugated with the π system of phenyl groups its reduction may
take place (25), Eq. (4.51):

$$H_5C_6 \diagdown \qquad C_6H_5 \diagup$$
$$\qquad\qquad C = C \qquad + \ 2e \ \xrightarrow[DMF]{Hg} \ (H_5C_6)_2 - CH - CH_2C_6H_5$$
$$H_5C_6 \diagup \qquad \diagdown Br$$

$$(4.51)$$

In the reduction of vinyl halides, stereochemical
changes, i.e., cis-trans isomerization may take place (28,
29).

$$H_5C_2 \diagdown \qquad I \diagup$$
$$\qquad C = C \qquad + 2e \ \xrightarrow{DMF}$$
$$H \diagup \qquad \diagdown C_2H_5$$

[71]

$$H_5C_2 \diagdown \qquad H \diagup$$
$$\qquad C = C$$
$$H \diagup \qquad \diagdown C_2H_5$$

[72]

$$H_5C_2 \diagdown \qquad C_2H_5 \diagup$$
$$\qquad C = C$$
$$H \diagup \qquad \diagdown H$$

[73]

$$H \diagdown \qquad Br \diagup$$
$$\qquad C = C \qquad + 2e \ \xrightarrow{Hg \atop C_2H_5OH}$$
$$H_5C_2 \diagup \qquad \diagdown CO_2H$$

[74]

$$H_5C_2 \diagdown \qquad H \diagup$$
$$\qquad C = C$$
$$H \diagup \qquad \diagdown C_2H_5$$

[75]

$$H \diagdown \qquad H \diagup$$
$$\qquad C = C$$
$$C_2H_5 \diagup \qquad \diagdown C_2H_5$$

[76]

70   :   30

$$H \diagdown \qquad Br \diagup$$
$$\qquad C = C \qquad + 2e \ \xrightarrow{Hg \atop C_2H_5OH}$$
$$CO_2H \diagup \qquad \diagdown CO_2H$$

[77]

$$H \diagdown \qquad H \diagup$$
$$\qquad C = C \qquad +$$
$$CO_2H \diagup \qquad \diagdown CO_2H$$

[78]

$$HO_2C \diagdown \qquad H \diagup$$
$$\qquad C = C \qquad +$$
$$H \diagup \qquad \diagdown CO_2H$$

[79]

$$HO_2C \diagdown \qquad CO_2H \diagup$$
$$\qquad CH - CH$$
$$HO_2C-CH_2 \diagup \qquad \diagdown CH_2CO_2H$$

[80]

The formation of the dimeric product [80], as well as
the stereochemical changes, are of interest and are discussed under the section on reaction mechanism below.

The reduction of aromatic halides generally leads to benzene. Thus, bromobenzene affords benzene in a quantitative yield (26). However, in the presence of carbon dioxide, benzoic acid has been isolated together with benzene (27).

Mechanism of Reduction. Vinyl and aromatic halides reduce at a more negative potential than their simple aliphatic analogs (see Table 4.3). Furthermore, because of the presence of unsaturation in the molecule, the mechanism of their reduction may differ from that of simple alkyl halides. Thus, it was suggested by Miller and co-workers (25) that, in the reduction of vinyl halides, an electron is added to the lowest unoccupied π molecular orbital to give the radical anion, which undergoes elimination of halide anion to form the vinyl radical. This in turn accepts another electron to give the anion which undergoes protonation, Eq. (4.52).

$$\begin{array}{ccc} \underset{/}{\overset{\backslash}{C}} = \underset{\backslash}{\overset{/}{C}} + 2e & \longrightarrow & \underset{/}{\overset{\backslash}{C}} \!\!-\!\! \underset{\underset{X}{\wedge}}{\overset{/}{C}^{\ominus}} \longrightarrow \underset{/}{\overset{\backslash}{C}} = \overset{/^{\ominus}}{C} \xrightarrow{+1e} \underset{/}{\overset{\backslash}{C}} = \overset{/^{\ominus}}{C} \\ [81] & & [82] \qquad\quad [83] \qquad\quad [84] \end{array}$$

(4.52)

It should be noted, however, that prior to the elimination of halide anion, rotation of the C-C single bond in compound [82] is possible, and this will allow the formation of the thermodynamically stable product. Alternatively, the vinyl anion [84] (or the radical [83], if it has enough stability) may equilibrate to yield the most stable olefin. Unfortunately, the question of equilibration of vinyl radicals and anions has not been settled (30-32).

The formation of dimeric products [80] from the reduction of monobromomaleic acid was taken as evidence for the formation of a vinyl radical intermediate (29). It should be noted, however, that the product may easily be explained

TABLE 4.3

Half-Wave Potential of Organic Halides[a]

Organic halide	Solution	$-E$[b]	Reference
CH_3-Cl	75% Dioxane	2.23	40
CH_3-Br	75% Dioxane	2.01	40
CH_3-I	75% Dioxane	1.63	40
CH_3-CH_2-Br	DMF	2.13	15
$CH_3-\underset{\underset{CH_3}{\vert}}{\overset{\overset{CH_3}{\vert}}{C}}-CH_2Br$	DMF	2.37	15
 △—Br	DMF	2.36	15
 □—Br	DMF	2.36	15
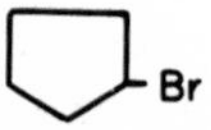 ⬠—Br	DMF	2.19	15
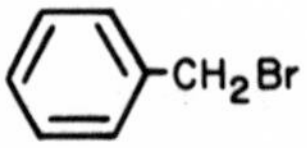 ⬡—CH_2Br	DMF	1.22	36
$Br-CH_2-CH_2-CH_2\overset{\oplus}{N}-Et_3$	DMF	1.34	6

Table 4.3 (continued)

Br	DMF	2.17 (Ag/Ag$^{\oplus}$)	<u>17</u>
	DMF	1.79 (Ag/Ag$^{\oplus}$)	<u>17</u>
Br	95% DMF	2.15	<u>38</u>
Br	DMF	2.05	<u>38</u>
CH$_2$=CH-Br	75% Dioxane	2.47	<u>40</u>
CH$_2$=CH-Cl	75% Dioxane	No wave	<u>40</u>
Br			
Cl	DMF	2.32	<u>36</u>
CH$_2$=CH-CH$_2$Br	DMF	2.54	<u>36</u>

Table 4.3 (continued)

$CH_2=CH-CH_2Br$	DMF	1.29	8
$CH_2=CH_2-CH_2Cl$	DMF	1.91	8
$CH_2=CH_2-CH_2I$	DMF	1.16	8
$Br-CH_2-CH_2Br$	95% DMF	1.23	38

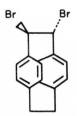

95% DMF	1.53	38
95% DMF	1.21	38
95% DMF	0.86	38
95% DMF	1.67	38
DMF	1.4	47

Table 4.3 (continued)

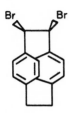

| | DMF | 1.12 | <u>47</u> |

Br-CH$_2$-CH$_2$-CH$_2$Cl | DMF | 2.04 | <u>47</u>

Br-CH$_2$-CH$_2$-CH$_2$Br | DMF | 1.91 | <u>6</u>

Br-CH$_2$+CH$_2$+$_2$CH$_2$Br | DMF | 1.99, 1.95^c | <u>6</u>

Br-CH$_2$+CH$_2$+$_3$CH$_2$Br | DMF | 2.14, 2.2^c | <u>6</u>

Br-CH$_2$+CH$_2$+$_4$-CH$_2$Br | DMF | 2.13 | <u>6</u>

(<u>cis</u>-<u>trans</u> isomers)

| DMF | 2.02 | <u>6</u> |

| DMF | 1.97 | <u>47</u> |

| DMF | 2.2 | <u>47</u> |

Table 4.3 (continued)

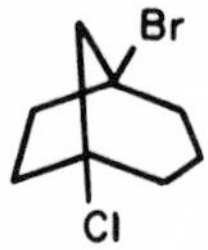	DMF	1.99	47
	DMF	2.28	47
CH_2Cl_2	75% Dioxane	2.23	40
CH_2Br_2	75% Dioxane	1.48	40
$CHBr_3$	75% Dioxane	0.49, 1.09, 1.5	40
$CHCl_3$	75% Dioxane	1.67	40
CBr_4	75% Dioxane	0.3, 0.75, 1.49	40
CCl_4	75% Dioxane	0.78, 1.71	40

[a]Unless otherwise specified, mercury was used as the cathode and calomel electrode was used as the reference.

[b]The reader should use the $E_{\frac{1}{2}}$ with caution when comparing values reported by different authors.

[c]Ascending part of wave is deformed by a sharp maximum.

by invoking the corresponding anionic intermediate [85]:

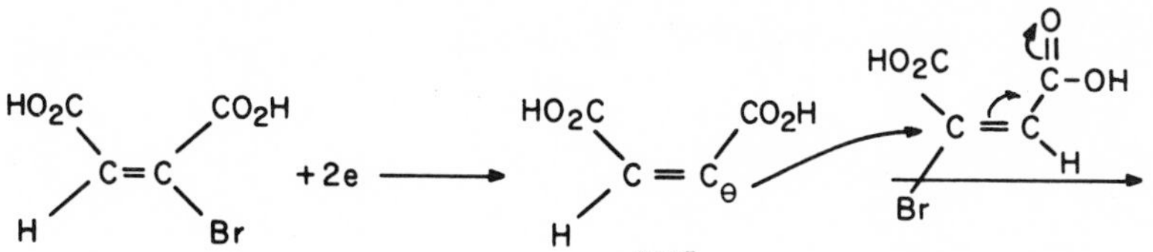

[85]

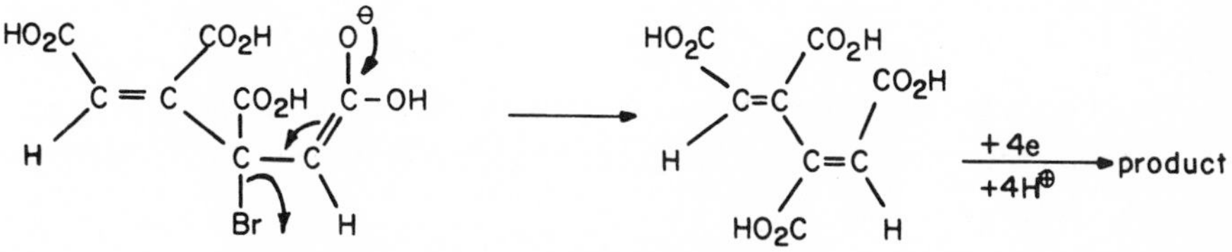

Similar anionic attacks on activated olefins have been reported, even in aqueous solutions (33).

Role in Organic Synthesis. The reduction of vinyl and aromatic halides has found limited use in organic synthesis. Most of the published work on this subject has dealt with the mechanism of the reduction rather than its synthetic utility.

4.3.3. POLYHALIDES

4.3.3.1. Gem Halides

The reduction of gem halides generally yields the corresponding hydrocarbons (34). However, this reduction has been used for the synthesis of polymers (8), as well as that of reactive intermediates (5), Eqs. (4.53)-(4.55):

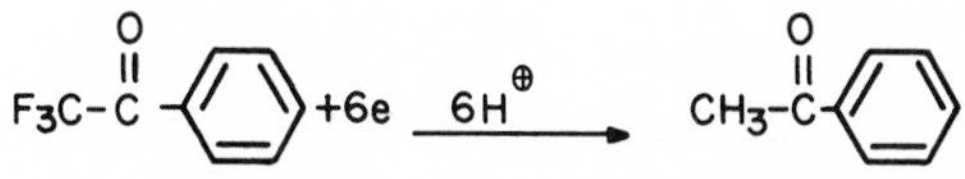

$$F_3C-\overset{O}{\overset{\|}{C}}-\langle\bigcirc\rangle + 6e \xrightarrow{6H^{\oplus}} CH_3-\overset{O}{\overset{\|}{C}}-\langle\bigcirc\rangle \qquad (4.53)$$

$$Cl_3C-\langle\bigcirc\rangle-CCl_3 + 2e \xrightarrow[\text{T.H.F.}]{-10°} Cl_2C=\langle\bigcirc\rangle=CCl_2 + Cl^-$$

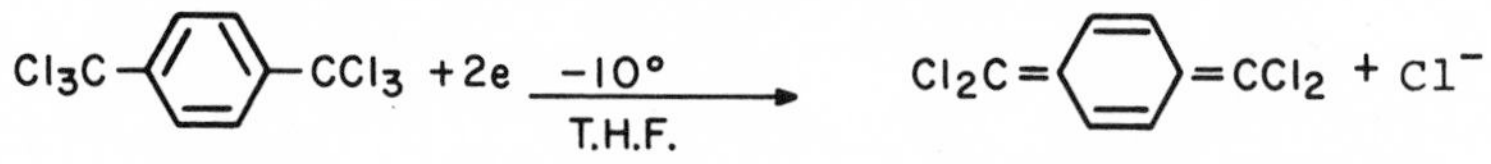

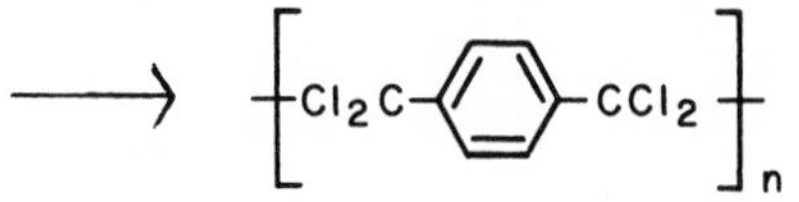

$$\longrightarrow \left[Cl_2C-\left\langle\bigcirc\right\rangle-CCl_2 \right]_n \qquad (4.54)$$

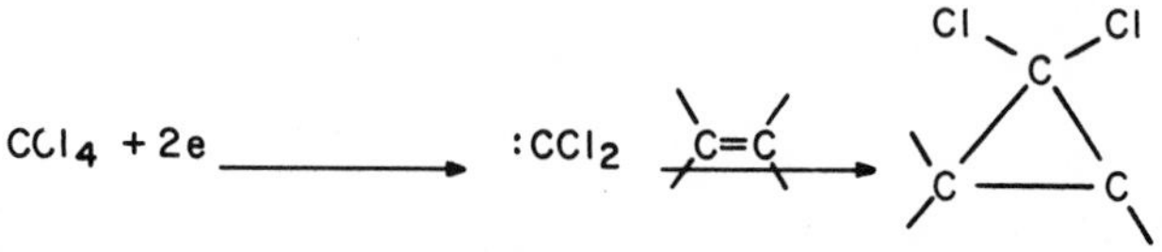

$$CCl_4 + 2e \longrightarrow :CCl_2 \quad \overset{}{\underset{}{C}}=\overset{}{\underset{}{C} \quad \longrightarrow \qquad (4.55)$$

The reduction of gem halides can be allowed to proceed in a stepwise manner by using controlled potential electrolysis, see Sec. 2.9 (35).

$$Cl_3C-\overset{O}{\overset{\|}{C}}-OH + 2e \xrightarrow{NH_4Cl} Cl_2-CH-\overset{O}{\overset{\|}{C}}-OH$$

100%

Mechanism of Reduction. Based on product analysis, the reduction of gem halides can best be explained by invoking anionic intermediates. Thus, it appears that this reduction proceeds in a similar manner described for monohalides.

Role in Organic Synthesis. The selective reduction of one of two gem halides, by using controlled potential electrolysis, can be a powerful tool in organic synthesis. Consider, for example, the reduction of 1,1,3,3-tetrachloro-2, 2,4,4-tetramethylocyclobutane [86] to prepare the corresponding bicyclobutane compound [87] (12).

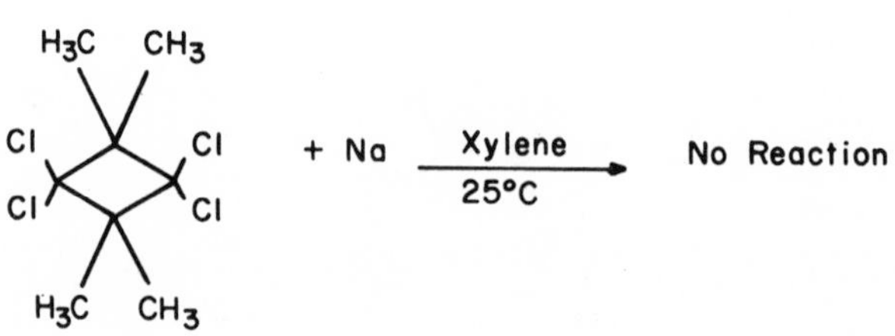

H₃C CH₃
Cl   Cl
Cl   Cl   + Na   Xylene   No Reaction
                                  25°C
H₃C CH₃
[86]

+ K _____"_____→   No Reaction

+ Na-K-alloy ____"____→ No Reaction

Na - K-alloy ____150°____→ Decomposition

Compound [86] exhibited two 2-electron polarographic waves
at -2.0 and -2.4V (s.c.e.). Its macroscale electrolysis at
the plateau of the first wave allowed the production of
the desired bicyclic product (12, 36).

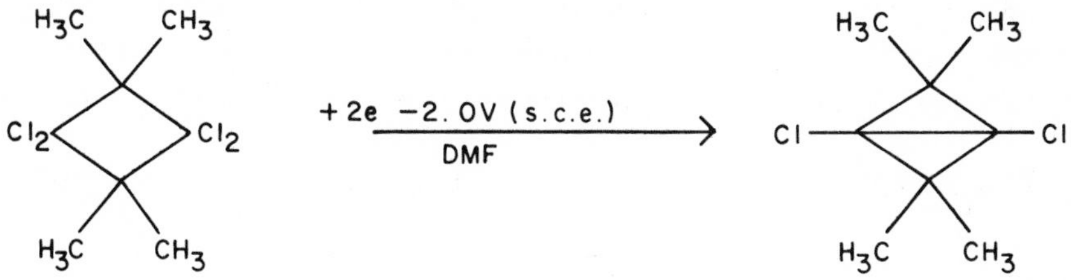

$$H_3C \quad CH_3 \qquad\qquad \xrightarrow[\text{DMF}]{+2e \ -2.0V \ (s.c.e.)} \qquad\qquad H_3C \quad CH_3$$

Cl$_2$ ⟨ ⟩ Cl$_2$ Cl — ⟨ ⟩ — Cl

H$_3$C   CH$_3$ H$_3$C   CH$_3$

[86]   [87]

The preparation of carbenes from gem halides was mentioned above.

4.3.3.2. Vicinal Halides

The electrochemical reduction of vicinal dihalides
resembles the reduction of the same dihalide with zinc or
sodium in that the product is, in general, the corresponding olefin. In some cases, however, reduction by electrolysis might be advantageous. This is mainly because the
controlled potential of the cathode can allow a discrimination in the reduction of one halide over another. For example, it is known that trans-dihalides, e.g., [88], reduce
at a more positive potential than their cis-isomers, e.g.,
[89] (9). Similarly, axial dihalides reduce at a more
positive potential than their equatorial analogs (38). Unfortunately, there have been no macroscale reduction

[88]   [89]

$EI/2 = -0.8V$ $-1.67V$

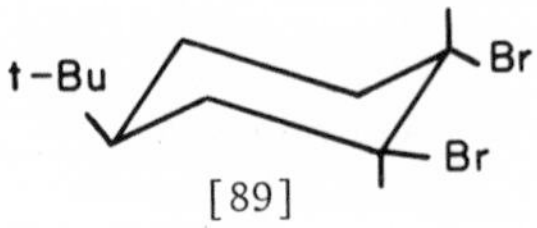

studies on the above dihalides by electrolysis or by conventional reducing agents.

The stereochemistry of <u>activated</u> dihalides, i.e.,
halides on carbon atoms having electron-withdrawing groups,
does not appear to be important since, in general, the olefin with the most thermodynamic stability is formed (<u>39</u>).
Thus, both the <u>meso</u>- and <u>racemic</u>-esters of α,α´-dibromo-
succinic acids [90] afford fumaric acid [91] upon reduction.

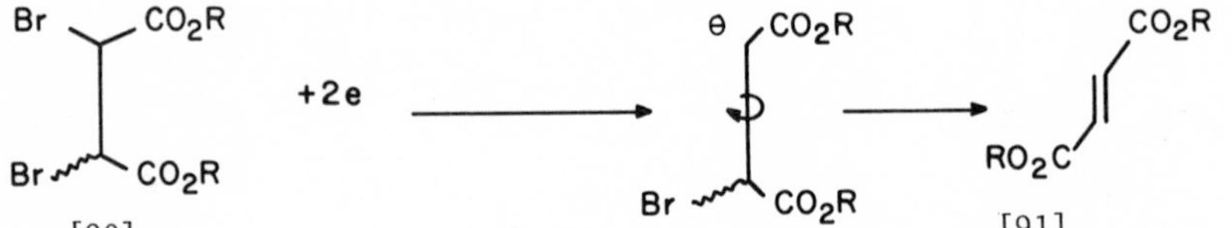

[90]

Rotation of bond or equilibration of anion

The reduction of aromatic vicinal dihalides affords
benzene; however, a benzyne intermediate has been trapped
in small yield (<u>4</u>).

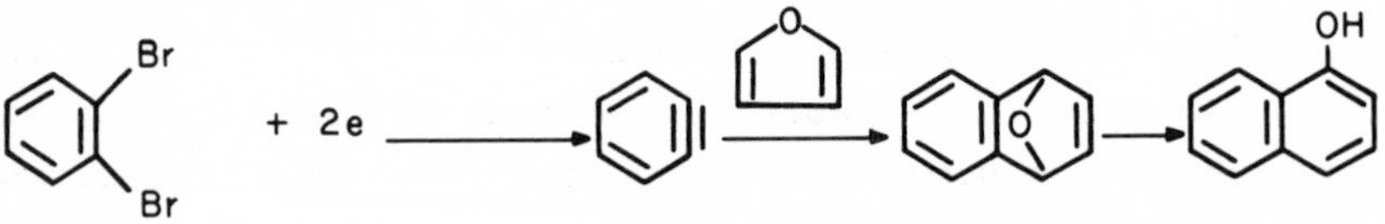

<u>Mechanism of Reduction.</u>   In principle, the mechanism
of reduction of vicinal dihalides resembles that of the
monohalide.   However, the presence of a second halogen imposes a new direction on this reduction, which may be utilized in organic synthesis.   For example, the great difference in $E_{\frac{1}{2}}$ (0.87 V ∿20 kcal/mole) in the axial and equatorial cyclohexane dibromides [88], [89] cannot be accounted

for by simple inductive effects. The fact that 1,2-dibromo-
ethane can be reduced in 75% aqueous dioxane to give ethylene
as the major product ($\underline{40}$) cannot be explained by invoking
anionic intermediates. Consequently, a concerted mechanism
in which the transition state [Eq. (4.66)] bears great resemblance to product (olefin) must be operative:

$$-\overset{\displaystyle X}{\underset{\displaystyle X}{\overset{\displaystyle |}{\underset{\displaystyle |}{C}}}}-\overset{\displaystyle /}{\underset{\displaystyle \backslash}{C}}-\ +2e\ \longrightarrow\ \left[\overset{\displaystyle \overset{\delta\ominus}{\overset{X}{/}}}{\underset{\displaystyle \underset{X}{\underset{\delta\ominus}{\diagdown}}}{C\ \cdots\ C}}\right]\ \longrightarrow\ \overset{\displaystyle /}{\underset{\displaystyle \backslash}}C=C\overset{\displaystyle \backslash}{\diagup}\ +\ 2\overset{\ominus}{X} \qquad (4.56)$$

transition state

Such a transition state includes the partial formation of
a bond. This bond would lower the activation energy of the
reaction. How well developed this bond is will determine
the $E_{\frac{1}{2}}$ (and hence the activation energy) of the reduction
of dihalides.

Role in Organic Synthesis. The reduction of vicinal
dihalides by electrolysis or by conventional reducing agents
will afford the corresponding olefin and/or saturated hydrocarbon. The advantage of electrolysis lies in the ability
of the controlled potential of the cathode to discriminate
between the reduction of different isomers (cis-trans, axial-
equatorial) of dihalides. While the utility of such a discrimination has not been fully explored, it is worthwhile
for the organic chemist to keep in mind.

4.3.4. α,ω-DIHALIDES

The electrochemical reduction of α,ω-dihalides is quite
analgous to that of monohalides. A variety of products,
however, can be obtained from such reductions ($\underline{6}$, $\underline{12}$):

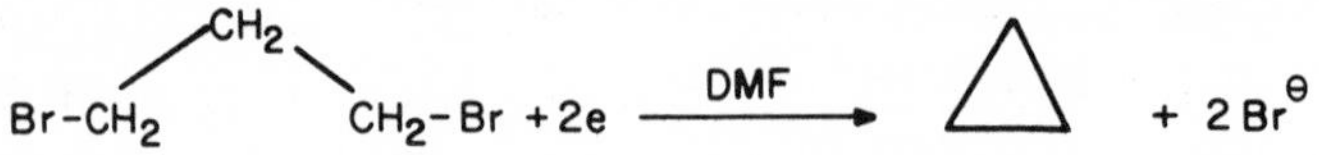

$$\text{Br-CH}_2\text{-CH}_2\text{-CH}_2\text{-Br} + 2e \xrightarrow{\text{DMF}} \triangle + 2\,\text{Br}^{\ominus} \qquad (4.57)$$

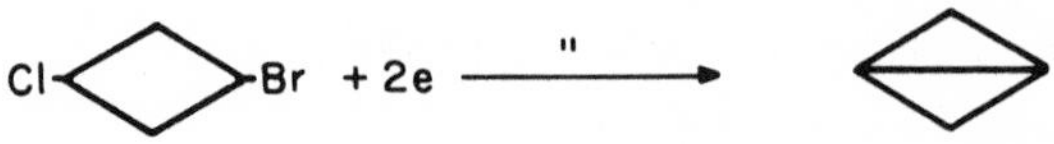

$$\text{Cl-}\diamondsuit\text{-Br} + 2e \xrightarrow{\quad''\quad} \diamondsuit \qquad (4.58)$$

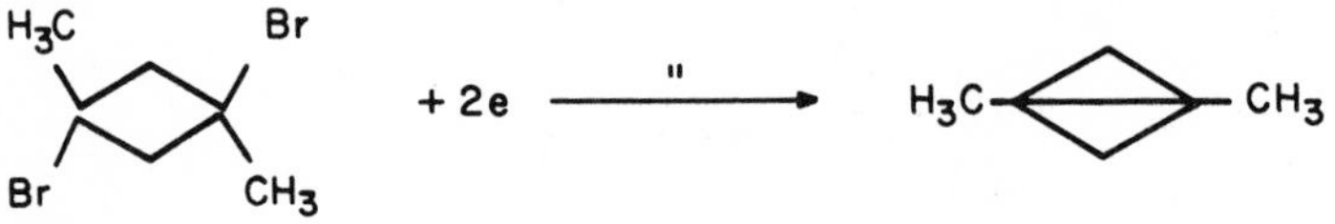

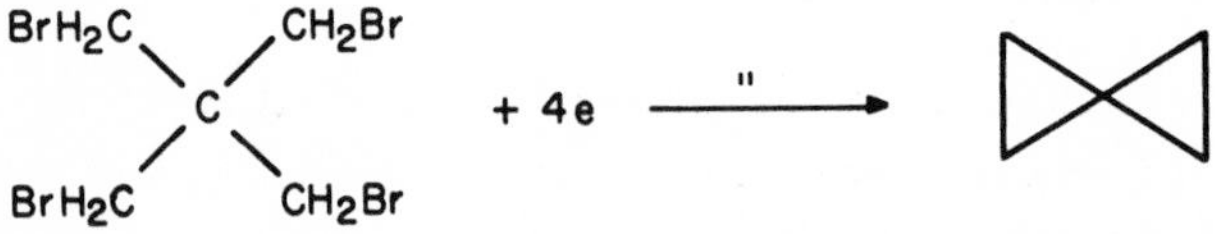

$$\underset{\text{Br}}{\overset{\text{H}_3\text{C}}{\diamondsuit}}\overset{\text{Br}}{\underset{\text{CH}_3}{}} + 2e \xrightarrow{\quad''\quad} \text{H}_3\text{C-}\diamondsuit\text{-CH}_3 \qquad (4.59)$$

$$\underset{\text{BrH}_2\text{C}}{\overset{\text{BrH}_2\text{C}}{}}\text{C}\overset{\text{CH}_2\text{Br}}{\underset{\text{CH}_2\text{Br}}{}} + 4e \xrightarrow{\quad''\quad} \bowtie \qquad (4.60)$$

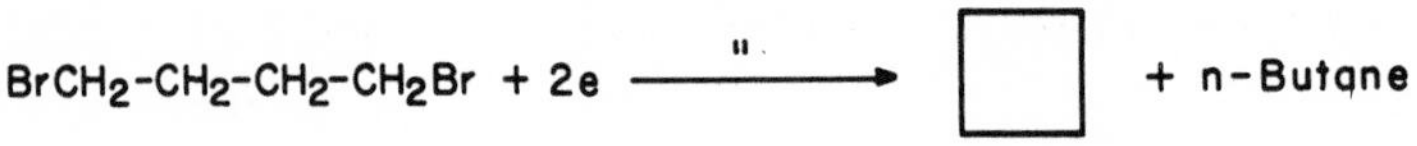

$$\text{BrCH}_2\text{-CH}_2\text{-CH}_2\text{-CH}_2\text{Br} + 2e \xrightarrow{\quad''\quad} \square + \text{n-Butane} \qquad (4.61)$$

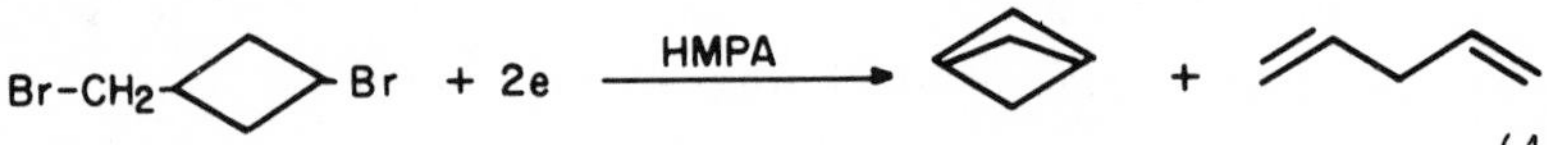

$$\text{Br-CH}_2\text{-}\diamondsuit\text{-Br} + 2e \xrightarrow{\text{HMPA}} \diamondsuit + \diagup\!\diagdown\!\diagup\!\diagdown \qquad (4.62)$$

The preparation of polymeric products has also been reported (8, 41).

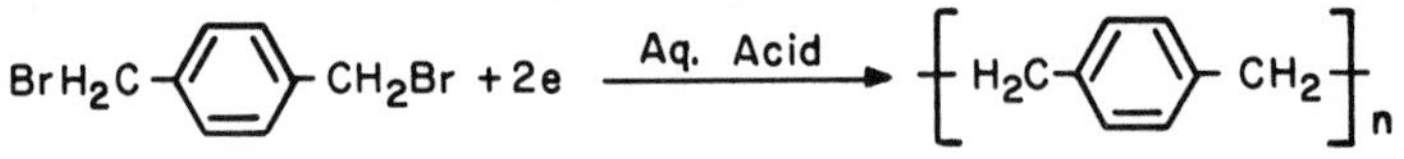

$$\text{BrH}_2\text{C-}\hexagon\text{-CH}_2\text{Br} + 2e \xrightarrow{\text{Aq. Acid}} \left[\text{H}_2\text{C-}\hexagon\text{-CH}_2\right]_n \qquad (4.63)$$

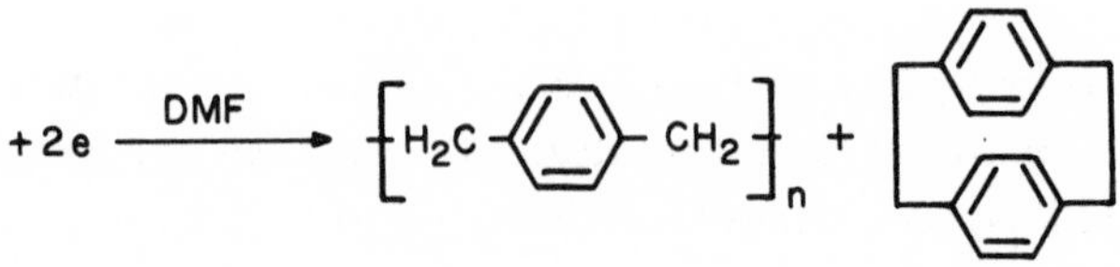

$$+ 2e \xrightarrow{\text{DMF}} \left[\text{H}_2\text{C-}\hexagon\text{-CH}_2\right]_n + \text{(cyclophane)} \qquad (4.64)$$

It has been found ($\underline{6}$) that in certain cases different
solvents in the reduction of α,ω-dihalides allow the formation of different products, Eqs. (4.65) and (4.66):

(4.65)

(unknown stereochemistry)

(4.66)

The explanation for this behavior is that the anionic
intermediate from the cleavage of the first carbon-bromine
bond abstracts a proton from DMF, but does not from hexamethylphosphoramide. Hence, the latter solvent is recommended for reactions involving anions whose protonation is
not desired.

     The reduction of aromatic α,ω-dihalides (i.e., $\underline{o}$- , $\underline{m}$-
and $\underline{p}$-) was studied by Wawzonek and co-workers ($\underline{4}$). The
$\underline{m}$-, and $\underline{p}$-dihalides afforded benzene, while the $\underline{o}$-compounds
afforded a small amount of benzyne, which was trapped with
furan to afford α-naphthol.

     Mechanism of Reduction. The reduction of α,ω-dihalides
in general proceeds in a stepwise manner through anionic
intermediates to afford the observed products. There are,
however, some exceptions. 1,3-Dibromopropane forms cyclopropane upon electrochemical reduction in over 80% yield,
even in the presence of excess water ($\underline{6}$), Eq. (4.57). The
product in Eq. (4.59), is decreased in the presence of
water. The reduction of α,α'-dibromo-$\underline{p}$-xylene ($\underline{8}$, $\underline{41}$) affords polymeric product, even in an acidic medium Eq.
(4.63). While all these observations may be explained by
anionic intermediates, which undergo cyclization faster
than protonation, it is more likely that these reactions
proceed in a concerted manner, with the following transition state:

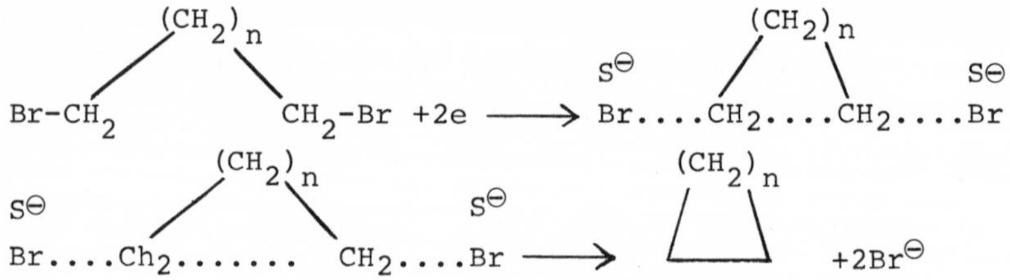

The above transition state was also proposed for the reduction of vicinal dihalides, and can be extended to the reduction of α,ω-dihalides. In reactions where the transition state greatly resembles the product, the central bond is well formed and the reduction may be considered to proceed in a "concerted" manner. Consequently, one observes a trend in the $E_{\frac{1}{2}}$ of α,ω-dihalides (see Table 4.3), which get more negative as the two halides are further separated. This may be an indication that the central bond gets weaker and then disappears as the distance between the two halides increases.

It must be kept in mind that the formation of the above transition state depends, to a great extent on the geometry of the starting material. (Thus, the authors found that compound [92] is easier to reduce than compound [94] by >7 kcal/mole,* which is difficult to explain on the basis of inductive effects.) The following interesting reduction was reported by Cristol and co-workers (37).

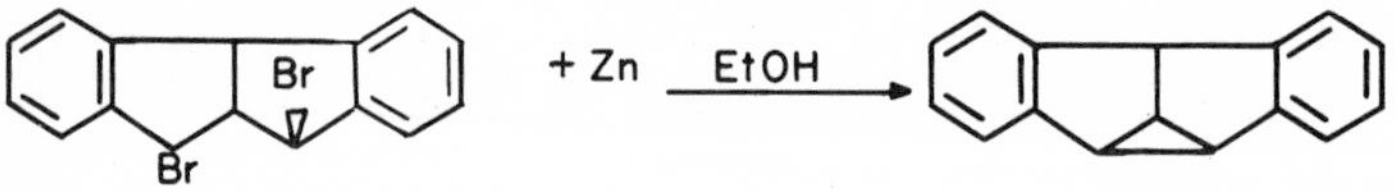

(4.67)

[92]   [93]

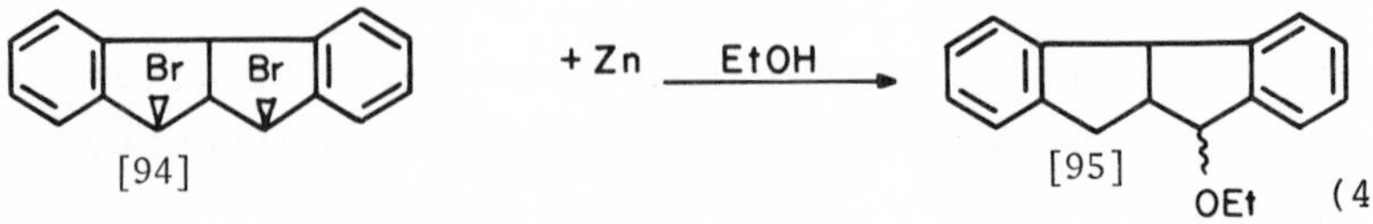

(4.68)

[94]   [95]

---

*This is about 0.3 V.

Role in Organic Synthesis.   Earlier it was mentioned
that the electrochemical reduction of α,ω-dihalides affords
a variety of products.   This type of reduction has a clear
advantage over conventional reducing agents.   For example,
the formation dichlorobicyclobutane [87] was discussed
under gem halides.

$$Cl_2 \diamondsuit Cl_2 + 2e \xrightarrow{\quad DMF \quad} Cl \diamondsuit Cl$$

$$[86] \qquad\qquad\qquad\qquad [87]$$

For reasons which are difficult to explain, the tetrachloride
[86] did not react with sodium and potassium at room temperature.   The reduction of 1,3-dibromo-2,2-bis(bromomethyl)-
propane is of interest.   Reduction of this compound with
conventional reducing agents yields a variety of products
which include the following (42-44):

$$BrH_2C \quad CH_2Br$$
$$\underset{BrH_2C}{\overset{}{\diagdown}} C \underset{CH_2Br}{\overset{}{\diagup}} + Zn \longrightarrow \bowtie + \square + CH_3-CH_2-C=CH_2$$
$$[96] \qquad\qquad [97] \quad [96] \qquad\qquad CH_3$$

$$CH_2 \text{ (} =CH_2\text{)}$$

$$(4.69)$$

The electrochemical reduction of the tetrabromide [96] under
uncontrolled potential gives spiropentane [97] in high yield
(12).   In both cases the reaction presumably proceeds in a
stepwise manner through the following intermediates [98]:

$$BrH_2C \quad CH_2Br$$
$$\underset{BrH_2C}{\overset{}{\diagdown}} C \underset{CH_2Br}{\overset{}{\diagup}} \xrightarrow{+2e} \bowtie \underset{CH_2Br}{\overset{CH_2Br}{}} \xrightarrow{+2e} \bowtie + \text{other products}$$

$$[96] \qquad\qquad\qquad [98]$$

It was the polarographic behavior of the tetrabromide [96]
that led to the conclusion that the isolation of the intermediate [98] was possible.   Compound [96] exhibits two 2-
electron waves (DMF/nBu$_4$NClO$_4$) at -1.8 and -2.3 V (s.c.e.).
Thus it was anticipated that its electrolysis at a controlled

potential <-1.8 V (s.c.e.) should lead to the formation of
the dibromide [98].   This was indeed realized (45).   Furthermore, the reduction of the dibromide 1,1-bis(bromomethyl)-
cyclopropane at -2.3 V (s.c.e.) afforded spiropentane [97]
in good yield.

    Controlled potential electrolysis can also be useful
in organic synthesis where the reduction of one of two halides in a molecule is desired (46), Eqs. (4.70) and (4.71):

(4.70)

$$\text{Br}-\!\!\left\langle\!\!\bigcirc\!\!\right\rangle\!\!-\text{I} \; + 2e \quad \xrightarrow{\text{DMF}} \quad \left\langle\!\!\bigcirc\!\!\right\rangle\!\!-\text{Br}$$

98 %

(4.71)

$$\text{Br}-\!\!\left\langle\!\!\bigcirc\!\!\right\rangle\!\!-\overset{\displaystyle\overset{O}{\|}}{C}CH_2-CH_2\,CH_2Cl + 2e \quad \xrightarrow{\text{DMF}} \quad \left\langle\!\!\bigcirc\!\!\right\rangle\!\!-\overset{\displaystyle\overset{O}{\|}}{C}CH_2-CH_2-CH_2Cl$$

96 %

4.3.5.   EXPERIMENTAL PROCEDURE

4.3.5.1.   Reduction to Hydrocarbons

$$RX + 2e \xrightarrow{\text{H}^{\oplus}} RH+ X^{\ominus}$$

    The experimental procedure for the reduction of organic
halides depends on the nature of the desired product.   For
reductions where R-X is transformed to R-H, reactions may
be carried out in a beaker in which a porcelain thimble is
suspended and which serves as the anode compartment (Fig.
4.2).
The solvent used for these reactions is generally aqueous
ethanol.   Enough ethanol is normally used to dissolve the
organic halide.   The supporting electrolyte may be an
inorganic acid or a tetraalkylammonium salt.

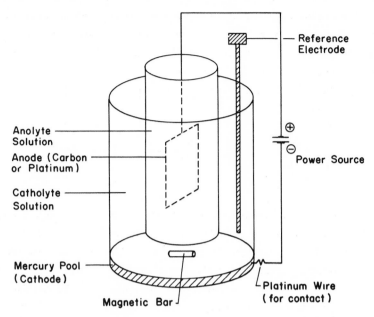

FIG. 4.2. Electrolysis cell for the reduction of organic halides.

4.3.5.2. Reduction to Olefins and Cyclopropanes

The above apparatus may also be used for the reduction of halides of the type:

$$X-\overset{|}{\underset{|}{C}}-\overset{|}{\underset{|}{C}}-X \quad + 2e \longrightarrow \;C{=}C\; + 2X^{\ominus}$$

$$X-\overset{|}{\underset{|}{C}}-\overset{|}{\underset{|}{C}}-\overset{|}{\underset{|}{C}}-X + 2e \longrightarrow \triangle + 2X^{\ominus}$$

These reactions are generally not affected by protic solvents. If it is feared that these reactions may be sensitive to aqueous solutions, then an aprotic solvent such as acetonitrile, DMF, hexamethylphosphoramide may be used.
For such reactions, inorganic (alkali halides) as well as organic salts (tetraalkylammonium halides) may be used as supporting electrolytes.

The reduction of dihalides to yield cyclic products
may also be carried out in a beaker, provided the cyclic
product is not volatile. For reactions which yield volatile'
products, the reader should consult Sec. 3.1.

Preparation 1,3-Dimethylbicyclobutane [100] (12)

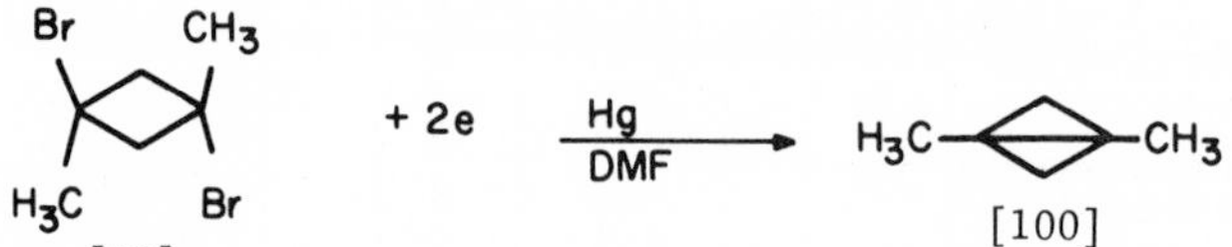

[99]   [100]

An H-type cell was used for this reaction and is described in Ref. 12. A solution of 50 g (0.2 mole) of 1,3-
dibromo-1,3-dimethylcyclobutane in 230 ml of 0.1 M LiBr/DMF
was introduced into the cathode chamber. This chamber was
connected to a dry ice-acetone trap in order to collect the
volatile products. A mercury pool was used as the cathode.
The anode chamber was charged with 150 ml of 0.1 M LiBr/DMF.
A platinum wire 6 in. long was used as the anode. A current
of 0.5 A was allowed to pass through the cell. The over-
all voltage was 40 V. Electrolysis was allowed to proceed
for 16 hr after which a colorless liquid was collected in
the trap. Distillation from repetitive runs at atmospheric
pressure afforded 9 to 15 g (55-94%) of 1,3-dimethylbicyclo-
butane.

Controlled Potential Electrolysis, Preparation of 1,1-bis
(bromomethyl)cyclopropane [98] (45)

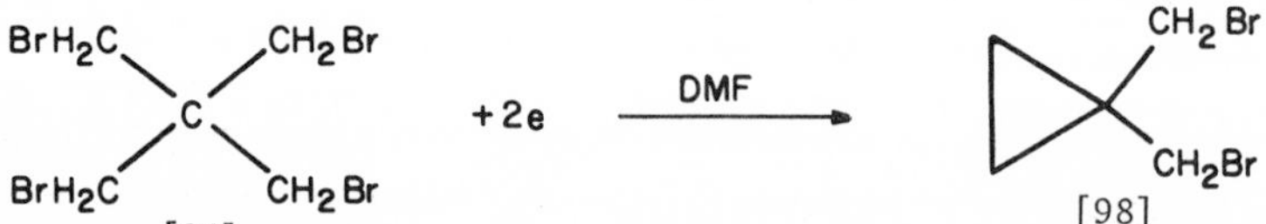

[96]   [98]

The apparatus used for this reaction is shown in Fig.
4.2. Twenty-five grams (0.06 mole) of 1,3-dibromo-2,2-bis

(bromomethyl)propane* was dissolved in 250 ml of 0.05 M
Et_4NBr/DMF and placed in the cathode compartment. The anode
compartment was charged with 0.05 M Et_4NBr/DMF so that the
liquid levels in the two compartments were equal. A calomel
reference electrode was suspended in the solution so that
it was approximately 6-7 mm above the surface of the mercury.** The potential of the mercury pool was registered
on an RCA vacuum tube voltmeter. A graphite cylinder 3 in.
long and ½ in. outer, diameter was used as the anode. The
over-all potential of the cell, 12 V, was adjusted so that the
potential of the cathode was -1.4 V (s.c.e.). This allowed
the passage of 0.3 A through the cell. During the course
of the reaction, samples from the cathode compartment (3
drops diluted to 20 ml with 0.05 M DMF/Et_4NBr) were withdrawn
and their polarograms recorded. When the first wave had
almost disappeared (13 hr) the reaction was terminated. At
this time the cathode solution was hydrolyzed with 250 ml
water and extracted into 500 ml pentane. Distillation of
the pentane at atmospheric pressure left a slightly yellowish
liquid 7.9 g (54% yield), which was distilled through a 1-ft
spinning band column, bp $63°/7$ mm, to give 7.1 g of the
desired product (48).

REFERENCES ON CARBON-HALOGEN BONDS

1. P. J. Elving, Record Chem. Progr. (Kresge-Hooker Sci.
    Lib.), 14, 99 (1953).

---

*This compound exhibits two 2-electron polarographic
waves at -1.8 and 2.3 V (s.c.e.) in DMF/Et_4NBr.

**This is approximately the same distance that this
electrode was stationed away from the dropping mercury electrode in the polarographic measurements. It is important
that this distance in the macroscale electrolysis be maintained as close as possible.

2. P. J. Elving and B. Pullman, in Advances in Chemical Physics, (I. Prigogine, ed.), Vol. 1, Interscience, New York, 1961, p. 7.

3. L. G. Feoktistov, Usp. Electrokhim. Org. Soedin., Akad. Nauk SSSR, Inst. Electrokhim, 135 (1966).

4. S. Wawzonek and R. C. Duty, J. Electrochem. Soc., 110, 420 (1963).

5. S. Wawzonek and R. C. Duty, J. Electrochem. Soc., 108, 1135 (1961).

6. M. R. Rifi, Tetrahedron Letters, 1969, 1043.

7. F. H. Covitz, J. Am. Chem. Soc., 89, 5403 (1967).

8. H. Gilch, J. Polymer Sci., 4, 1351 (1966).

9. W. Jura and R. Gaul, J. Am. Chem. Soc., 80, 5402 (1958).

10. P. J. Elving and T. Leone, J. Am. Chem. Soc., 79, 1546 (1957).

11. G. Klopman, Helv. Chim. Acta, 44, 1908 (1964).

12. M. R. Rifi, J. Am. Chem. Soc., 89, 4442 (1967).

13. A. P. Tomilov and Yu. D. Simrnov, Zh. Vses. Khim. Obschestva im. D. I. Mendeleeva, 10(1), 101 (1965); C. A., 63, 3897 (1965).

14. S. Wawzonek, R. C. Duty, and J. H. Wagenknecht, J. Electrochem. Soc., 111, 74 (1969).

15. F. Lambert and K. Kobagaski, J. Am. Chem. Soc., 82, 5324 (1929).

16. J. Pearson, Trans. Faraday Soc., 44, 683 (1948).

17. J. W. Sease, P. Chang, and J. Groth, J. Am. Chem. Soc., 86, 3154 (1969).

18. P. J. Elving and B. Pullman, Advan. Chem. Phy., 3, 1 (1961).

19. A. Streitwieser and C. Perrin, J. Am. Chem. Soc., 86, 4938 (1964).

20. J. W. Sease, F. G. Burton, and S. L. Nickol, J. Am. Chem. Soc., 90, 2595 (1968).

21. F. L. Lambert, J. Org. Chem., 31, 4184 (1966).

22. R. Annino, R. E. Erickson, J. Michalovic, and B. McKay, J. Am. Chem. Soc., 88, 4424 (1966).

23. C. L. Mann, J. L. Webb, and E. M. Walborsky, Tetrahedron Letters, 1966, 2249.

24. H. E. Ulery, J. Electrochem. Soc., 116, 1201 (1969).

25. L. L. Miller and E. Riekena, J. Org. Chem., 34, 3359 (1969).

26. J. W. Sease and R. C. Reed, Electrochemical Society Meeting, New York City, May 4-9, 1969.

27. M. R. Rifi, unpublished results.

28. A. J. Fry and M. A. Mitnick, J. Am. Chem. Soc., 91, 6207 (1969).

29. P. J. Elving, I. Rosenthall, J. R. Hayes, and A. J. Martin, Anal. Chem., 33, 330 (1961).

30. P. S. Skell and R. G. Allen, J. Am. Chem. Soc., 80, 5997 (1958).

31. A. A. Oswald, K. Griesbaum, B. B. Hudson, Jr., and J. M. Bregman, J. Am. Chem. Soc., 8b, 2877 (1964).

32. J. A. Kampeier and G. Chen, J. Am. Chem. Soc., 87, 2608 (1965).

33. J. P. Petrovich, M. M. Baizer, and M. R. Ort, J. Electrochem. Soc., 116, 749 (1969).

34. S. Stocker and R. Jenevein, Chem. Commun., 934 (1968).

35. T. Meites and L. Meites, Anal. Chem., 27, 1531 (1955).

36. M. R. Rifi, unpublished results.

37. S. Cristol, A. Dahl, and W. Lim, J. Am. Chem. Soc., 92, 2670 (1970).

38. J. Zavada, J. Krupicka, and J. Sicker, Collection Czech. Chem. Commun., 28, 1664 (1963).

39. P. J. Elving, I. Rosenthal, and A. Martin, J. Am. Chem. Soc., 77, 5218 (1955).

40. Von M. v. Stakelberg and W. Strake, J. Electrochem. Soc., 53, 118 (1949).

41. F. H. Covitz, J. Am. Chem. Soc., 89, 540 (1967).

42. D. E. Applequist, J. Org. Chem., 23, 1715 (1958).

43. V. A. Slobey, J. Am. Chem. Soc., 68, 1335 (1946).

44. H. O. House, R. C. Lord, and H. S. Rao, J. Org. Chem., 21, 1487 (1956).

45. M. R. Rifi, J. Org. Chem., 36, 2017 (1971).

46. A. J. Fry, M. A. Mitnick, and R. G. Reed, J. Org. Chem., 35, 1232 (1970).

47. M. R. Rifi, 21st Meeting, C.I.C.I., Prague, Czechoslovakia, Sept. 28 to Oct. 2, 1970.

48. M. Slobadin and I. N. Shokhar, J. Gen. Chem. USSR, 21, 223 (1951).

4.4.   UNSATURATED COMPOUNDS

4.4.1.   GENERAL CONSIDERATIONS

In many respects, the electrochemical reduction of unsaturated compounds, such as olefins and benzene derivatives, resembles the reductions of these compounds with conventional reducing agents. However, it is the ability to control the potential of the electrode as well as the choice of the reaction conditions that make reduction by electrolysis a versatile tool in organic synthesis. The work of Baizer ($\underline{1}$) on the dimerization of activated olefins has opened up the door for the synthesis of a variety of cyclic compounds from simple molecules:

$$CH_2=CHX + 2e \longrightarrow X-CH_2-CH_2-CH_2-CH_2X$$

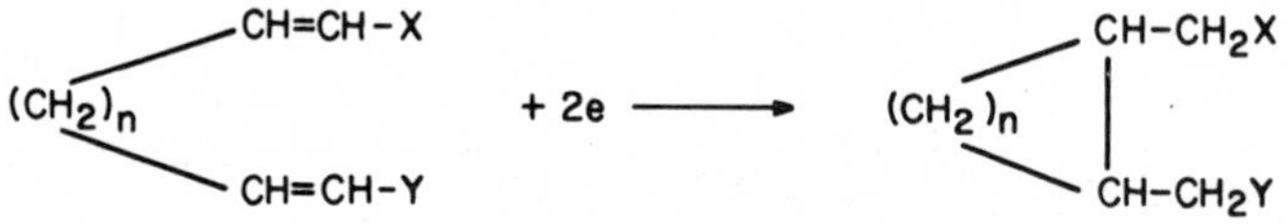

$$(CH_2)_n \Big\langle {CH=CH-X \atop CH=CH-Y} + 2e \longrightarrow (CH_2)_n \Big\langle {CH-CH_2X \atop CH-CH_2Y}$$

(X and Y are electron-withdrawing nonelectroactive groups). The dimerization of acrylonitrile, which is now carried out on a commercial basis,* is one of the best demonstrations that electrolysis has a significant role in organic synthesis.

The work of Bankeser ($\underline{2}$, $\underline{3}$) which describes the "electrochemical Birch reduction," teaches how a simple change in the design of the electrolysis cell can lead to formation of a variety of products, Eq. (4.72):

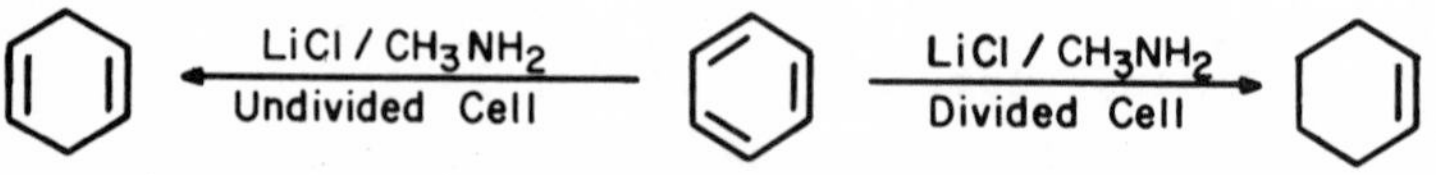

$$\text{(benzene)} \xleftarrow[\text{Undivided Cell}]{\text{LiCl / CH}_3\text{NH}_2} \text{(benzene)} \xrightarrow[\text{Divided Cell}]{\text{LiCl / CH}_3\text{NH}_2} \text{(cyclohexadiene)} \quad (4.72)$$

---

*Monsanto Co., Chem. Eng., Nov. 8, 1965, p. 238.

In this reaction, the reducing agent is lithium metal,
which is generated electrochemically from the supporting
electrolyte. Even in polymerization reactions, electrolyses may yet reveal distinct advantages in the polymerization of olefins over conventional initiating reagents.
(This topic is dealt with in a separate chapter of this
book.)

4.4.2. EFFECT OF REACTION CONDITIONS

The Solvated Electron. Reaction conditions, i.e.,
nature of electrode, pH of medium, etc., do not play a major
role in determining the course of reduction of unsaturated
compounds. There are indications, however, that certain
solvents, e.g., hexamethylphosphoramide in ethanol, have
some influence on the mode of reduction. In such a medium
it has been postulated (4, 5) that electrons from the cathode
migrate to the bulk of the solution and are solvated (solvated electrons).

$$e \quad cathode + S \rightleftharpoons e_s \quad (where \ s \ is \ the \ solvent).$$

Reduction of the substrate can then take place away from the
cathode. Thus, benzene was reduced to cyclohexane, cyclohexadiene, and cyclohexene. On the other hand, electrolysis
of benzene in ethylenediamine (6) gave cyclohexene with a
trace of cyclohexadiene and no detectable amount of cyclohexane. These results were explained as due to the faster
rate of reduction of cyclohexene to cyclohexane (6). In
ethanol/hexamethylphosphoramide, the relative amounts of
cyclohexadiene, cyclohexene, and cyclohexane were found to
depend on the concentration of ethanol, current density,
and the temperature of the reaction (7). Furthermore, the
absence of hexamethylphosphoramide in this reduction reduced
the current efficiency from 95 to less than 1%.

The nature of the supporting electrolyte and cathodes
may also play a role in determining the course of the re-

action. For example, the reduction of benzene to unconjugated dienes, which do not undergo further reduction, was
attributed to the use of tetrabutylammonium cation as the
supporting electrolyte and mercury as the cathode (9).
Tetramethyl and tetraethylammonium cations were found to
be unsuitable for such reductions.

Electrochemical reduction via the "solvated electron"
is obviously similar to the very well-known Birch reduction,
which is carried out with alkali metals in liquid ammonia
and other amine solvents (8). However, it can be seen from
the above discussion that the electrochemical method appears
to be more sensitive to reaction conditions. This characteristic may be an asset in the area of organic synthesis
and is discussed below.

4.4.3. REDUCTION OF AROMATIC COMPOUNDS

Reduction Products. Reduction of aromatic compounds
generally affords the dihydro derivative which may be reduced further at a more negative potential. Benzene, which
does not exhibit a polarographic wave, can be reduced to a
variety of products depending on the reaction conditions,
Eqs. (4.73)-(4.76):

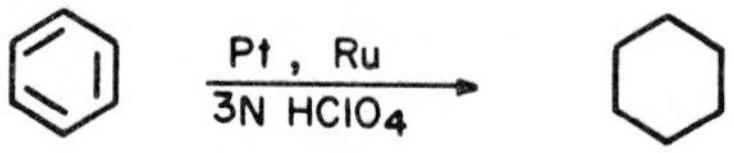

$$\text{benzene} \xrightarrow[\text{3N } HClO_4]{\text{Pt , Ru}} \text{cyclohexane} \qquad (4.73)$$

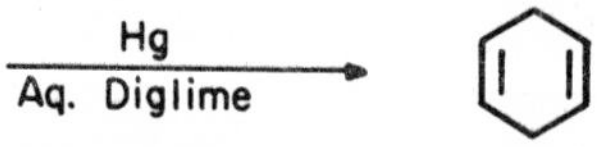

Ref. 10

$$\text{benzene} \xrightarrow[\text{Aq. Diglime}]{\text{Hg}} \text{cyclohexadiene} \qquad (4.74)$$

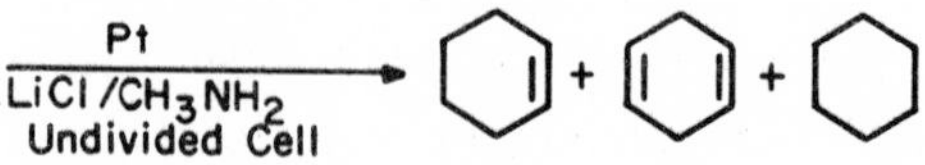

Ref. 9

$$\text{benzene} \xrightarrow[\substack{LiCl/CH_3NH_2 \\ \text{Undivided Cell}}]{\text{Pt}} \text{cyclohexene} + \text{cyclohexadiene} + \text{cyclohexane} \qquad (4.75)$$

Ref. 2

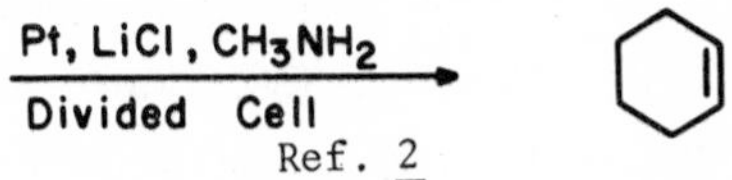

$$\xrightarrow[\substack{\text{Divided Cell} \\ \text{Ref. } \underline{2}}]{\text{Pt, LiCl, CH}_3\text{NH}_2}$$

(4.76)

The reduction of aromatic compounds under the appropriate reaction conditions can lead to their functionalization (<u>11</u>-<u>13</u>), Eqs. (4.77) and (4.78):

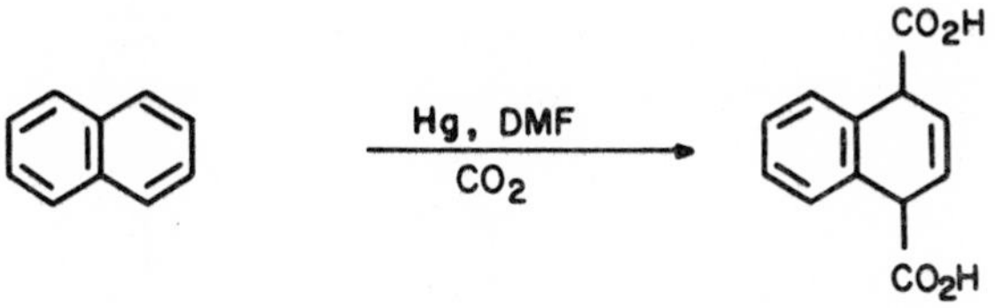

$$\xrightarrow[\text{CO}_2]{\text{Hg, DMF}}$$

(4.77)

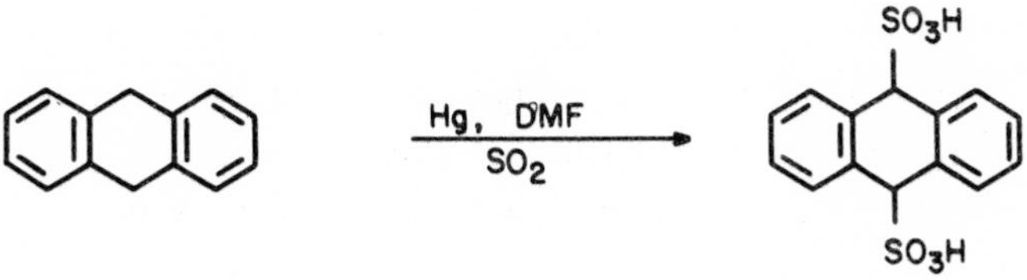

$$\xrightarrow[\text{SO}_2]{\text{Hg, DMF}}$$

(4.78)

$$\xrightarrow[\text{H}_2\text{SO}_4]{\text{Pt}}$$

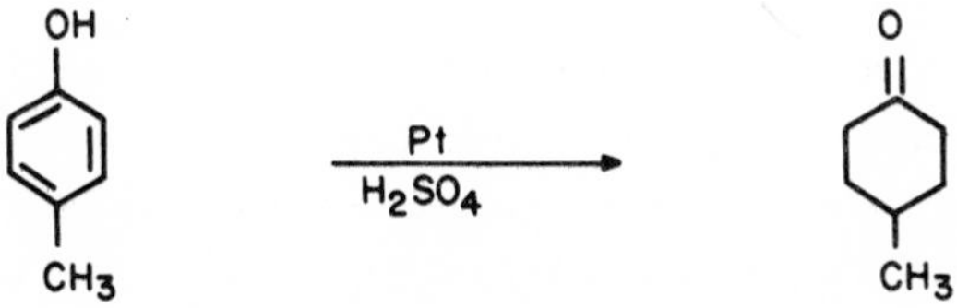

(4.79)

Some aromatic hydrocarbons give dimeric products upon reduction as in the case of phenanthrene (<u>11</u>).

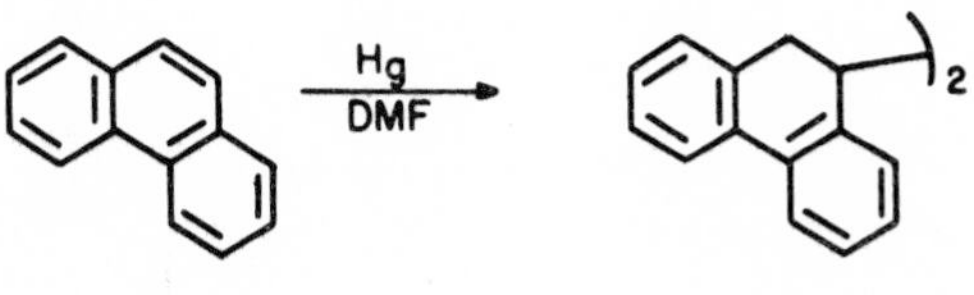

$$\xrightarrow[\text{DMF}]{\text{Hg}}$$

Mechanism of Reduction. The mechanism of the electrochemical reduction of aromatic compounds has been recently
discussed (14). Consequently, only a brief discussion of
it is presented in this section.

In either protic or aprotic media, aromatic hydrocarbons are first reduced to the radical anion [101]. The potential of reduction depends on the stability of this radical anion. Thus, it is not surprising that benzene reduces
at a higher potential than anthracene.

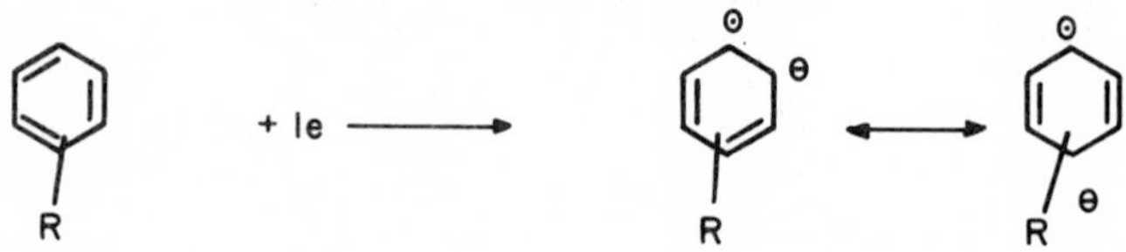

[101]

The radical anion abstracts a proton from the solvent
to give the radical, which may dimerize or undergo further
reduction and protonation to the dihydro product.

The mechanism for the electrochemical Birch-type reduction is of interest (2, 3):

When the reaction is carried out in a divided cell, the
radical anion formed [101] abstracts a proton from the
amine to give the amide ion and the diene radical [102].
The diene radical [102] undergoes further reduction and
protonation to afford the nonconjugated diene [103]. However,
this diene is isomerized by the amide anion to the conjugated
species [104], which undergoes further reduction to cyclohexene. Thus, the over-all reaction may be described as
follows:

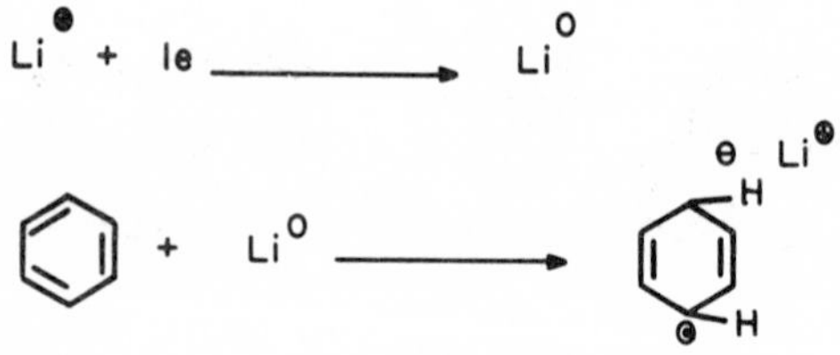

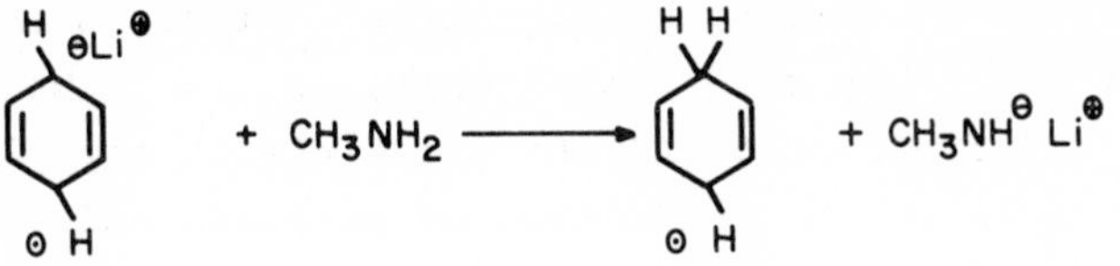

[101] + CH$_3$NH$_2$ $\longrightarrow$ [102] + CH$_3$NH$^\ominus$ Li$^\oplus$

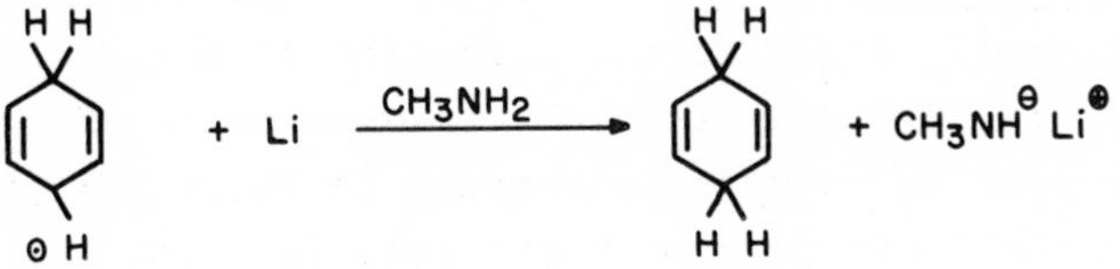

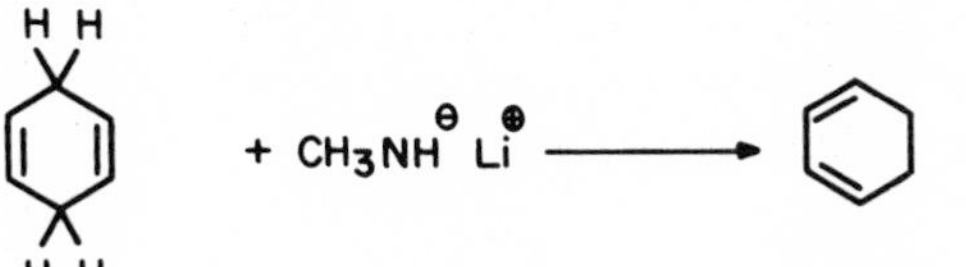

[102] + Li $\xrightarrow{\text{CH}_3\text{NH}_2}$ [103] + CH$_3$NH$^\ominus$ Li$^\oplus$

[103] + CH$_3$NH$^\ominus$ Li$^\oplus$ $\longrightarrow$ [104]

[104] + 2 Li + 2 CH$_3$NH$_2$ $\longrightarrow$ + 2 CH$_3$NH$^\ominus$ Li$^\oplus$

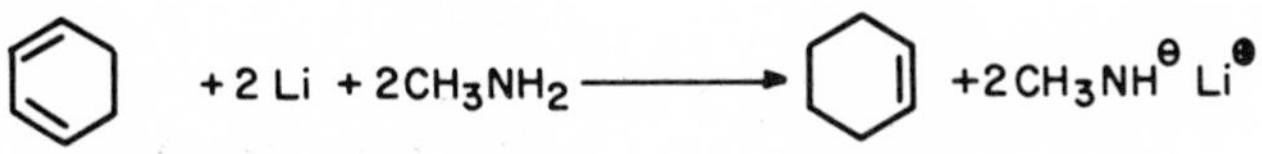

When the reaction is carried out in an undivided cell,
species formed at the anode may interact with those at the
cathode. Thus, methylamine is oxidized (at the anode) to
form methylamine hydrochloride. This species neutralizes
the methylamide formed as shown above, and prevents the
isomerization of the unconjugated diene, which resists
further reduction and is isolated as the final product.

4.4.4. REDUCTION OF OLEFINS

The electrochemical reduction of olefins is one of the
few areas where electrolysis has begun to show its place in
organic synthesis. Thus, the electrochemical reduction of

acrylonitrile is already being carried out on an industrial
scale (1) and it is anticipated that, in time, the polymerization of olefinic compounds by electrolysis may play an
important role in the production of plastics and other products. (For further details on polymerization reactions
see Chap. 6.)

Reduction Products. The reduction of olefins in protic
media generally leads to the formation of the corresponding
hydrocarbons (probably through the generation of hydrogen).
Thus, propene is reduced to propane in about 90% yield on a
platinum electrode in phosphoric acid solution (15). Activated olefins can also undergo dimerization reactions (16-18),
Eqs. (4.80) and (4.81):

$$CH_2=CH-CN \xrightarrow[\substack{R_4N^{\oplus}X^{\ominus} \\ \text{Aq. solution}}]{Hg} NC-CH_2(CH_2)_2-CH_2CH \qquad (4.80)$$

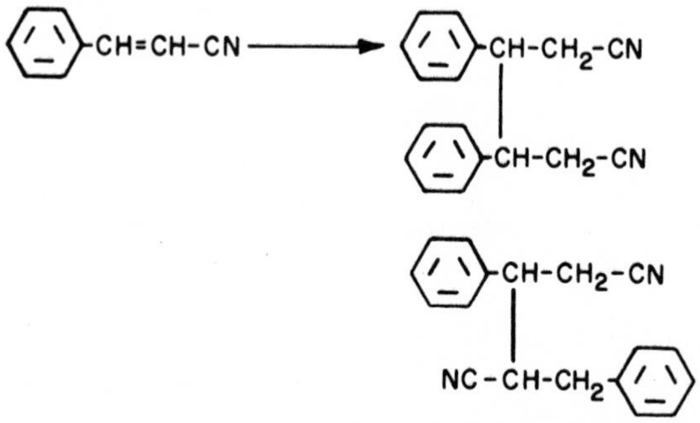

(4.81)

An interesting reaction developed by Baizer and co-
workers (1) involves an intradimerization of activated
dienes [105] to give cyclic products [106]:

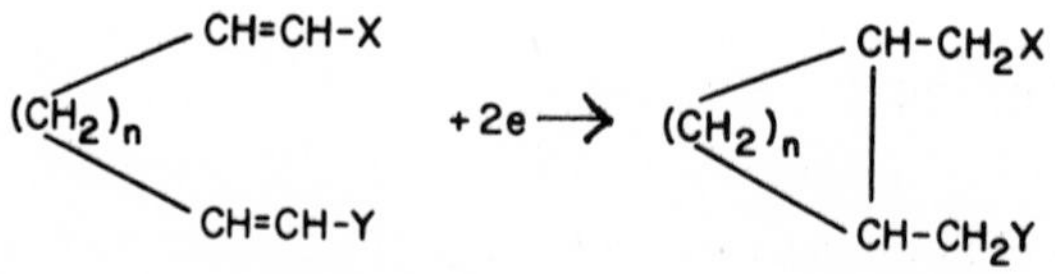

[105]   [106]

where X and Y are nonelectroactive electron-withdrawing groups. For n = 1-4 the yields of the cyclic product were fairly high; for n = 5, the yield was quite low, Eqs. (4.82)-(4.85):

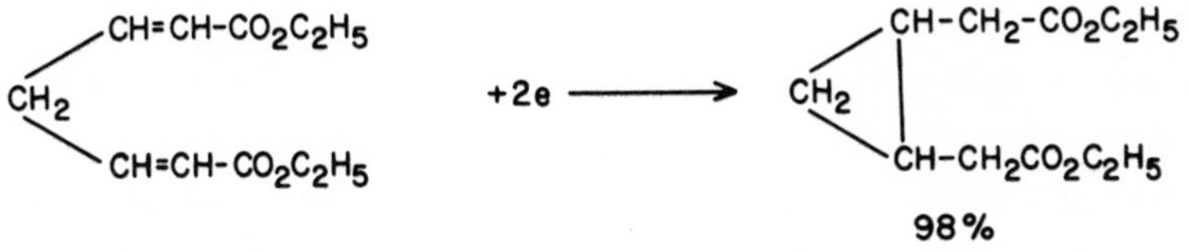

$$
\begin{array}{c}
\text{CH=CH-CO}_2\text{C}_2\text{H}_5 \\
\text{CH}_2 \qquad\qquad \xrightarrow{+2e} \qquad \text{CH}_2 \qquad \text{CH-CH}_2\text{-CO}_2\text{C}_2\text{H}_5 \\
\text{CH=CH-CO}_2\text{C}_2\text{H}_5 \qquad\qquad \text{CH-CH}_2\text{CO}_2\text{C}_2\text{H}_5
\end{array}
\qquad (4.82)
$$

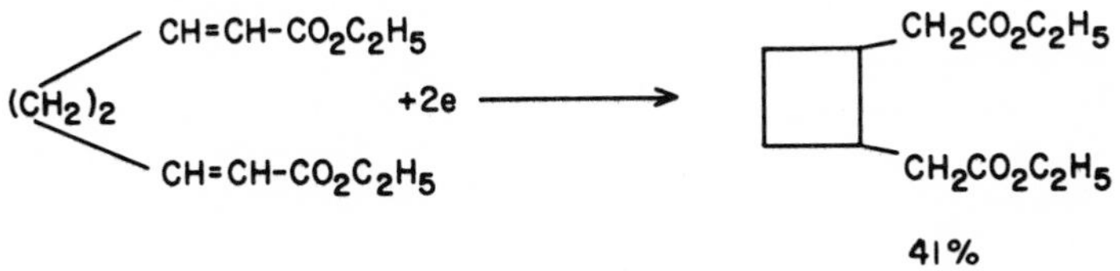

98%

$$
\begin{array}{c}
\text{CH=CH-CO}_2\text{C}_2\text{H}_5 \\
\text{(CH}_2)_2 \qquad\qquad \xrightarrow{+2e} \qquad\qquad \text{CH}_2\text{CO}_2\text{C}_2\text{H}_5 \\
\text{CH=CH-CO}_2\text{C}_2\text{H}_5 \qquad\qquad\qquad \text{CH}_2\text{CO}_2\text{C}_2\text{H}_5
\end{array}
\qquad (4.83)
$$

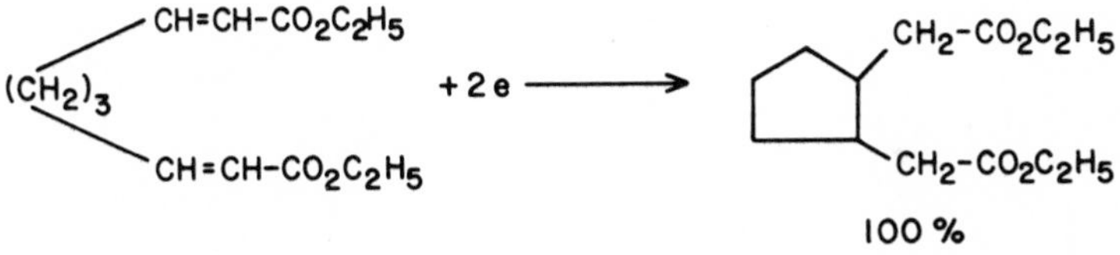

41%

$$
\begin{array}{c}
\text{CH=CH-CO}_2\text{C}_2\text{H}_5 \\
\text{(CH}_2)_3 \qquad\qquad \xrightarrow{+2e} \qquad\qquad \text{CH}_2\text{-CO}_2\text{C}_2\text{H}_5 \\
\text{CH=CH-CO}_2\text{C}_2\text{H}_5 \qquad\qquad\qquad \text{CH}_2\text{-CO}_2\text{C}_2\text{H}_5
\end{array}
\qquad (4.84)
$$

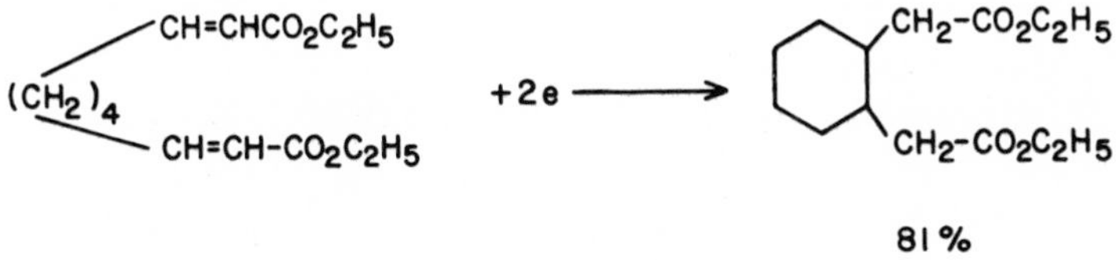

100%

$$
\begin{array}{c}
\text{CH=CHCO}_2\text{C}_2\text{H}_5 \\
\text{(CH}_2)_4 \qquad\qquad \xrightarrow{+2e} \qquad\qquad \text{CH}_2\text{-CO}_2\text{C}_2\text{H}_5 \\
\text{CH=CH-CO}_2\text{C}_2\text{H}_5 \qquad\qquad\qquad \text{CH}_2\text{-CO}_2\text{C}_2\text{H}_5
\end{array}
\qquad (4.85)
$$

81%

In aprotic solvents, activated olefins undergo poly- merization reactions ([19]-[22]):

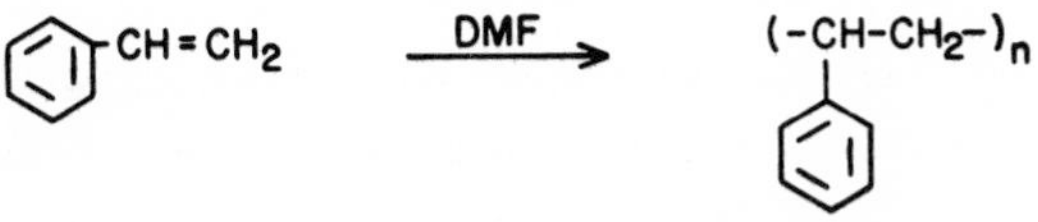

$$
\text{C}_6\text{H}_5\text{-CH=CH}_2 \xrightarrow{\text{DMF}} (\text{-CH-CH}_2\text{-})_n
$$

$$CH_2{=}CH\text{-}CN \xrightarrow{DMF} (\text{-}CH_2\text{-}\underset{\underset{\displaystyle CN}{|}}{CH}\text{-})_n$$

This topic is dealt with in some detail in the chapter on electroinitiated polymerization.

<u>Mechanism of Reduction</u>. The mechanism of the electrochemical reduction of olefins resembled that of aromatic hydrocarbons. The addition of one electron affords the radical anion [107], which may give the observed products as follows:

$$RCH{=}CHR' + 1e \longrightarrow \overset{\odot}{R}CH\text{-}\overset{\ominus}{C}H\text{-}R' \longleftrightarrow R\overset{\ominus}{C}H\text{-}\overset{\odot}{C}HR'$$

[107]

$$R\overset{\odot}{C}H\text{-}\overset{\ominus}{C}HR' + H^{\oplus} \longrightarrow R\overset{\odot}{C}H\text{-}CH_2R'$$

$$2R\overset{\odot}{C}H\text{-}CH_2R' \longrightarrow \begin{array}{l} RCH\text{-}CH_2R' \\ \;\;| \\ RCH\text{-}CH_2R' \end{array}$$

$$R\overset{\odot}{C}H\text{-}CH_2R' \xrightarrow[+H\oplus]{+1e} RCH_2\text{-}CH_2R$$

$$2R\overset{\odot}{C}H\text{-}\overset{\ominus}{C}HR' \longrightarrow \begin{array}{l} R\text{-}CH\text{-}\overset{\ominus}{C}H\text{-}R' \\ \;\;\;\;\;| \\ R\text{-}CH\text{-}CH^{\ominus}R' \end{array} \xrightarrow{RCH{=}CHR'} \text{Polymer}$$

Polymerization reactions may occur via a radical or anionic mechanism, although indications are that the radical anion dimerizes to give the dianion, which attacks the starting material to afford polymers.

The internal dimerization of dienes to give cyclic products was postulated to proceed via a concerted mechanism as follows (<u>1</u>), Eqs. (4.86) and (4.87):

$$
\text{(-)} \xrightarrow{e} \quad \underset{\text{CATHODE}}{}
\begin{array}{c}
X \\ | \\ CH \\ \| \\ CH \\ \diagdown (CH_2)_n \\ CH \\ \| \\ CH \\ | \\ Y
\end{array}
\quad \longrightarrow \quad
X-\overset{}{CH}-CH \underset{}{\overset{(CH_2)_n}{\diagup \diagdown}} CH-CH^{\ominus}-Y \qquad (4.86)
$$

$$
X-\overset{\ominus}{CH}-CH \underset{}{\overset{(CH_2)_n}{\diagup \diagdown}} CH-\overset{\ominus}{CH}-Y \xrightarrow[+2H^{\oplus}]{+1e} X-CH_2 CH \underset{}{\overset{(CH_2)_n}{\diagup \diagdown}} CH-CH_2 Y \qquad (4.87)
$$

This mechanism was based, among other observations, on the fact that dienes which afforded cyclic products exhibited polarographic waves at a more positive half-wave potential than those which gave alycyclic products.

4.4.5. ROLE IN ORGANIC SYNTHESIS

The electrochemical reduction of unsaturated compounds has an important role in organic synthesis, since it allows the investigator in this field some control over the nature of the isolated products. For example, it was mentioned that in the reduction of benzene a simple change in the design of the electrolysis cell (i.e., divided vs undivided) can afford cyclohexane or cyclohexadiene (2). The formation of the "solvated electron" under different reaction conditions appears to favor the formation of certain products to others (9). When applied to the reduction of olefins, solvated electrons, generated electrochemically, behave in a different manner than those generated from alkali metals. For example, 2,3-dimethyl-2-butene and cyclohexene are both reduced electrochemically in the same reaction in ethanol/hexamethylphosphoramide (6). However, reaction of these compounds with lithium in ethylamine did not allow the reduction of the sterically hindered butene (23). It would be of interest to compare the same reduction with lithium in ethanol/hexamethylphosphoramide.

The importance of electrolysis in organic synthesis is also shown in the selective reduction of aromatic nucleic [108] in the presence of isolated double bonds (26):

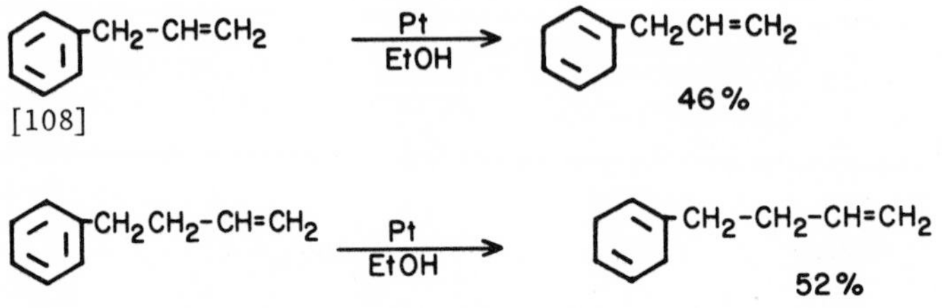

$$\text{Ph–CH}_2\text{–CH=CH}_2 \xrightarrow[\text{EtOH}]{\text{Pt}} \text{Ph–CH}_2\text{CH=CH}_2 \quad 46\%$$

[108]

$$\text{Ph–CH}_2\text{CH}_2\text{–CH=CH}_2 \xrightarrow[\text{EtOH}]{\text{Pt}} \text{Ph–CH}_2\text{–CH}_2\text{–CH=CH}_2 \quad 52\%$$

By catalytic methods (e.g., Raney nickel and hydrogen), the reverse would be true, since the double bond would reduce preferentially.

One of the potential advantages of the electrochemical reduction of unsaturated compounds lies in the formation of polymers. Thus, it has been possible to control the molecular weight (and perhaps the molecular weight distribution) of the polymeric chain by monitoring the impressed current into the reaction solution (21).

4.4.6. ACETYLENES

Compared to olefins, the reduction of acetylenes is more difficult to accomplish. Consider the following polarographic observations (27):

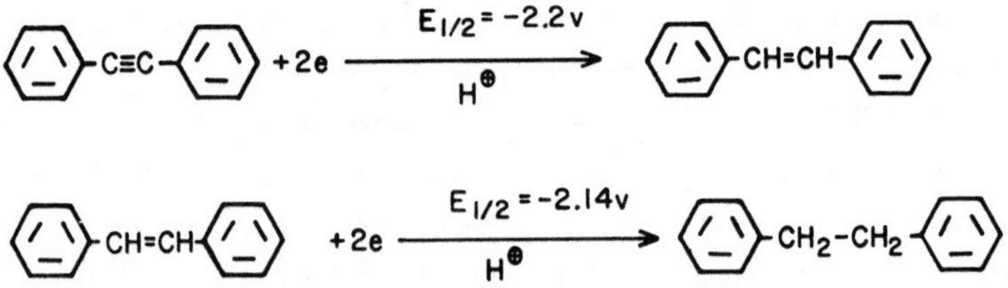

$$\text{Ph–C≡C–Ph} + 2e \xrightarrow[\text{H}^{\oplus}]{E_{1/2}=-2.2\text{v}} \text{Ph–CH=CH–Ph}$$

$$\text{Ph–CH=CH–Ph} + 2e \xrightarrow[\text{H}^{\oplus}]{E_{1/2}=-2.14\text{v}} \text{Ph–CH}_2\text{–CH}_2\text{–Ph}$$

It was observed that the height of the wave for diphenyl-acetylene is twice that for stilbene. This would indicate that, in general, the reduction of acetylenes will lead to the formation of the corresponding saturated product.

4.4.7. EXPERIMENTAL PROCEDURE

4.4.7.1. Reduction of Olefins

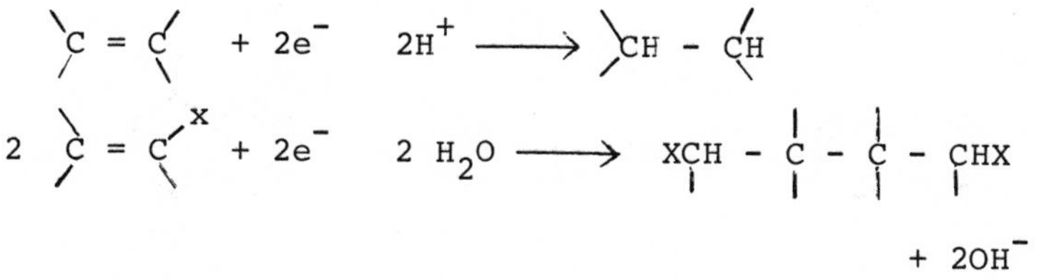

$$\text{C} = \text{C} \ + 2e^- \quad 2H^+ \longrightarrow \ \text{CH} - \text{CH}$$

$$2 \ \text{C} = \text{C}^X \ + 2e^- \quad 2\ H_2O \longrightarrow \ XCH - \overset{|}{\underset{|}{C}} - \overset{|}{\underset{|}{C}} - \underset{|}{C}HX$$

$$+ \ 2OH^-$$

where X is an electron-withdrawing group, e.g., CN.

Reduction of simple aliphatic olefins is carried out in a straightforward manner. A divided cell is recommended, but with suitable precautions (absence of halides, inertness of olefin toward oxidation) a nondivided cell may be satisfactory. Electrode materials are not critical, although low overvoltage metals may be desirable because of the resemblance of this reaction to catalytic hydrogenation. Tetraalkylammonium salts are recommended for use as supporting electrolytes. The solvent should contain a proton donor such as water or alcohol.

With activated olefins, reductive coupling product may be favored under the following conditions:
1. High concentration of starting material.
2. Neutral solution or low proton availability.
3. High overvoltage cathode (Sec. 3.1.3).

4.4.7.2. Reduction of Aromatic Compounds

The course of reduction of this class of compounds is highly dependent on the reaction conditions
where E is an electrophile, e.g., CO_2.

$$\text{(benzene)} \ + \ 6e \ \xrightarrow{\ 6H^{\oplus}\ } \ \text{(cyclohexane)} \qquad (1)$$

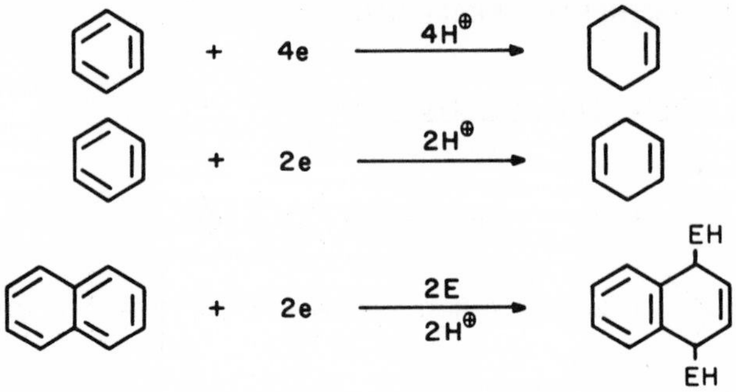

$$\text{(benzene)} + 4e \xrightarrow{4H^{\oplus}} \text{(cyclohexene)} \qquad (2)$$

$$\text{(benzene)} + 2e \xrightarrow{2H^{\oplus}} \text{(cyclohexadiene)} \qquad (3)$$

$$\text{(naphthalene)} + 2e \xrightarrow[2H^{\oplus}]{2E} \text{(1,4-dihydronaphthalene, EH)} \qquad (4)$$

Reaction (1) is favored by the use of a low hydrogen overvoltage cathode (e.g., Pt) in an aqueous acidic medium. Reaction (2) is normally carried out in a divided cell using ammonia or amines as solvents and lithium salts as the supporting electrolyte. On the other hand, reaction (3) is favored by using the same solvent and supporting electrolyte as (2) in an undivided cell. Platinum is the recommended cathode for reactions (2) and (3). Reaction (4) is in essence a reductive substitution which is normally carried out under similar conditions to reaction (3) in the presence of an electrophile.

4.4.7.3. Reduction of Acetylenes

$$-C \equiv C- + 2e + 2H^{\oplus} \longrightarrow \overset{\diagdown}{/}C = C\overset{\diagup}{\diagdown}$$

The design of the cell is not critical for this type of reaction. Thus, a divided or undivided cell may be used. The solvent and supporting electrolyte are essentially the same as those described for reactions (2) and (3) under aromatic compounds.

4.4.7.4. Reduction of Cinnamic Acid (24)

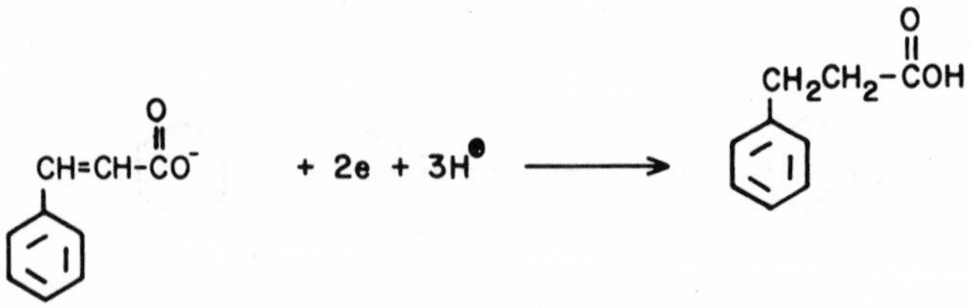

$$\underset{\text{Ph}}{\text{CH=CH-}\overset{\overset{\textstyle O}{\|}}{C}\text{O}^{-}} + 2e + 3H^{\bullet} \longrightarrow \underset{\text{Ph}}{\text{CH}_2\text{CH}_2\text{-}\overset{\overset{\textstyle O}{\|}}{C}\text{OH}}$$

The electrolysis cell may be a beaker into which a
porous cup is suspended. The cup serves as the anode chamber. External cooling can be provided by immersing the
beaker into a water or ice-water container. For the above
reaction mercury was used as the cathode and was placed at
the bottom of the beaker. Connection of cathode to power
source was made by means of an insulated copper wire, except
at one end, which was immersed in the mercury. The anode,
a heavy sheet of lead, was placed in the porous cup (which
should be located close to the mercury surface). A magnetic
bar may be placed on the surface of the mercury to provide
agitation.

Into the cathode compartment, 2 liters of 7-8% Na_2SO_4
(concentration not critical) were introduced. The porous
cup was also filled with this solution so that the levels
of the liquids in both compartments were equal. During
electrolysis the anode solution is kept alkaline by adding
aqueous sodium hydroxide ($\sim$2.7 moles NaOH). To the cathode
solution, 200 g (1.35 moles) of cinnamic acid is added, followed by a solution of 35 g (0.88 mole) of sodium hydroxide
in 150 ml water. The latter solution is added slowly to
avoid the formation of lumps. Electrolysis is conducted
at 5-10 A. During electrolysis some of the product adheres
to the wall of the beaker and is washed down with water
using a wash bottle. At the conclusion of the reaction
(76-80 A hr) the cathode solution is decanted and acidified
with excess sulfuric acid. The hydrocinnamic acid separates
as an oil and solidifies upon cooling. The yield of the
acidic product is 180-200 g. The product, boiling point
194-197°/75 mm, is colorless and melts at 47.5-48. The yield
(distilled) is 160-180 g, 80-90%, depending on the quality
of cinnamic acid used.

4.4.7.5. Reduction of Acrylonitrile

The reduction of acrylonitrile is described in detail,
including a diagram for continuous electrolysis, by Baizer.*

*M. Baizer, J. Electrochem. Soc., 111, 215 (1964).

4.4.7.6. The Reduction of Acetylenes (25)

A simple, home-made cell for the reduction of acetylenes
was described by Benkeser (2, 3). This includes two chambers divided by a sheet of asbestos and each fitted with a
condenser and an electrode. The electrodes were connected
by means of a wire which was sealed through the glass.
While this cell is simple, any H-type cell (described
previously) or divided cell is sufficient, and the reader
may design the one which is most applicable for his system.
It should be noted that the reduction of a number of acetylenes may be carried out in an undivided cell. Thus, for
these reductions, a 3-neck flask may be used. Two of the
necks are used to suspend the electrodes and the third is
equipped with a condenser. A magnetic bar may be used for
stirring.

General procedure. The desired electrolysis cell is
charged with 34 g (0.8 mole) lithium chloride, 900 ml of
anhydrous methylamine, and 0.1 mole of the acetylene to
be reduced. Both electrodes were made out of platinum.
In case a divided cell is used the same ingredient without
the acetylene in the anode compartment may be used. Electrolysis is carried out at a current of 2 A for 2 hr and 40
min (19,300 C). The solvent was evaporated (or distilled)
and the residue hydrolyzed with 300 ml water and extracted
into ether. The ether was dried ($CaSO_4$) and evaporated.
The residue may be distilled or analyzed by vapor phase
chromatography.

Using an undivided cell, 3-octyne was reduced as described above to give 3-octene (58%), which was 98% in the
trans form.*

_____

*The reader should compare these results with the re-
duction of 4-octyne reported by Lampbell and co-workers,
J. Am. Chem. Soc., 65, 965 (1943).

REFERENCES ON UNSATURATED COMPOUNDS

1. J. P. Petrovich, J. D. Anderson, and M. M. Baizer, J. Org. Chem., 31, 3897 (1966).

2. R. A. Benkeser and E. M. Kaiser, J. Am. Chem. Soc., 85, 2858 (1963).

3. R. A. Benkeser, E. Kaiser, and R. Lambert, J. Am. Chem. Soc., 86, 527 (1964).

4. H. W. Sternberg, R. E. Markby, I. Wander, and M. Mohliner, J. Electrochem. Soc., 113, 1060 (1966).

5. H. W. Sternberg, R. E. Markby, I. Wender, and D. M. Mohilner, J. Am. Chem. Soc., 89, 186 (1967)

6. H. Sternberg, The Synthetic and Mechanistic Aspects of Electroorganic Chemistry, U. S. Army Research Office, Durham, N. C., Oct. 14-16, 1968.

7. H. W. Sternberg, R. E. Markby, I. Wender, and D. M. Mohilner, J. Am. Chem. Soc., 91, 4191 (1969).

8. A. J. Birch, Nature, 158, 60 (1946).

9. T. Osa, T. Yamagishi, T. Kodama, and A. Misono, The Synthetic and Mechanistic Aspects of Electroorganic Chemistry, U. S. Army Research Office, Durham, N. C., Oct. 14-16, 1968.

10. S. H. Langer, J. Electrochem. Soc., 116, 1228 (1969).

11. S. Wawzonek and D. Wearing, J. Am. Chem. Soc., 81, 2067 (1959).

12. W. C. Neikam, U. S. Pat. 3,344,047 (1967).

13. M. Sitaraman and V. Raman, Current Sci., 16, 23 (1947); C. A., 41, 4468 (1947).

14. M. Peover, Electroanalytical Chemistry (A. Bard, ed.), Dekker, New York, 1967, p.1.

15. H. J. Barger, Jr., J. Org. Chem., 34, 1489 (1969).

16. Monsanto Co., U. S. Pat. 553,851; Netherland Pat. 67,07472 (1967).

17. J. P. Petrovich, M. Baizer, and M. Ort, J. Electrochem. Soc., 116, 749 (1969).

18. I. Knunyants and N. Vyazanki, Dokl. Akad. Nauk, S.S.S.R., 113, 112 (1957); C. A., 51, 14637 (1957).

19. B. L. Funt and F. D. Williams, J. Polymer Sci., A, 2, 865 (1964).

20. D. Goering and H. Janas, West German Pat. 937, 919 (1956); C. A., 53, 3950 (1959).

21. M. R. Rifi, unpublished results.

22. B. L. Funt, S. Bhadani, and D. Richardson, Polymer
     Reprints, 7, 153 (1966).

23. A. P. Krapcho and M. E. Nadel, J. Am. Chem. Soc., 86,
     1096 (1964).

24. A. W. Ingersoll, Org. Syn. Coll., 1, 311 (1941).

25. R. A. Benkeser and C. A. Tincher, J. Org. Chem., 33,
     2727 (1968).

26. R. A. Benkeser, The Synthetic and Mechanistic Aspects
     of Electroorganic Chemistry, U. S. Army Research Office,
     Durham, N. C., Oct. 14-16, 1968.

27. H. A. Laitinen and S. Wawzonek, J. Am. Chem. Soc., 64,
     1765 (1942).

4.5. CARBON-NITROGEN GROUPS

4.5.1. NITRILES

General Considerations. The transfer of electrons to
unsubstituted alkylnitriles occurs at a high negative potential. This is why acetonitrile is commonly used as solvent
in polarographic as well as macroscale electrolytic studies.
On the other hand, aromatic nitriles are reduced reversibly
to give the radical anion whose ESR spectrum has been observed (1). Some of the substituted radical anions have a
considerably long half-life. This stability may be utilized
for the synthesis of interesting compounds, as is discussed
below.

Effect of Reaction Conditions on Course of Reaction.
While a direct transfer of electrons to aliphatic nitriles
is difficult to accomplish, their reduction can be achieved
under the appropriate reaction conditions. For example, the
use of a low hydrogen over-voltage cathode (e.g., Pt) in
aqueous acid solution affects the quantitative transformation of acetonitrile to the corresponding amine (2), probably
via hydrogenation reaction, Eq. (4.88):

$$CH_3CN \xrightarrow[\text{Aq. HCl}]{\text{Pt}} CH_3-CH_2NH_2 \qquad\qquad (4.88)$$

$$(100\%)$$

On the other hand, the use of a mercury cathode in an aprotic medium produces low molecular weight polymer from the
reduction of acetonitrile ($\underline{3}$) Eq. (4.89):

$$CH_3\text{-}C\equiv N \xrightarrow[\substack{(Neat) \\ R_4NX}]{Hg} (-\underset{\underset{CH_3}{|}}{C}=N-) \qquad (4.89)$$

The use of appropriate solvents may determine the nature of
products formed from the reduction of nitriles. Thus, while
in aqueous media, aromatic nitriles are transformed to the
corresponding amines ($\underline{4}$), their reduction in DMF affords
dimeric products ($\underline{1}$) Eqs. (4.90) and (4.91):

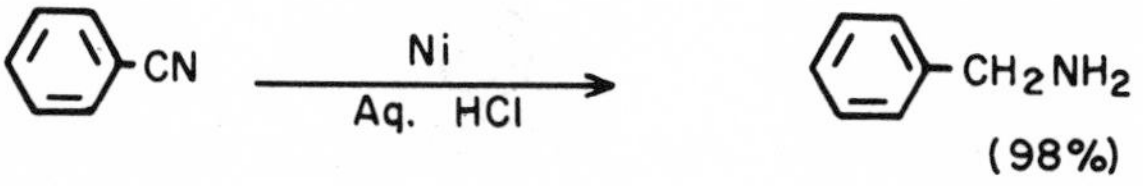

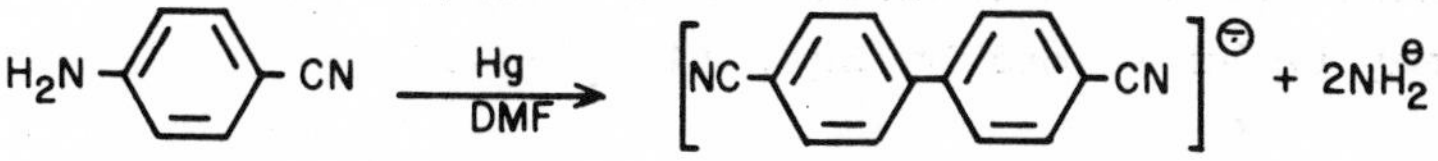

Mechanism of Reduction. The mechanism of the reduction
of nitriles depends on the reaction conditions. Thus, with
a low hydrogen voltage electrode (e.g., Pt) in aqueous solutions, the electrochemical reduction involves the generation
of hydrogen, which is responsible for the conversion of nitriles to amines. In nonaqueous media, direct transfer of
electrons to the carbon-nitrogen π bonds may take place to
afford the radical anion ($\underline{1}$, $\underline{5}$) [109]:

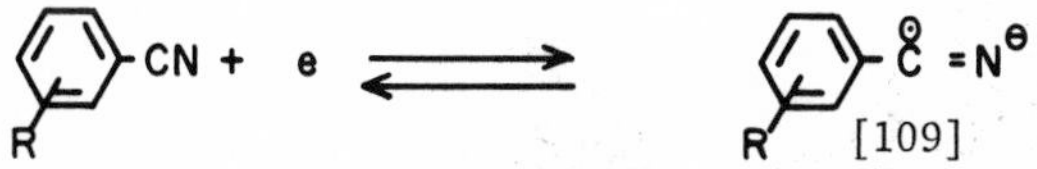

In DMF, benzonitrile exhibits one one-electron polarographic wave ($\underline{1}$) and its reduction in this medium affords

a red-orange product with a well-resolved ESR spectrum.
The radical anion [110] can lead to dimeric (**1**) or polymeric
products (**3**), [111, 112]:

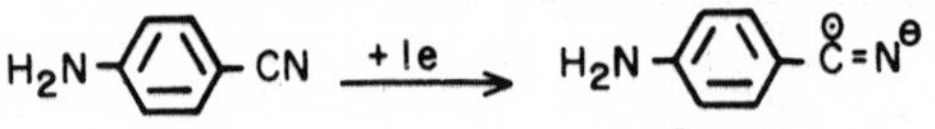

$$H_2N-\langle\ \rangle-CN \xrightarrow{+1e} H_2N-\langle\ \rangle-\overset{\ominus}{\underset{\parallel}{C}}=N^{\ominus}$$

[110]

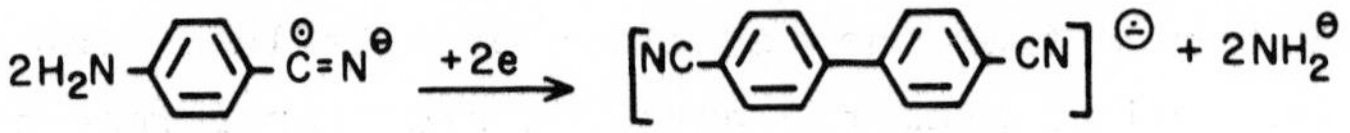

$$2H_2N-\langle\ \rangle-\overset{\ominus}{\underset{\parallel}{C}}=N^{\ominus} \xrightarrow{+2e} \left[NC-\langle\ \rangle-\langle\ \rangle-CN\right]^{\ominus} + 2NH_2^{\ominus}$$

[111]

$$\langle\ \rangle-CN \longrightarrow \left[-C=N-\right]_n$$

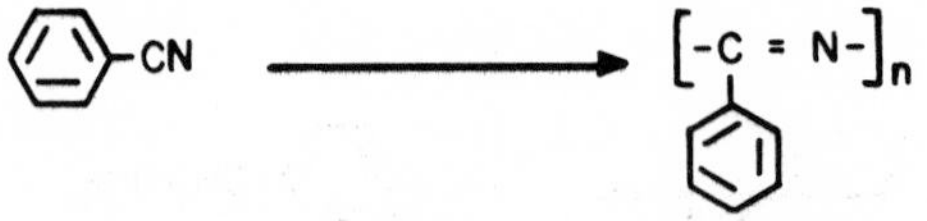

[112]

The formation of low molecular weight polymer from acetonitrile may be explained by Eqs. (4.92)-(4.95):

Initiation

$$CH_3-C \equiv N + 1e \longrightarrow CH_3-\overset{\ominus}{C}=N^{\ominus} \tag{4.92}$$

$$2CH_3\overset{\ominus}{C}=N^{\ominus} \longrightarrow {}^{\ominus}N=\underset{\underset{CH_3}{|}}{C}-\underset{\underset{CH_3}{|}}{C}-N^{\ominus} \tag{4.93}$$

Propagation

$$Polymer \xleftarrow{N\equiv C-CH_3} {}^{\ominus}N=\underset{\underset{CH_3}{|}}{C}-\underset{\underset{CH_3}{|}}{\overset{\ominus}{C}}=N \xrightarrow{CH_3-C\equiv N} Polymer \tag{4.94}$$

Termination

$$(-\underset{\underset{CH_3}{|}}{C}=N)_n-\underset{\underset{CH_3}{|}}{C}=N^{\ominus} \xrightarrow{CH_3CN} (-\underset{\underset{CH_3}{|}}{C}=N)_n-\underset{\underset{CH_3}{|}}{C}=NH + {}^{\ominus}CH_2-CN \tag{4.95}$$

The above mechanism is quite analogous to electrochemical
polymerization of styrene (6). In order to determine whether
the above termination reaction, Eq. (4.95), is responsible
for the production of low molecular weight polymer, benzonitrile was reduced under the same conditions. This afforded
a product with a high degree of polymerization and whose
structure [112] was identified from the spectral properties
of the product (3):

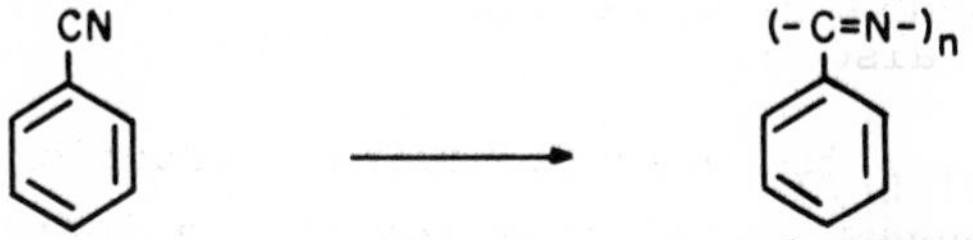

[112]

Role in Organic Synthesis. In general, the reduction
of nitriles by electrochemical or conventional methods gives
the corresponding amine. Consequently, electrochemical reduction (aside from its possible ease of handling) has not
yet offered an advantage over the use of conventional reducing agents (7).

4.5.2. IMINES AND OXIMES

General Considerations. The electrochemical reduction
of carbon-nitrogen double bonds was recently reviewed by
Lund (8). Consequently, only a brief discussion of this
subject, with emphasis on imines and oximes, is described
here.

Interest in the electrochemical reductions of imines
began with the study of the polarographic behavior of some
ketones which reduced at a high negative potential. Such
ketones [113] could be easily transformed to the corresponding imines [114], whose reduction occurred at a moderately
negative potential (9, 10).

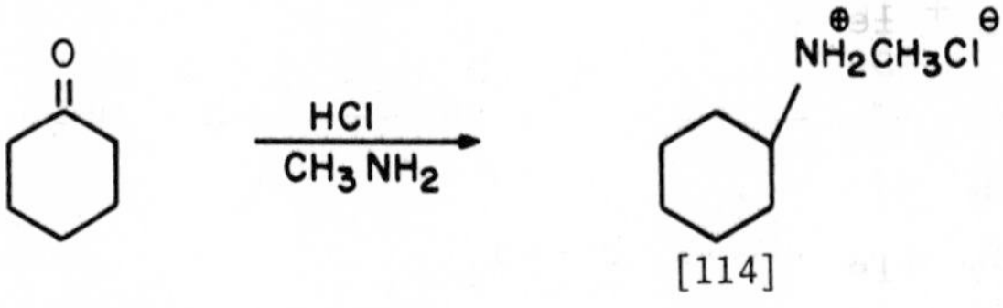

[114]

Thus, in many respects, the mechanism of reduction of imines, which affords the corresponding amines, resembles that of carbonyl compounds. Since imines are intermediates in the reduction of oximes (see mechanism below), the reduction of the latter also resembles the reduction of carbonyl compounds.

4.5.3. EFFECT OF REACTION CONDITIONS ON COURSE OF REDUCTION

The ease of the electrochemical reduction of imines and oximes leaves little room for the variation of reaction conditions in order to control the course of the reaction. In general, reduction under acidic conditions is easier, indicating that the reduced species is the protonated starting material (11, 12).

Mechanism of Reduction. Under acidic conditions, imines exhibit two one-electron waves which merge at higher pH values (9). The half-wave potential of the first wave is pH dependent, while that of the second wave is not. In aprotic media, imines exhibit a well-developed, diffusion controlled, polarographic wave. Thus, the over-all mechanism may be described by Eqs. (4.96)-(4.99):

Low pH:

$$\text{C}=NH + H_3O^{\oplus} \rightleftharpoons \text{C}=\overset{\oplus}{N}H_2 + H_2O \qquad (4.96)$$

$$\text{C}=\overset{\oplus}{N}H_2 + 1e \xrightarrow[\text{wave}]{\text{1st}} \overset{\ominus}{\text{C}}-NH_2 \qquad (4.97)$$

$$\text{C}^{\ominus}-NH_2 + 1e \xrightarrow[\text{2nd wave}]{H^{\oplus}} \text{C}H-NH_2 \qquad (4.98)$$

High pH:

$$\text{C}=NH + 1e \longrightarrow \overset{\ominus}{\text{C}}=\overset{\ominus}{N}H \xrightarrow{H^{\oplus}} \overset{\ominus}{\text{C}}-NH_2 \qquad (4.98a)$$

$$\text{C}^{\ominus}-NH_2 + 1e \xrightarrow{H^{\oplus}} \text{C}H-NH_2 \qquad (4.99)$$

The reduction of oximes consumes four electrons to afford the corresponding amines. Their reduction occurs at more positive potentials with a decrease in the pH values of the medium, indicating that the reduced species is the protonated oxime, Eq. (4.100), (11). In 5-20% sulfuric acid, oximes are converted to their conjugate acids and their reduction becomes pH independent (12). Under acidic conditions imines, and not hydroxylamines, are believed to be the intermediates in the reduction of oximes, since the reduction of hydroxylamine would occur at a more negative potential. At pH values above 8, a second polarographic wave appears whose height develops at the expense of the first wave. Thus, the over-all reduction may be described by Eqs. (4.100)-(4.104):

Low pH:

$$\text{\textbackslash C=NOH} + H^{\oplus} \longrightarrow \text{\textbackslash C=N-OH}_2^{\oplus} \tag{4.100}$$

$$\text{\textbackslash C=NOH}_2^{\oplus} + 2e + H^{\ominus} \longrightarrow \text{\textbackslash C=NH} + H_2O \tag{4.101}$$

$$\text{\textbackslash C=NH} + H^{\oplus} \rightleftharpoons \text{\textbackslash C=NH}_2^{\oplus} \xrightarrow[+H^{\oplus}]{+2e} CH-NH_2 \tag{4.102}$$

High pH:

$$\text{\textbackslash C=NOH} \rightleftharpoons \text{\textbackslash C=NO}^{\ominus} + H^{\oplus} \tag{4.103}$$
$$\text{(difficult to reduce)}$$

$$\text{\textbackslash C} = NH \xrightarrow[+2H^{\oplus}]{+2e} \text{\textbackslash CH-NH}_2 \tag{4.104}$$

The intervention of imines in the reduction of oximes is consistent with the observation described for the reduction of benzaldehyde oxime (9). In a buffer solution using a mercury cathode, benzyl amine was obtained. Under similar

conditions the reduction of benzylhydroxylamine was not
possible.

Role in Organic Synthesis. The reduction of imines
and oximes may be accomplished electrochemically or with
conventional reducing agents such as sodium metal or hydrogen on platinum. However, in some cases, reduction by the
two methods affords different products (13) Eqs. (4.105) and
(4.106):

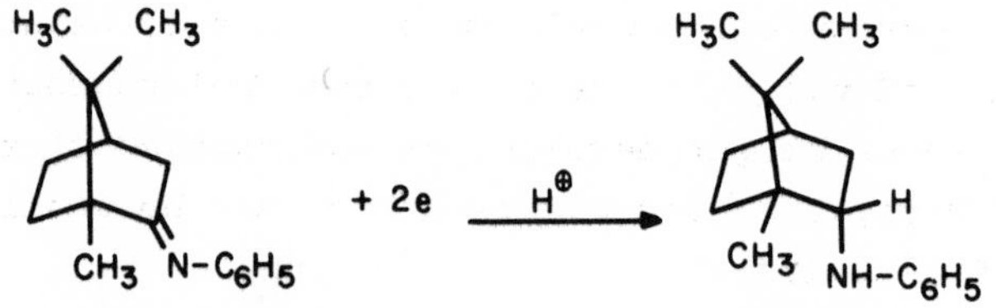

$$+ 2e \quad \overset{H^{\oplus}}{\longrightarrow} \qquad (4.105)$$

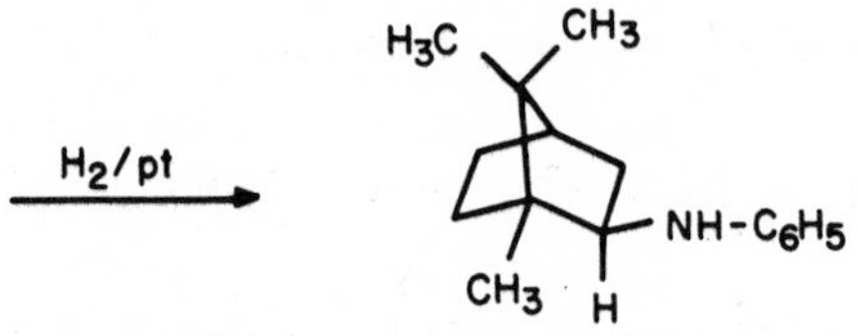

$$(4.106)$$

$$\overset{H_2/pt}{\longrightarrow}$$

Thus, the electrochemical reduction of imines (and perhaps
oximes) may play an important role in organic synthesis
when a product with a certain stereochemistry is desired.

Under the appropriate reaction conditions, the syn-
and anti- forms of oximes exhibit different reduction behavior (9). Thus, under basic conditions, only the syn-
isomer exhibits a polarographic wave. Therefore, it may
be possible to separate the syn- and anti-isomers by electrochemical reduction.

4.5.4. EXPERIMENTAL PROCEDURES

Reduction of Nitriles (2). The electrolysis cell used
for the reduction of nitriles is described in Fig. 4.3.

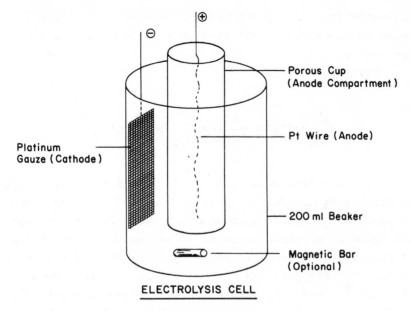

ELECTROLYSIS CELL

FIG. 4.3.

4.5.4.1. General Procedure

Preparation of Cathode. A mixture of 50 ml water, 10
ml of concentrated hydrochloric acid, and a solution of
0.5-1.0 g palladium chloride in a small amount of hydrochloric acid was used as the catholyte. The anolyte was
dilute hydrochloric acid. Electrolysis was carried out at
room temperature at a current density of 0.02 A/cm^2. When
the color of palladium chloride had disappeared (indicating
the deposition of palladium metal on the platinum cathode)
the catholyte was removed and replaced with dilute hydrochloric acid.

Electrolytic Reaction. In cases where the nitrile was
sparingly soluble in aqueous hydrochloric acid, a small
amount of glacial acetic acid was added. An ice bath was
used to cool the reaction solution. The addition of small
amounts of palladium chloride was found to be helpful in
the reduction. At the end of the electrolysis, the catholyte

was evaporated to dryness under vacuum and the residual
amine hydrochloride identified as such or as a suitable
derivative.

   Examples:

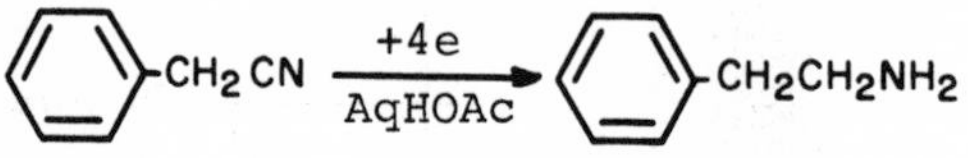

   The catholyte consisted of 1 g (0.008 mole) of benzyl-
nitrile in 30 ml glacial acetic acid, 10 ml of water, and
10 ml of concentrated hydrochloric acid. The anolyte was
10% aqueous hydrochloric acid. (The amount of anolyte should
be such that its level is about equal to that of the catholyte.) Electrolysis was carried out at a current density
of 2 A/dm^2. The temperature of the medium was about 10°.
At the end of 2 hr the current was turned off and the catholyte was evaporated to dryness under vacuum. This afforded
1 g, 75% of phenylethylamine hydrochloride.

   Other nitriles reduced according to the above procedure
are shown in Table 4.4.

TABLE 4.4

Reduction of Nitriles (2)

Nitrile	Product	Yield (%)
CH_3CN	$CH_3CH_2NH_2$	100
$NC-CH_2-CH_2CN$	$H_2N-CH_2-(CH_2)_2-CH_2NH_2$	70
$NC-(CH_2)_4CN$	$H_2N-(CH_2)_6-NH_2$	40
$NC-CH_2-CO_2H$	$H_2NCH_2CH_2CO_2H$	40

Reduction of Imines and Oximes to Amines

$$\begin{array}{l} R \\ \diagdown \\ C = NH + 2H^{+} + 2e^{-} \longrightarrow \\ \diagup \\ R' \end{array} \quad \begin{array}{l} R \\ \diagdown \\ CH - NH_2 \\ \diagup \\ R' \end{array}$$

$$\begin{array}{l} R \\ \diagdown \\ C= NOH + 4H^{+} + 4e^{-} \longrightarrow \\ \diagup \\ R' \end{array} \quad \begin{array}{l} R \\ \diagdown \\ CH-NH_2 + H_2O \\ \diagup \\ R' \end{array}$$

The electrolysis cell for these reductions should be
divided, since the starting materials and products are susceptible to oxidation. Aqueous acid is the ideal solvent/
electrolyte for this type of reduction, but mixed solvents
may be employed when required for solubility. Although
platinized platinum has generally been used as the cathode
material, other metals will probably work as well.

Reduction of Cyclohexanone Imines

The reduction of these compounds may be carried out in
the apparatus shown in Fig. 4.3 for the reduction of nitriles*.

The catholyte consisted of a solution of 1 ml of
freshly distilled cyclohexanone in 40 ml of 33% aqueous
methylamine, 50 ml of 4 $\underline{N}$ hydrochloric acid, and 100 ml of
water. The anolyte** was 10% aqueous hydrochloric acid.
Reduction was carried out at a controlled potential of -1.65
V (s.c.e.). After 2 hr of electrolysis, which allowed the
passage of 2 electrons per mole cyclohexanone, the reaction
mixture was extracted 3 times into ether, which was washed

*Although a slightly different apparatus is described
by Lund, Acta Chem. Scand., 11, 283 (1957).

**The nature of the anolyte was not mentioned in Ref.
9, consequently the above composition is recommended.

3 times with dilute sodium hydroxide. After drying, the
ether was evaporated, together with traces of methylamine.
This gave a residue which was redissolved in ether. Dry
hydrogen chloride was bubbled into the ether, thus causing
the precipitation of 925 mg of N-methylcyclohexalamine
hydrochloride.

Reduction of Oximes:

$$\text{C}_6\text{H}_5\text{-CH=N-OH} \xrightarrow[\text{HCl}]{+4e} \text{C}_6\text{H}_5\text{-CH}_2\overset{\oplus}{\text{N}}\text{H}_3\text{Cl}^{\ominus}$$

     The anti-isomer of benzaldoxime was dissolved in a
solution of 30% peroxide-free tetrahydrofuran and 70% aqueous buffer (sodium borate). The anolyte was the same as
that described in the first example above. Reduction was
carried out at a potential of -1.7 V (s.c.e.). After the
addition of 4 electrons per molecule, the catholyte was extracted 3 times into ether and the extracts combined and
dried. Addition of dry hydrogen chloride to the ether extracts caused the precipitation of 700 mg solid, which was
identified as benzylamine hydrochloride.

REFERENCES ON CARBON-NITROGEN GROUPS

1. P. H. Rieger, I. Bernal, W. H. Reinmuth, and G. Fraenkel,
    J. Am. Chem. Soc., 85, 683 (1963).

2. M. Ohta, Bull. Chem. Soc. (Japan), 17, 485 (1942).

3. M. R. Rifi, unpublished results.

4. S. Sawa, Japan Pat. 5019 (1951); C. A., 47, 2615 (1953).

5. I. G. Sevastyanova and A. P. Tomilov, Zh. Obshch. Khim
    (Engl. Ed.), 33, 2741 (1963).

6. B. L. Funt, in Macromolecular Reviews, Vol. 1, Inter-
    science, New York, 1967, p. 35.

7. C. R. Noller, in Chemistry of Organic Compounds, Saunders,
    1951, p. 248.

8. H. Lund, in Chemistry of the Carbon-Nitrogen Double Bond,
    Interscience, New York, 1970, p. 505.

9.   H. Lund, Acta Chem. Scand., 13, 249 (1959).

10. P. Zuman, Nature, 165, 485 (1950).

11. P. Souchay and S. Ser, J. Chim. Phys., 49, C172 (1952).

12. H. J. Gardner and W. P. Georgans, J. Chem. Soc., 1956
     4180.

13. A. J. Fry and R. Reed, J. Am. Chem. Soc., 91, 6448
     (1969).

4.6.   MISCELLANEOUS REDUCTIONS

So far, we have discussed the reduction of organic
compounds that have been studied in some detail. Other
classes of compounds that have not received as much attention are briefly discussed in this section.

4.6.1.   ORGANOSULFUR COMPOUNDS

The electrochemical reduction of organosulfur compounds
has not been used to a great extent in organic synthesis.
In general, the reduction of sulfur-oxygen double bond is
easier than the reduction of the carbon-oxygen analog.
Thus, sulfoxides are easily reduced to the corresponding
mercaptans (1-3).

$$R-\overset{\overset{O}{\|}}{S}-R + 2e \xrightarrow{2H^{\oplus}} R-S-R + H_2O$$

This reaction is of interest in that it is polarographically
irreversible, yet the oxidation of the mercaptan gives the
corresponding sulfoxide. Carbon-sulfur double bonds are
also easier to reduce than the corresponding carbon-oxygen
analog (7, 8).

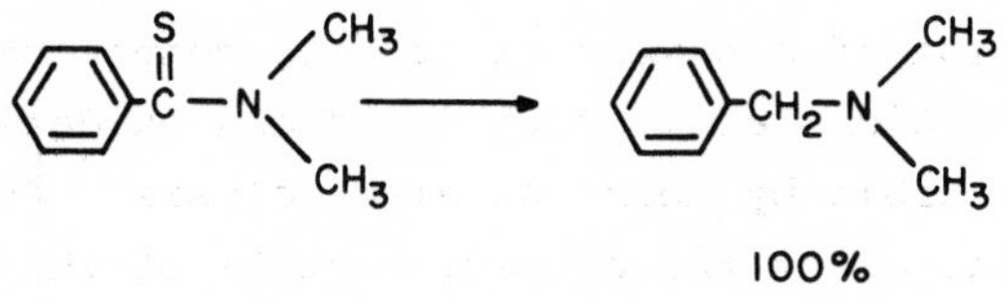

100%

The reduction of sulfur-sulfur single bonds is simple and
gives the corresponding mercaptans (4, 5). On the other
hand, the reduction

$$R-S-S-R + 2e \xrightarrow{2H^{\oplus}} 2RSH$$

of carbon-sulfur single bond is difficult, except in some
aromatic thiocyanates (6).

The mechanism of the electrochemical reduction of organosulfur compounds has been studied in some detail (1,
10-12). For those who are interested in this topic, the
suggested references at the end of this section should be
consulted. Unfortunately, the facile reaction of organosulfur compounds with mercury has complicated the study of
their polarographic behavior (13).

4.6.2. ORGANOMETALLIC COMPOUNDS

The polarographic behavior of organometallic compounds
has been studied in detail (9, 14-19). Although such behavior
has not been fully utilized in organic synthesis, the impli-
cation is that one could do so. Consider, for example, the
formation of metal-metal bonds (16, 20, 24), Eqs. (4.107-
4.109):

$$2R_3SnCl + 2e \longrightarrow R_3Sn-SnR_3 + 2Cl^{\ominus} \qquad (4.107)$$

$$2(H_5C_2)_2TlBr + 2e \longrightarrow (H_5C_2)_2Tl-Tl(C_2H_5)_2 + 2 Br^{\ominus} \qquad (4.108)$$

$$2(H_5C_2)_3 PbCl + 2e \longrightarrow (H_5C_2)_3Pb-Pb(C_2H_5)_3 \qquad (4.109)$$

The preparation of a number of organic compounds, which
include transition metals with a certain oxidation state,
should be feasible by means of electrolysis. Furthermore,
the determination of the oxidation state of these metals

in organic reactions could be easily done by means of polarography, which can detect concentration in about 10^{-4} M.
Wilkinson (21-23) reported the polarographic behavior of
ferrocene, bis-cyclopentadienyl compounds of rhodium, cobalt,
titanium, zirconium, and other metals. From such studies,
it was found that the ferrocene-ferrocinium ion couple is
thermodynamically reversible with an E_o = -0.56 V relative
to the normal hydrogen electrode.

4.6.3. PEROXIDES

The reduction of peroxides is irreversible and proceeds in a two-electron transfer to afford the corresponding alcohols (25).

$$R-O-O-R + 2e \xrightarrow{2H^{\oplus}} 2\ ROH$$

The ease of reduction of peroxides is in the following
order (26, 27): $RCO_3H = (RCO_2)_2 > RO_2H > RCO_3-\underline{t}-Bu > \underline{t}-BuO_2H > RO_2R > (\underline{t}-BuO)_2$. A comprehensive list of $E_{\frac{1}{2}}$ of
peroxides was reported by Martin (25).

The electrochemical reduction of peroxides is not of
synthetic utility, however, the reduction of molecular oxygen to form superoxide ion has recently received a considerable amount of attention (28, 29). In aprotic solvents
containing quaternary ammonium salts as supporting electrolytes, the reduction of oxygen gives (O_2^-), which is somewhat
stable at room temperature but eventually decays in the following manner, Eqs. (4.110 and 4.111):

$$2O_2^{\ominus} + Et_4N^{\oplus} \longrightarrow HO_2^{\ominus} + O_2 + Et_3N + \underset{H}{\overset{H}{\diagdown}}C=C\underset{H}{\overset{H}{\diagup}} \qquad (4.110)$$

$$2HO_2^{\ominus} \longrightarrow O_2 + 2OH^{\ominus} \qquad (4.111)$$

The superoxide $O_2^{\ominus}$ is obviously a powerful oxidizing
agent. Thus, when fluorene is oxidized in the presence of

oxygen it gives fluorenone readily (29), while in the absence of oxygen no oxidation took place (29). The superoxide is also a nucleophile and, as such, may be of use in organic synthesis (29, 30), Eqs. (4.112)-(4.114):

$$O_2^{\ominus} + RX \longrightarrow RO_2^{\cdot} + X^{\ominus} \qquad\qquad (4.112)$$

$$O_2^{\ominus} + RO_2^{\ominus} \longrightarrow RO_2^{\ominus} + O_2 \qquad\qquad (4.113)$$

$$RO_2^{\ominus} + RX \longrightarrow R_2O_2 + X^{\ominus} \qquad\qquad (4.114$$

It can be seen from the above equations that peroxides may be conveniently prepared from alkyl halides and oxygen.

4.6.4. REDUCTION OF SOME SINGLE (σ) BONDS

Carbon-Oxygen Bond. This reduction is similar to that of carbon-halogen bonds, however, what makes the reduction of the latter much easier lies in the leaving ability of the halide anions (31, 32). The ease of reduction depends greatly on the nature of the substituent, i.e., R=CH$_3$ < C$_2$H$_5$ < Ph and brings about the replacement of the oxygen with a hydrogen:

$$R_3'\text{-}C\text{-}O\text{-}R + 2e \xrightarrow[H^{\oplus}]{solvent} R_3CH + R'O^{\ominus}$$

Since there are two available positions for carbon oxygen bond cleavage, the group which can best accommodate a carbon anion (or a radical) will undergo cleavage.

Carbon-Nitrogen Bonds. This reduction can only occur in certain molecular structures such as α-activating groups [115] (34).

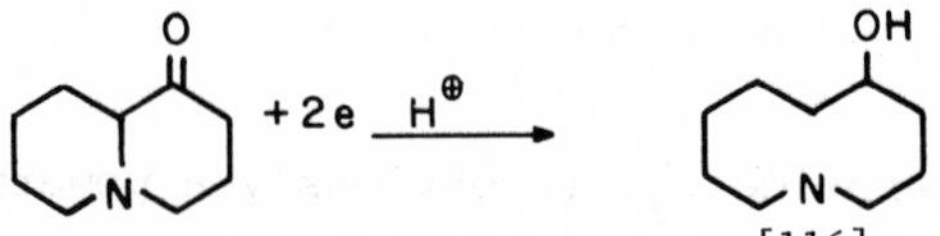

[115]   [116]

Quaternary ammonium salts [117] can be reduced to the corresponding amines (35-38).

$$R_4N^\oplus X^- + 2e \xrightarrow{\text{H}^\oplus} R_3N + RH$$

[117]

The mechanism of this reduction is reported to proceed via radical intermediates (36), thus, the group which can best support such an intermediate would undergo reduction.

4.6.5. OTHER REDUCTIONS

Scattered reports have appeared in the literature describing the electrochemical reduction of a variety of organic compounds which are of insufficient general utility to be discussed in this book. For interested readers the following references are suggested:

For information on carbon-sulfur single bonds see Refs. 33, 39-42; carbon-carbon single bonds see Refs. 43, 44; for carbonium ions see Refs. 45-47; and for phosphonium and arsonium salts see Ref. 48.

REFERENCES ON MISCELLANEOUS REDUCTIONS

1.   R. Bowers and H. Russell, Anal. Chem., 32, 405 (1960).

2.   H. Drushel and J. Miller, Anal. Chem., 29, 1459 (1956).

3.   M. Nicholson, J. Am. Chem. Soc., 76, 2539 (1954).

4.   I. M. Kolthoff, W. Stricks, and N. Tanaka, J. Am. Chem. Soc., 77, 4739 (1955).

5.   M. J. Allen and H. G. Steiman, J. Am. Chem. Soc., 74, 3932 (1952).

6.   K. Schwabe and J. Voight, Z. Elektrochem., 56, 44 (1952).

7.   K. Kindler, Arch. Pharm., 265, 390 (1927).

8.   K. Kindler, Ann., 431, 187 (1923).

9.   S. J. Leach, Aut. J. Chem., 13, 520 (1960).

10. R. Elofson, F. F. Gadallah, and L. A. Gadallah, Can. J. Chem., 47, 3979 (1969).

11. D. Bernard, M. E. Evans, G. M. C. Higgins, and J. F. Smith, Chem. Ind., 20 (1961).

12. C. W. Johnson, C. G. Overberger, and W. J. Seagers, J. Am. Chem. Soc., 75, 1495 (1953).

13. J. Donahue and J. W. Oliver, Anal. Chem., 41, 753 (1969).

14. A. Kirrman and E. Kleine-Peter, Bull. Soc. Ch. (France), 894 (1957).

15. N. S. Hush and K. B. Oldman, J. Electroanal. Chem., 6, 34 (1963).

16. R. E. Dessy, W. Kitching, and T. Chivers, J. Am. Chem. Soc., 88, 453 (1966).

17. R. E. Dessy, W. Kitching, T. Psarras, R. Salinger, A. Chen, and T. Chivers, J. Am. Chem. Soc., 88, 460 (1966).

18. R. E. Dessy, T. Chivers, and W. Kitching, J. Am. Chem. Soc., 88, 467 (1966).

19. R. E. Dessy, F. E. Stary, R. B. King, and M. Waldrop, J. Am. Chem. Soc., 88, 471 (1966).

20. G. Costa, Ann. Chim. (Rome), 40, 559 (1950); C. A., 45, 6509 (1951).

21. J. A. Page and G. Wilkinson, J. Am. Chem. Soc., 74, 6149 (1952).

22. F. A. Cotton, R. O. Whipple, and G. Wilkinson, J. Am. Chem. Soc., 75, 3586 (1953).

23. G. Wilkinson, P. L. Pauson, and F. A. Cotton, J. Am. Chem. Soc., 76, 1970, 4231 (1954).

24. G. Costa, Ann. Chim. (Rome), 40, 541 (1950); C. A., 45, 6507 (1951).

25. A. J. Martin, Org. Analysis, 4, 1 (1960).

26. W. Lewis and F. Quackenbush, J. Am. Oil Chemist's Soc., 26, 53 (1940).

27. W. Lewis, F. Quackenbush, and T. DeVries, Anal. Chem., 21, 762 (1949).

28. M. E. Peover and B. S. White, Electrochim. Acta, 11, 1061 (1966).

29. R. Dietz, M. E. Peover, and H. P. Rothbaum, Synthetic and Mechanistic Aspects of Electroorganic Chemistry, U. S. Army Research Office, Durham, N. C., Oct. 14-16, 1968.

30. A. L. Berre and Y. Berger, Bull. Soc. Chim. (France), 2363, 2368 (1966).

31. S. Wawzonek, H. Laitinen, and S. Kwiatkowski, J. Am. Chem. Soc., 66, 827 (1944).

32. R. Pasternak, _Helv. Chim. Acta_, _31_, 753 (1948).

33. H. Lund, _Acta Chem. Scand._, _14_, 1927 (1960).

34. N. J. Leonard, S. Swann, Jr., and J. Figueras, _J. Am._
     Chem. Soc., _74_, 4620 (1952).

35. V. Horak and P. Zuman, _Collection Czech. Chem. Commun._,
     _26_, 173 (1961).

36. J. Mayell and A. Bard, _J. Am. Chem. Soc._, _85_, 421 (1963).

37. M. Finkelstein, R. Petersen, and S. Ross, _J. Am. Chem._
     Soc., _81_, 2361 (1959).

38. S. Ross, M. Finkelstein, and R. Peterson, _J. Am. Chem._
     Soc., _92_, 6003 (1970).

39. E. Colichman and D. L. Love, _J. Org. Chem._, _18_, 40
     (1953).

40. S. Tang and P. Zuman, _Collection Czech. Chem. Commun._,
     _28_, 829, 1524 (1963).

41. J. H. Wagenkneckt and M. M. Baizer, _J. Electrochem._
     Soc., _114_, 1095 (1967).

42. H. Lund, _Acta Chem. Scand._, _17_, 2325 (1963).

43. J. Volke and J. Holubek, _Collection Czech. Chem. Commun._,
     _28_, 1597 (1963).

44. P. J. Elving and J. M. Markowitz, _J. Phys. Chem._, _65_,
     686 (1961).

45. S. Wawzonek, R. Berkey, and D. Thomson, _J. Electrochem._
     Soc., _103_, 513 (1956).

46. R. Breslow, W. Bahary, and W. Reinmuth, _J. Am. Chem._
     Soc., _83_, 1763 (1961).

47. L. Horner, H. Fuchs, H. Winkler, and A. Rapp,
     Tetrahedron Letters, _1963_ 965.

48. L. Horner and H. Fuchs, _Tetrahedron Letters_, _1963_ 1573

5.1. THE KOLBE REACTION

5.1.1. GENERAL CONSIDERATIONS

The Kolbe electrolysis is probably one of the earliest reactions applied in organic synthesis, namely in the formation of dimeric products from the oxidation of carboxylic acid salts:

$$RCO_2^\ominus \; -1e \longrightarrow RCO_2^\ominus \xrightarrow{\;-CO_2\;} R^\bullet + CO_2$$

$$2R^\bullet \longrightarrow R-R$$

Since its early conception, its application in organic chemistry has been described in several papers (1-9), and the whole topic has been discussed in review articles (10-13).

The early work of this reaction centered on the preparation of dimeric products from the radical intermediates obtained from the oxidation of carboxylic acid salts. However, it later became apparent that other intermediates, e.g., carbonium ions, can also be formed. This expanded the utility of the Kolbe reaction in organic synthesis and confronted researchers with a challenge to control this oxidation reaction in order to obtain the desired intermediate. It is with this point in mind that we wish to describe the Kolbe reaction.

5.1.2. EFFECT OF REACTION CONDITIONS ON COURSE OF REACTION

Nature of Electrode. In general, the reason why different electrodes have different effects on the course of reactions is not always clear. Several explanations, which are mostly related to the surface of the electrodes (which may offer specific adsorption, catalysis, field effect, etc.)

have been advanced. In the oxidation of sodium acetate in
the presence of aromatic compounds, the use of a graphite
anode produced mainly methylated products with some acetoxylation, while the use of a platinum anode afforded the acetoxy-
lated products exclusively (9). Koehl (14) has shown that
the use of graphite as anodes in the oxidation of simple
aliphatic acids failed to produce dimeric products. This
is due, presumably, to the formation of reactive carbonium
ions, which react immediately to form olefins or undergo
chemical rearrangement. Thus, while there is no rule of
thumb as to which anode will give a specific intermediate,
the use of carbon generally affords carbonium ions, while
the use of platinum gives the radical intermediate.

Nature of Solvent. Depending on the intermediate
formed in the Kolbe reaction, the solvent plays an important
role in determining the nature of the products obtained.
For example, the oxidation of β-phenylpropionic acid on a
platinum anode in glacial acetic acid affords n-propyl benzene and 2-phenylethylacetate (15). On the other hand, the
oxidation of phenylacetic acid on a platinum anode in DMF
affords 1,2-diphenylethane in 88% yield (16). When carbonium ions are intermediates in the Kolbe reactions, the use
of water vs alcohols as solvents will obviously produce
different products.

Structure of Acid. It was shown by Fichter and co-
workers (17) that the Kolbe synthesis does not take place
in aqueous solution when a double bond is too close to the
carbonyl group. In fact, appreciable coupling was not found
until the double bond was in the γ-position or further.
Similarly, no coupling product was observed from benzoic
and phenylacetic acid. However, the oxidation of phenylacetic acid in DMF was shown to afford 1,2-diphenylethane
in 88% yield (16).

The presence of substituents α- to the carboxyl group
has a large influence in suppressing the formation of coupling products. Thus, little or no such products were ob-

served for most α-alkylmethoxy-, hydroxy-, halogen, keto-,
cyano-, and amino- groups.

5.1.3. MECHANISM OF REACTION

In the early work on the Kolbe reaction, a dimeric
product would easily be explained as arising from radical
intermediates. However, based on the early data, the nature
of the intermediates was unknown. Below is a brief description of the evolution of the mechanism of the Kolbe electrolysis.

Formation of Acyl Peroxides. The oxidation of acids
was originally thought to proceed through acyl peroxides
(18):

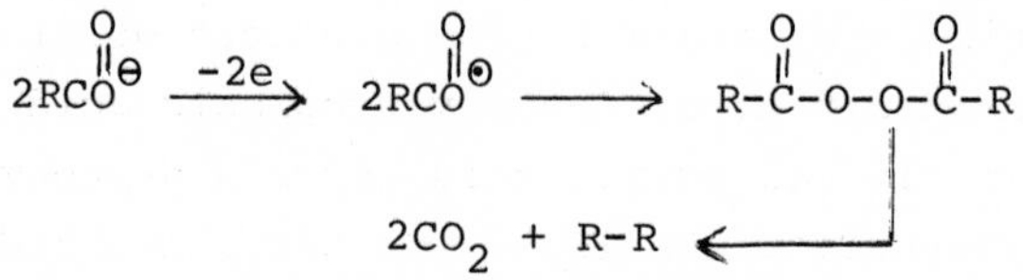

$$2RCO_2^{\ominus} \xrightarrow{-2e} 2RCO_2^{\odot} \longrightarrow R-\overset{O}{\underset{\|}{C}}-O-O-\overset{O}{\underset{\|}{C}}-R$$

$$2CO_2 + R-R \longleftarrow$$

This suggests that under special conditions (e.g., low temperatures) the isolation of the peroxide might be possible.
The failure to detect any appreciable amounts of such peroxides has shed some doubt on the validity of this theory.

Formation of Hydrogen Peroxide. Glasstone and co-
workers (19, 20) advanced another mechanism for the Kolbe
reaction which involves the formation of hydrogen peroxide
in aqueous solutions.

$$OH^{\ominus} \xrightarrow{-e} OH^{\odot} \longrightarrow H_2O_2$$

$$2CH_3-\overset{O}{\underset{\|}{C}}-O^{\ominus} + H_2O_2 \longrightarrow 2CH_3-\overset{O}{\underset{\|}{C}}O^{\odot} + 2\ OH^{\ominus}$$

$$2CH_3\overset{O}{\underset{\|}{C}}-O^{\odot} \longrightarrow CH_3-\overset{O}{\underset{\|}{C}}-O-O-\overset{O}{\underset{\|}{C}}-CH_3 \longrightarrow C_2H_6 + 2CO_2$$

The following observations, however, make it doubtful
whether this mechanism is operative to a great extent:

1.   Attempts to detect any amounts of hydrogen peroxide
during the reaction were not successful (10).

2.   The addition of hydrogen peroxide to acetate solutions produces only trace amounts of hydrocarbons (19).

3.   The Kolbe synthesis is quite efficient in nonaqueous media, such as DMF (16, 21).

Formation of Radicals.   The formation of radical intermediates in the Kolbe reaction was first advanced by Crum,
Brown and Walker and was later developed by Walker and co-
workers (22, 23).

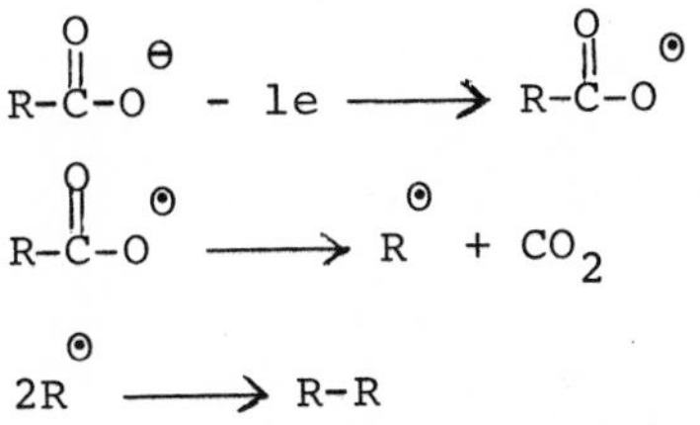

$$R-\overset{\overset{O}{\|}}{C}-O^{\ominus} \quad - \ 1e \longrightarrow R-\overset{\overset{O}{\|}}{C}-O^{\odot}$$

$$R-\overset{\overset{O}{\|}}{C}-O^{\odot} \longrightarrow R^{\odot} + CO_2$$

$$2R^{\odot} \longrightarrow R-R$$

While evidence for the formation of radical intermediates in the Kolbe reaction has been advanced (24, 25),
very little is known about the nature of these radicals.
For example, there is no ESR evidence to support their exis-
tence.   This, however, need not be regarded as evidence for
their absence, since it is possible that these radicals are
adsorbed on the surface of the electrode.   For further discussion on radical intermediates see Ref. 26.

Formation of Carbonium Ions.   While the use of graphite
anodes is reported to afford products that can best be explained by carbonium ion intermediates, Corey and co-workers
(1, 2) have shown that extensive oxidation at platinum electrodes also favors the formation of these ions, Eqs. (5.1)
and (5.2):

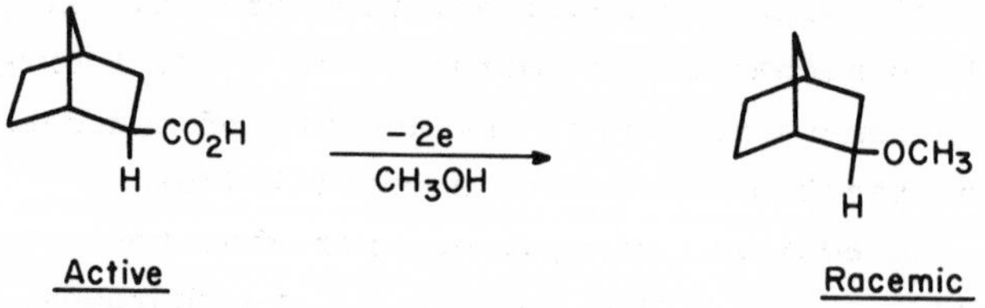

Active   **Racemic**

$$\text{(5.1)}$$

Corey has also shown that the formation of carbonium ions can be a useful tool in organic synthesis:

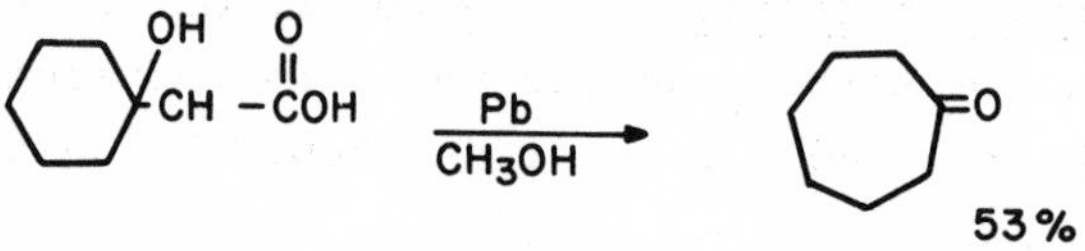

53%

$$\text{(5.2)}$$

The formation of carbonium ions can take place according to one (or more) of the following routes:

From an investigation of the electrolysis of cyclopropane carboxylic acids, Shono and co-workers (27) favored the formation of the acylonium ion from the oxidation of the acyl radical. The electrolysis of cyclopropanol carboxylic acid in the presence of water afforded allyl alcohol, presumably through cyclopropanol (38). Similarly, cyclohexane carboxylic acid was converted to cyclohexanol (39).

For a discussion of reaction conditions favoring the formation of radicals and carbonium ions, see experimental procedures (Sec. 5.1.5).

5.1.4. ROLE IN ORGANIC SYNTHESIS

The versatility, as well as the simplicity, of the Kolbe reaction are two of the important factors that have

led to its popularity. Table 5.1 illustrates some of the
types of products that can be prepared by this reaction.

TABLE 5.1

Types of Compounds Prepared Via the Kolbe Electrolysis

Starting Material	Reaction Conditions		Product(s)	Reference
	Anode	Solution		
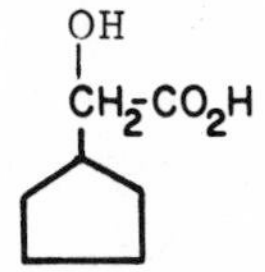	Pt	CH_3OH		2
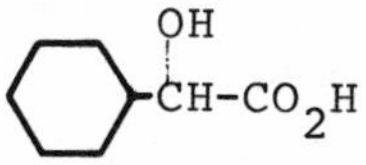	Pt	CH_3OH	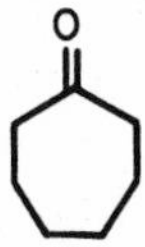	2
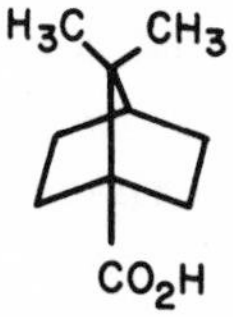	Pt	CH_3OH	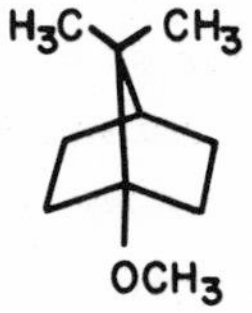	2
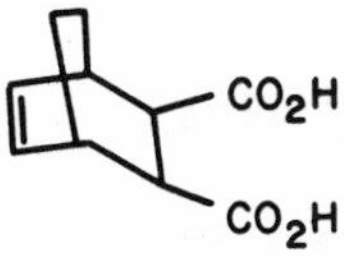	Pt	H_2O	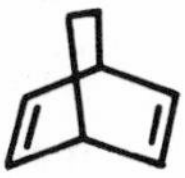	6
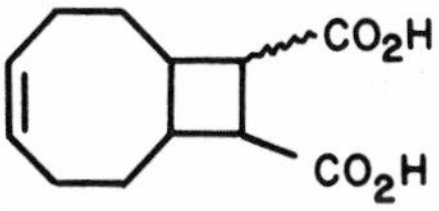	Pt	$(CH_3)_3N$ pyridine	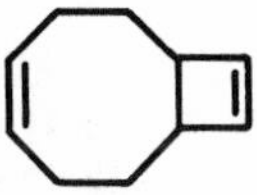	28
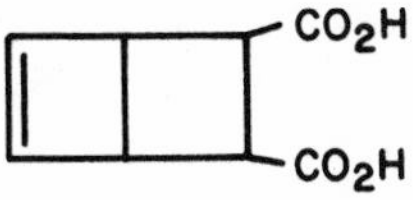	Pt	10% Aq. Py. $(CH_3)_3N$	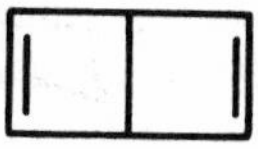	28

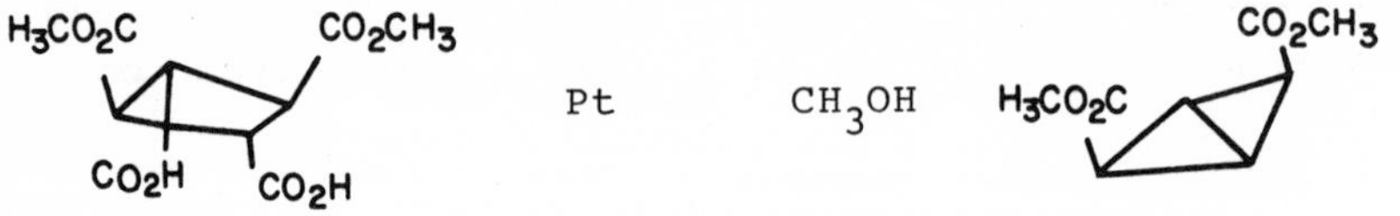

29

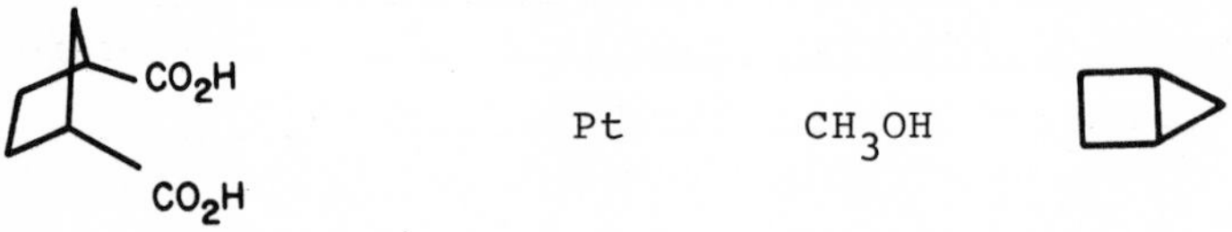

30

The preparation of the compounds in Table 5.1 by conventional chemical methods may not always be possible. For
example, the chemical synthesis of bicyclo[2.2.2]octadiene
could not be prepared from the appropriate bicyclo[2.2.2]-
octene (31). Only multistep chemical processes afforded
practical yields (6).

The facile decarboxylation of norbornyl-type carboxylic
acids allows the formation of nonclassical carbonium ions.
Corey (2) studied the anodic decarboxylation of several nor-
bornyl carboxylic acids and reported the following reactions,
Eqs. (5.3)-(5.5):

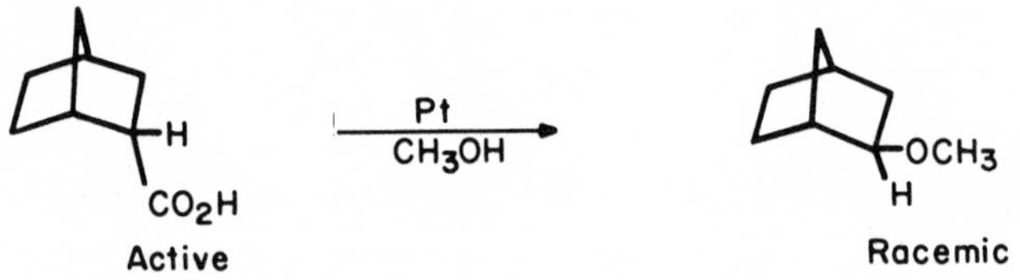

(5.3)

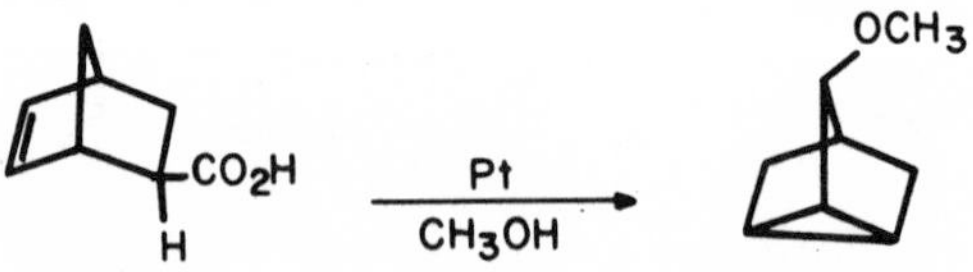

(5.4)

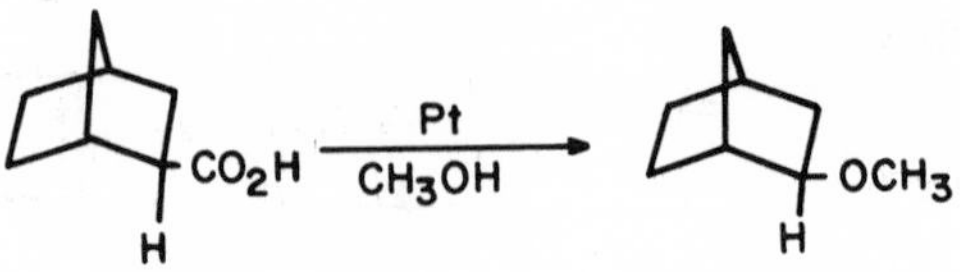

(5.5)

The products obtained in those reactions are quite analogous to those obtained via the bridged ions [118] shown below, in solvolytic reactions ($\underline{32}$, $\underline{33}$).

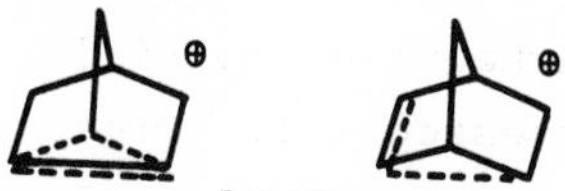

[118]

Gassman and co-workers ($\underline{34}$) performed the anodic oxidation on the epimeric isomers of bicyclo[3.1.0]hexane-3-carboxylic acids [119] in pyridine-water solution. The cis-isomer gave the following products:

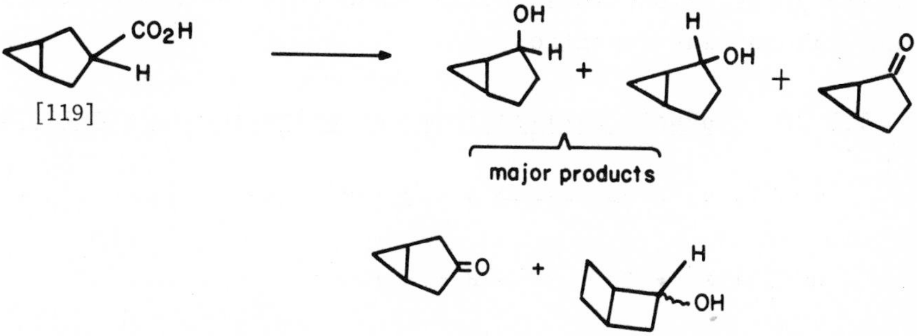

[119]

major products

These reactions were compared with the acetolysis of the bicyclo[3.1.0]hexyl-3-tosylates ($\underline{35}$) and with the deamination of bicyclo[3.1.0]-hexyl-3-amines ($\underline{36}$). The cis- and trans-tosylates afforded the 3-acetates upon solvolysis with the cis-isomer as the predominant product. On the other hand, deaminations, such as in anodic decarboxylation, produced a variety of products.

5.1.5. EXPERIMENTAL PROCEDURES

5.1.5.1. Products Derived from Radical Intermediates

Coupling products resulting from dimerization of radical intermediates are favored by the use of a platinum anode. A nondivided cell may be utilized in general since neither the starting material nor the product is sensitive to reduction. The use of aqueous or partially aqueous media is recommended but not essential. Organic solvents which are stable toward carbon radicals can also be used, such as DMF or acetonitrile. The carboxylate salt can serve as its own electrolyte, although perchlorate or fluoroborate salts can be used to achieve higher conductivity. Since the reaction involves coupling of a radical intermediate, higher yields may be obtained by using as high a concentration of carboxylate as possible and by conducting the electrolysis at high current densities.

5.1.5.2. Products Derived from Carbonium Ion Intermediates

$$RCO_2^{\ominus} + SH - 2e \longrightarrow R - S + H^{\oplus} + CO_2$$

$$RCO_2^{\ominus} + S^- - 2e \longrightarrow R - S + CO_2$$

where SH or S^- are nucleophiles, e.g., alcohols, acetates, or water.

In general, a graphite anode should be used, since it favors the formation of carbonium ions. However, if the resulting carbonium ion is sufficiently stabilized, it will also form on platinum anodes. Undivided cells can generally be utilized unless the nucleophile is sensitive to reduction. If possible, the desired nucleophile should be used as the solvent; for example, in the formation of ethers, the corresponding alcohol could be used. When alcohols are desired, water should be used as solvent, possibly together with a nonnucleophilic solvent such as DMF. As before, the carboxylate salt, or the nucleophile salt (e.g., acetate or alkoxide), can function as electrolyte, and should be utilized if possible. To discourage self-condensation to esters,

high concentrations of carboxylate and high current densities should be avoided.

<u>Dimerization Reaction</u> (<u>37</u>)

$$ROC\text{-}(CH_2)_8\text{-}C\text{-}O^{\ominus} \longrightarrow R\text{-}O\text{-}C\text{-}(CH_2)_{16}\text{-}C\text{-}OR$$

[120]   [121]

The electrolysis cell is a tall beaker (see Fig. 5.1)
which is fitted with a rubber stopper to accommodate a condenser and three copper wires connected to the electrodes.
The reaction solution is stirred by means of a magnetic
stirring bar. Three electrodes, made out of platinum wire,
were used and were spaced equidistantly from each other.
The anode was 12 cm^2 in area while the cathodes were 5 cm^2
in area.

The beaker was charged with 500 ml absolute methanol
and 1.1 g (0.05 g atom) sodium. When the sodium had dis-

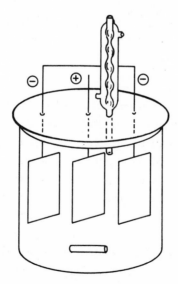

FIG. 5.1. Electrolysis cell for a Kolbe-type
electrolysis.

solved, 216 g (1.0 mole) of methyl hydrogen sebacate was
added and allowed to dissolve in the sodium methide solution.   An over-all potential of 50 V was applied across the
cell and allowed the passage of 1-2 A.   This caused a mild
reflux, which could be regulated by varying the applied
voltage.   (Towards the end of the reaction the voltage was
about 120 V.)   The reaction is completed within 30-40 hr
(longer reaction time may cause the formation of polymeric
materials) as indicated when a few drops of the solution
show an alkaline reaction to phenolphthalein.   Upon completion of the reaction, the electrodes are promptly removed
from the solution, which solidifies after cooling.   The
product is acidified with acetic acid and the solvent removed under reduced pressure.   The product [121] is isolated and identified by conventional methods.   The yield
varied between 67 and 74%.

Formation of Olefins (1)

[122]

meso-or dl- isomer

[123]

trans - isomer

A similar apparatus to that described above, Fig. 5.1,
may be used.   A solution of 540.6 mg (2.0 mmoles) of meso-2.
3-diphenylsuccinic acid and 1.12 ml (8.0 mmoles) of tri-
ethylamine in 10 ml of 90% pyridine-water was maintained
at reflux (external heating) under nitrogen.   The cathode
and anode were made from smooth platinum and the solution
was stirred by means of a magnetic stirrer.   Electrolysis
was performed at 134 V for 550 min during which time the
current fell from 0.14 A to 0.10 A.   The dark reaction mixture was evaporated on a steam bath under 10 mm vacuum,
treated with 50 ml ether and extracted with 2 x 20 ml of
N-hydrochloric acid.   This was followed with extraction

with 2 x 20 ml of N-sodium hydroxide and 1 x 10 ml of saturated salt. The pale yellow ether solution was dried ($MgSO_4$)
and evaporated, leaving an amber-colored oily, crystalline
solid (810 mg) after drying under high vacuum. This material
was identified as compound [123]. The yield was 68%.

Note: The meso- and dl-isomers of the above acid can
also be oxidized with lead tetraacetate to afford trans-
stilbene. This is really not surprising, since it appears
that both oxidation reactions (electrochemical and with lead
tetraacetate) afford the same stable carbonium ion, which
gives the thermodynamically favored product. Furthermore,
the authors feel that more electrochemical work needs to be
done in order to determine the most favorable conditions for
the generation of carbonium or radical intermediates.

REFERENCES ON THE KOLBE REACTION

1. E. J. Corey and J. Casanova, Jr., J. Am. Chem. Soc.,
    85, 165 (1963).

2. E. J. Corey, N. L. Bauld, R. T. LaLonde, J. Casanova,
    Jr., and E. T. Kaiser, J. Am. Chem. Soc., 82, 2645 (1960).

3. W. Paudler, R. E. Herbener, and A. G. Zeiler, Chem. Ind.
    (London), 46, 1909 (1965).

4. L. Riccobini, Gazz. Chim. Ital., 70, 747 (1940).

5. M. Finkelstein and R. C. Peterson, J. Org. Chem., 25,
    136 (1960).

6. H. H. Westberg and H. J. Dauben, Jr., Tetrahedron
    Letters, 1968 5123.

7. R. G. Woolford and W. S. Lin, Can. J. Chem., 44, 2783
    (1966).

8. R. Brettle and G. B. Cox, J. Chem. Soc., (c), 1969 1227.

9. V. D. Parker, Chem. Comm., 1164 (1968).

10. C. L. Weedon, Quart. Rev. (London), 6, 380 (1952).

11. A. P. Tomilov and M. Ya. Fioshin, Russ. Chem. Rev.,
    32, 30 (1963).

12. L. Eberson, Chemistry of the Carboxyl Group, Wiley,
    New York, 1969.

13. A. K. Vijh and B. E. Conway, Chem. Rev., 67, 623 (1967).

14. W. J. Koehl, Jr., J. Am. Chem. Soc., 91, 1227 (1969).

15. W. A. Bonner and F. D. Mango, J. Org. Chem., 29, 430 (1964).

16. L. Rand and A. F. Mohar, J. Org. Chem., 30, 3885 (1965).

17. F. Fichter and T. Holbro, Helv. Chim. Acta, 20, 333 (1937).

18. F. Fichter, Trans. Electrochem. Soc., 75, 309 (1939).

19. S. Glasstone and A. Hickling, Chem. Rev., 25, 407 (1939).

20. A. Hickling, Quart. Rev. (London), 3, 95 (1949).

21. B. E. Conway and M. Dzieciuch, Can. J. Chem., 41, 21, 38, 55 (1964).

22. S. Shukla and J. Walker, Trans. Faraday Soc., 27, 35 (1931).

23. S. Shukla and J. Walker, Trans. Faraday Soc., 28, 547 (1932).

24. K. Clusius and W. Schanzer, Z. Physik. Chem., 192A, 273 (1943).

25. C. L. Wilson and W. Lippincott, J. Am. Chem. Soc., 78, 4290 (1956); J. Electrochem. Soc., 103, 672 (1956).

26. L. Eberson, Acta Chem. Scand., 17, 2004 (1963).

27. T. Shono, I. Nishiguchi, Shoji Yamane, and R. Oda, Tetrahedron Letters, 1969 1965.

28. P. Radlick, R. Klem, S. Spurlock, J. Sims, E. van Tamelen, and T. Whiteside, Tetrahedron Letters, 1968 5117.

29. A. F. Velturo and G. Griffin, J. Org. Chem., 31, 2241 (1966).

30. T. Campbell, A. Velturo, and G. Griffin, Chem. Ind., 1235 (1969).

31. J. Hine, J. A. Brown, L. H. Zalkow, W. E. Gardner, and M. Hine, J. Am. Chem. Soc., 77, 694 (1955).

32. S. Winstein and P. Trifan, J. Am. Chem. Soc., 74, 1147, 1154 (1952).

33. J. D. Roberts, E. R. Trumbull, W. Bennett, and R. Armstrong, J. Am. Chem. Soc., 72, 3116 (1950).

34. P. G. Gassman and F. V. Zalar, J. Am. Chem. Soc., 88, 2252 (1966).

35. S. Winstein and J. Sonnenberg, J. Am. Chem. Soc., 83, 3235, 3244 (1961).

36. E. J. Corey and R. L. Dawson, J. Am. Chem. Soc., 85, 1782 (1963).

37. S. Swann, Jr., and W. E. Carrison, Jr., Org. Syn., 41, 33 (1961).

38. F. Fichter and O. Reeb, Helv. Chim. Acta, 6, 450 (1923); N. J. Demjanoff, *F. Russ. Phys. Chem. Soc.*, 61, 1861 (1929).

39. N. J. Demjanoff, *F. Russ. Phys. Chem. Soc.*, 36, 314 (1904).

5.2. OXIDATION OF UNSATURATED COMPOUNDS

5.2.1. ANODIC SUBSTITUTION

5.2.1.1. General Considerations

The electrochemical oxidation of aromatic compounds represents one of the most useful electrochemical reactions in the area of organic synthesis. Thus, a simple change in solvent and/or supporting electrolyte allows the preparation of a variety of compounds. Consider, for example, the reactions in Table 5.2.

The reactions in Table 5.2 may be compared to the well-documented Friedel-Crafts electrophilic substitution reactions. While anodic substitution can be of great utility in organic synthesis, in practice, experimental procedures have been somewhat limited because of the few anodes, such as platinum carbon and certain oxides of lead, that can be employed. Nevertheless, it has been the continued persistence of a number of researchers that has aroused the interest of the organic, as well as the physical organic, chemists in this field.

A detailed discussion of anodic substitution would involve a discussion of a tremendous amount of data which would make this chapter (and the whole book) longer than originally intended. Consequently, this topic is dealt with in general terms. However, for those who wish to dwell on the subject in some detail, the following recent key works are recommended: Refs. 7-10.

There has been some controversy concerning the mechanism of anodic substitution on the aromatic nucleus. Thus,

TABLE 5.2

Oxidation of Aromatic Compounds

Aromatic Compound	Reaction Conditions		Products	Ref.
	Electrode	Solution		
	PbO$_2$	Aq. H$_2$SO$_4$	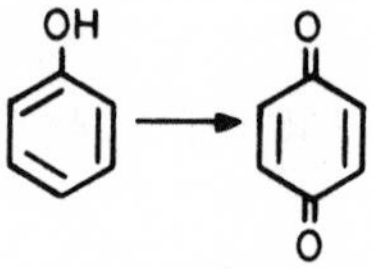	1
	Pt	I$_2$,LiClO$_4$,CH$_3$CN	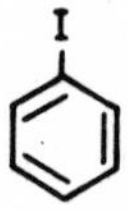	2
	Pt	KOAC, HOAC	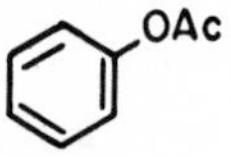	3
	Graphite	Aq. NaNO$_3$	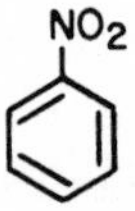	4
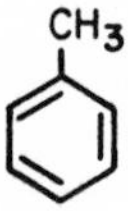	Pt	CH$_3$CN	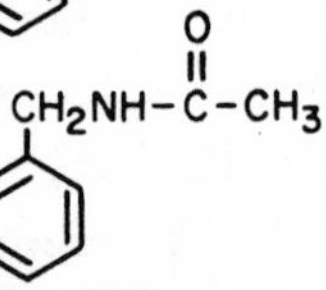 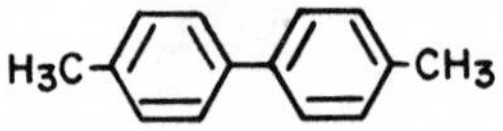	5
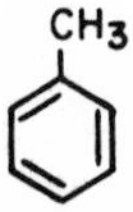	Pt	HCN, NaCN, CH$_3$OH	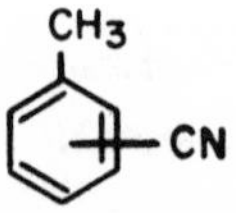	6

while some prefer a direct electron transfer from the aromatic species to the anode to form the radical cation (7),
other investigators prefer an attack of an anodically generated substrate, e.g., methoxy radical, bromine, etc., on

$$Ar - 1e \longrightarrow Ar^{\oplus}$$

the hydrocarbon (11, 12). A third group has presented some
rather convincing data, pointing to the involvement of the
anode surface, in the form of oxides, in the oxidation reactions (13, 14). The mechanism of the oxidation of substituted aromatic compounds has also been the subject of some
controversy. Eberson advocates the oxidation of the side
chain (15), while Parker and co-workers invoke the oxidation
of the solvent and/or supporting electrolyte, which then
abstract a hydrogen from the side chain of the aromatic compound (16). Whatever the mechanism of anodic substitution
may be, it is hoped that the discussion below will guide
the reader in choosing the appropriate reaction conditions
in order to obtain the desired products.

5.2.1.2. Anodic Alkoxylation

A significant amount of this work has centered on the
alkoxylation of furans. Thus, furan was oxidized on a platinum electrode in methanolic solution containing ammonium
bromide and gave the dimethoxy product [124], Ref. 17.

In this reaction, it was believed that bromine is formed
at the anode while hydrogen and ammonia are formed at the
cathode. Bromine reacts with furan in methanol to afford
2,5-dibromodihydrofuran, which then undergoes nucleophilic
substitution by methanol to give the observed 2,5-dimethoxy-
dihydrofuran and hydrogen bromide. The latter reacts with

ammonia to regenerate ammonium bromide. Thus, the only ingredients that are consumed are electricity and methanol.
The in situ generation of bromine in the electrochemical
method obviates its potentially hazardous handling, which
is needed in the analagous conventional method for the preparation of the dimethoxydihydrofuran (18). More recently,
the methoxylation of furans was carried out by using other
supporting electrolytes, such as $NaOCH_3$, and H_2SO_4 (19).
That bromine is involved in the methoxylation of furans was
shown by Ross and co-workers (20), who obtained the same
isomer ratio, 50:50 cis:trans, from the electrochemical methoxylation of 2,5-dimethylfuran as well as methoxylation with
Br_2/CH_3OH.

The electrochemical methoxylation of benzene has not
yet been accomplished. In contrast, however, substituted
benzenes [125] can be methoxylated under the appropriate
conditions (21). This type of reaction is difficult to accomplish by conventional chemical methods.

The alkoxylation of aromatic side chains has also been
reported (22). The mechanism for this reaction probably involves the abstraction of hydrogen atom from the side chain
by a radical, e.g., alkoxy radical, which is generated at
the anode.

5.2.1.3. Anodic Acetoxylation

Acetoxylation reactions in homogeneous solutions have
not yet been reported. This may be attributed to the facile
decarboxylation of acetoxy radicals (rate constant $\sim10^9$
sec^{-1}) within the solvent cage, thus preventing any contact
with the surrounding substrates (23). For example, the ther-

mal decomposition of diacetyl peroxides in the presence of
aromatic compounds does not result in acetoxylation of the
aromatic species (24). In anodic acetoxylation, i.e., oxidation of acetate ions (as opposed to acetic acid) in the
presence of aromatic hydrocarbons, acetoxy radicals may be
formed at the surface of the electrode (as was pointed out
in the discussion of the Kolbe reaction) in the presence of
aromatic hydrocarbons, and thus affect their acetoxylation.
This explains the formation of methylated products in the
acetoxylation of naphthalene and biphenyls (27): $CH_3CO_2^{\ominus}$
$\longrightarrow CH_3^{\cdot}+CO_2$). However, since aromatic hydrocarbons oxy-
dize at a lower potential than acetate anions, the interven-
tion of acetoxy radicals in such reactions was deemed unlikely (25). In fact, negatively substituted aromatics, whose
oxidation potential is more positive than that of acetate
anions, did not undergo acetoxylation. Eberson has found
that the distribution of o-, m-, and p-isomers in anodic
acetoxylation is similar to that found in electrophilic
aromatic substitution reactions (26). Table 5.3 shows some
examples of this.

The potential of the anode may play an important role
in the acetoxylation of aromatic compounds. Thus, it was
found (27) that the yield of acetoxyanisoles from anisole
can be increased from 27 to 40% by carrying the electrolysis at a controlled potential of 1.50 V (s.c.e.) and by
passing 50% of the theoretical amount of electricity through
the solution. The presence of acetate ion is also of importance in determining the course of anodic acetoxylation.
Thus, nuclear acetoxylation of anisole occurs only in the
presence of acetate ion. No reaction occurred when tosylate
or perchlorate were used in acetic acid.* The role of acetate ion was also studied in the acetoxylation of amides
(28, 29). Thus, with nitrate, rather than with acetate,

---

*This observation was explained as due to a two-electron
oxidation of the substrate followed by a concerted attack by
acetate ion (26).

TABLE 5.3

Oxidation of Aromatic Compounds in HOAC/NaOAC
on a Platinum Anode.

Aromatic Compound	Acetate-Isomers (%)			
	o-	m-	p-	α-side chain
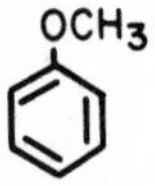 OCH₃	67.4	3.5	29.1	--
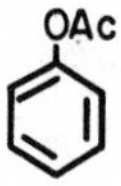 OAc	40.1	5.0	54.9	--
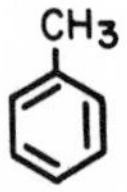 CH₃	43.2	11.1	45.7	28.6
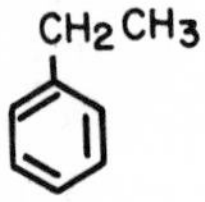 CH₂CH₃	43.8	10.2	46.0	50.5
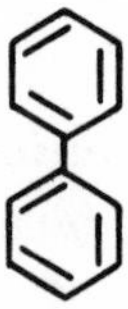 (biphenyl)	30.7	0.9	68.4	--

as the supporting electrolyte, an increase in N-alkyl ace-toxylation was observed. The oxidation of mesitylene was also sensitive to the supporting electrolyte (30). Thus, with NH_4NO_3/HOAC, only side chain acetoxylation occurred. However, with KOAC, a considerable amount of nuclear ace-toxylation occurred.

5.2.1.4. Anodic Cyanation

The mechanism of aromatic cyanation is quite similar to that of anodic acetoxylation. Thus, the aromatic sub-strate is oxidized in the presence of cyanide ion** and leads to the formation of the observed product.

$$ArH \xrightarrow{-1e} ArH^{\oplus} \xrightarrow{-1e} Ar^{\oplus} + H^{\oplus}$$

$$Ar^{\oplus} + CN^{\ominus} \longrightarrow ArCN$$

Table 5.4 describes the o-, m-, and p-isomer distributions of cyanation and acetoxylation reactions (26, 33). Unlike acetoxylation reactions, cyanation occurs almost exclusively at the aromatic nucleus.

Anodic cyanation was studied by Andreades and co-workers (35), who developed a procedure which affords high yield of products. These investigators proposed the following

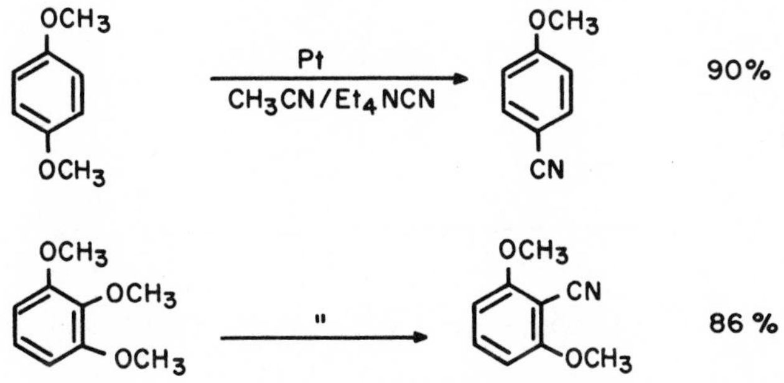

**In spite of the fact that the oxidation potential of cyanide anion occurs at a less positive potential than aro-matic compounds, no cyanation reaction takes place until the aromatic compound is oxidized (34).

TABLE 5.4

Comparison of o-, m-, and p-, Isomer Distribution
in Anodic Acetoxylation and Cyanation (26, 33)

Organic Compound	Acetoxylation (%)			Cyanation (%)		
	o-	m	p-	o-	m-	p-
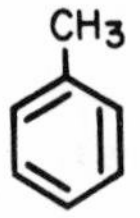 CH₃ (toluene)	43.2	11.1	45.7	40	8	52
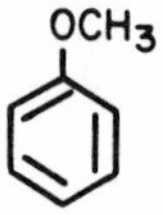 OCH₃ (anisole)	67.4	3.5	29	53	<0.1	47
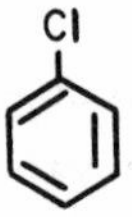 Cl (chlorobenzene)	36.8	5.5	57.7	50	<0.5	50
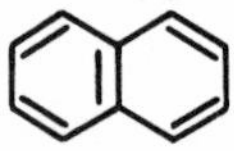 (naphthalene)	96.	3.9		90	10	

mechanism for the above reaction, Eqs. (5.6) and (5.7).

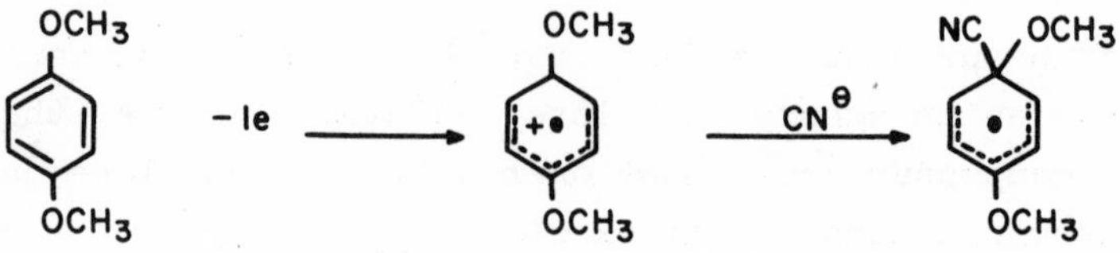

$$\text{(OCH}_3\text{ / OCH}_3\text{ benzene)} \xrightarrow{-1e} \text{(cation radical)} \xrightarrow{CN^{\ominus}} \text{(NC, OCH}_3\text{ product)} \qquad (5.6)$$

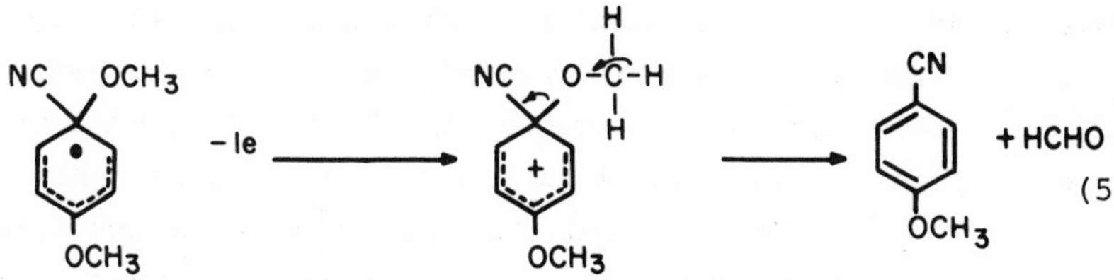

(5.7)

The site of substitution by cyanide ion occurred at the carbon with the highest spin-density which could be predicted from ESR studies.

5.2.1.5. Anodic Acetamidation

Anodic acetamidation is quite similar to the well-known Ritter reaction, which involves the reaction of a carbonium ion with acetonitrile.

$$ArH \quad -1e \longrightarrow ARH^{\dot{+}} \xrightarrow{-1e} Ar^{\oplus} + H^{\oplus}$$

$$Ar^{\oplus} + CH_3-C\equiv N \longrightarrow Ar-N=C^{\oplus}-CH_3$$

$$Ar-N=C^{\oplus}-CH_3 + H_2O \longrightarrow Ar-NH-\overset{\overset{\displaystyle O}{\|}}{C}-CH_3$$

The above reaction applies to aromatic as well as aliphatic compounds, In the aliphatic series, the carbonium ion may be formed from a Kolbe-type reaction (31, 36), Eq. (5.8):

(5.8)

The oxidation of toluene in acetonitrile [Eq. (5.9)] is of interest. If the amount of water in reaction medium is less than 0.1mM, acetamidation occurs to give [126] in almost quantitative yield.

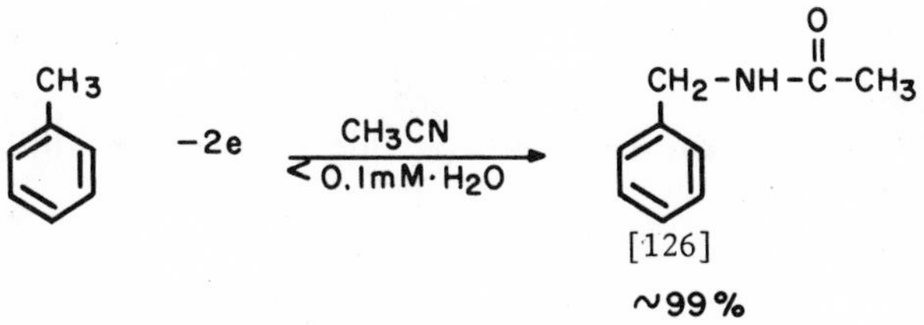

(5.9)

[126]

~99%

However, when the oxidation was carried out in wet acetonitrile >.1mM water, benzyl alcohol, benzaldehyde, benzoic
acid, and bibenzyl were isolated together with benzylacet-
amide.   There has been some controversy concerning the
mechanism of acetamidation of toluene in wet and dry acetonitrile.   For a discussion on this subject and the role of
water in this reaction, Refs. 6, 31, 36 are recommended.

5.2.2.   OXIDATION OF AROMATIC ALCOHOLS AND AMINES

Oxidation of Phenols.   Fichter was among the early investigators to study the electrochemical oxidation of phenol
(37) and reported the following products upon oxidation under
acidic solutions:

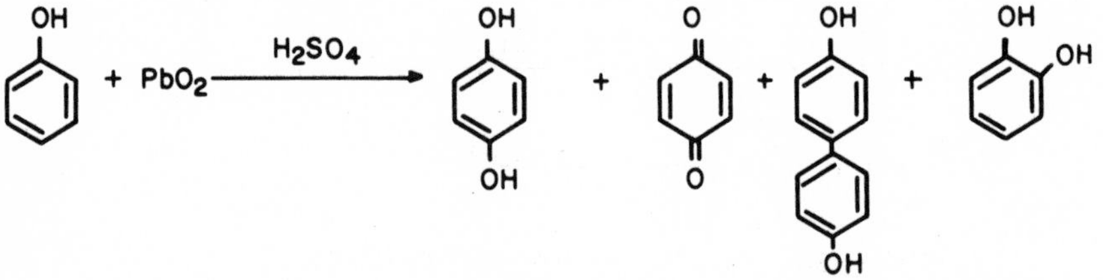

A detailed study of this reaction in the author's laboratories (38) showed that this reaction can be controlled to
afford hydroquinone in over 90% yield.   The mechanism of the
oxidation is proposed by Eqs. (5.10)-(5.12):

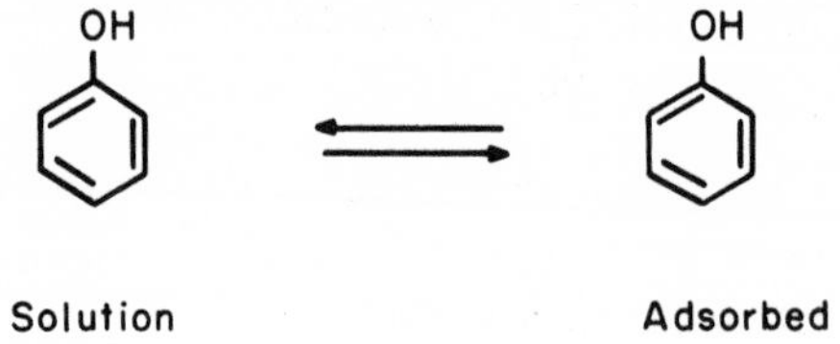

(5.10)

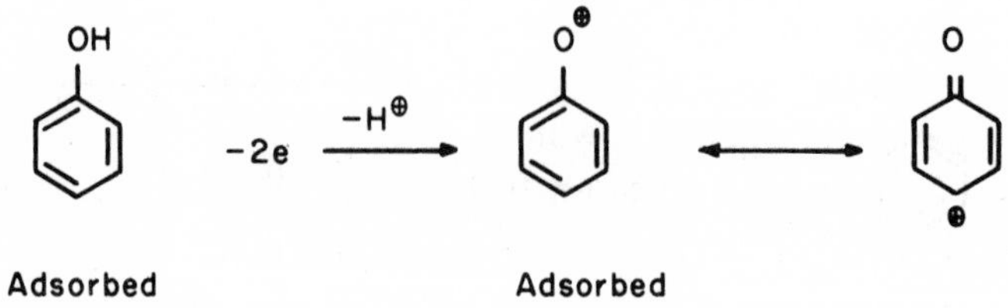

(5.11)

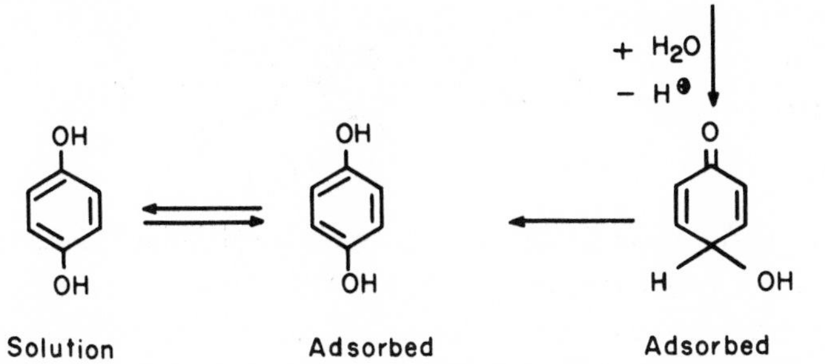

(5.12)

Solution   Adsorbed   Adsorbed

An interesting speculation concerning the high yield of p-hydroquinone is that it is due to the field effect of the anode, which keeps the positive charges far apart due to coulombic repulsion, Eq. (5.13):

[127]   versus   [128]   (5.13)

At 10^7 V/cm the energy difference is ~2.5 kcal/mole.

The above cation [127] has been trapped in other reactions, Ref. 39, Eq. (5.14):

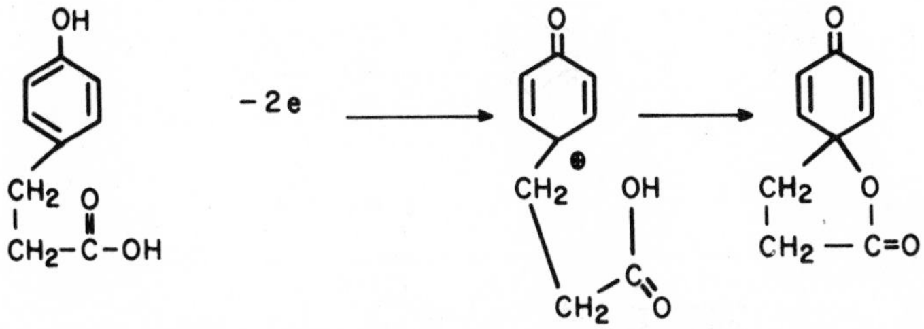

(5.14)

The oxidation of phenol has been used in the area of electrocoating, since polymeric material can be deposited on several anodic substrates (40, 41). This reaction is discussed in more detail in Chap. 6. The dimerization of phenols and its derivatives may be of utility in organic synthesis. Thus, 2,6-di-t-butylphenol [129] afforded the corresponding biphenol [130] in good yield (42).

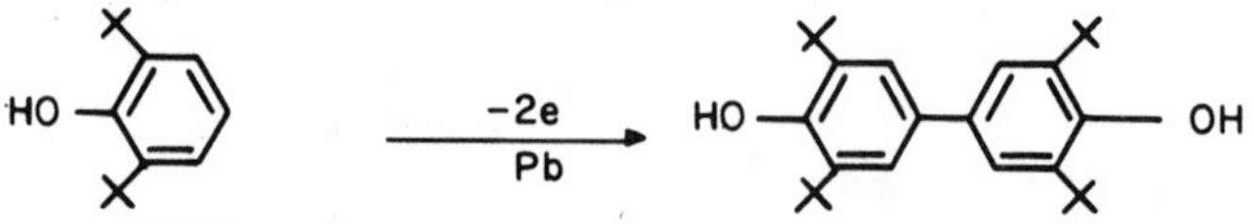

$$\text{[129]} \quad \xrightarrow[\text{Pb}]{-2e} \quad \text{[130]}$$

Under similar conditions, no 4,4'-dihydroxydiphenyl could be obtained from the oxidation of phenol. The product(s) was mainly polymeric in nature. For further discussion on the oxidation of phenols Refs. 43–48 are recommended.

Oxidation of Anilines. The electrochemical oxidation of anilines is perhaps of more synthetic utility than the oxidation of phenols, since a variety of products may be obtained from this reaction, Eqs. (5.15)-(5.17):

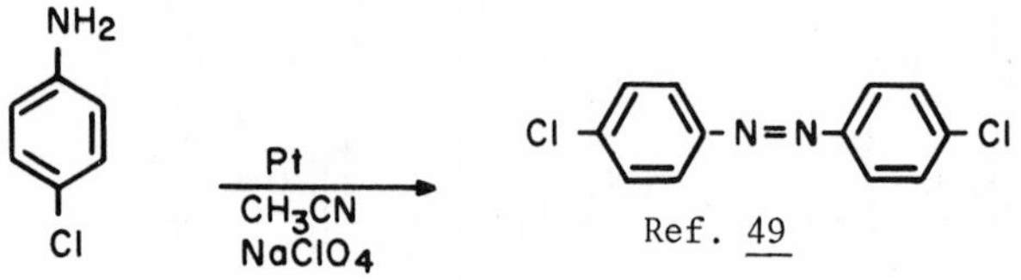

$$\xrightarrow[\substack{CH_3CN \\ NaClO_4}]{Pt} \quad Cl-\langle\rangle-N=N-\langle\rangle-Cl \qquad (5.15)$$

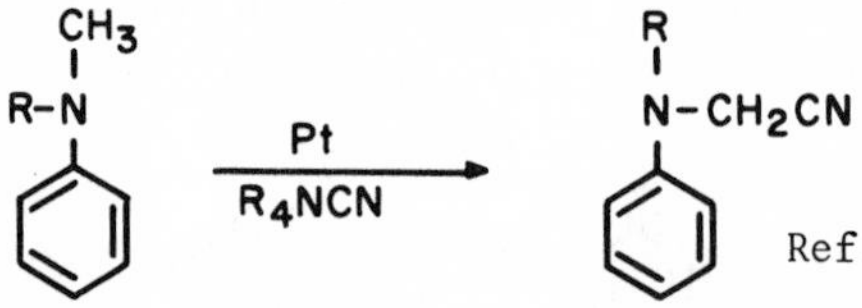

Ref. 49

$$\xrightarrow[R_4NCN]{Pt} \qquad (5.16)$$

Ref. 50

$$\xrightarrow[\substack{Aq.C_2H_5OH \\ HCl, NH_4SCN}]{C} \qquad Ref. 51 \qquad (5.17)$$

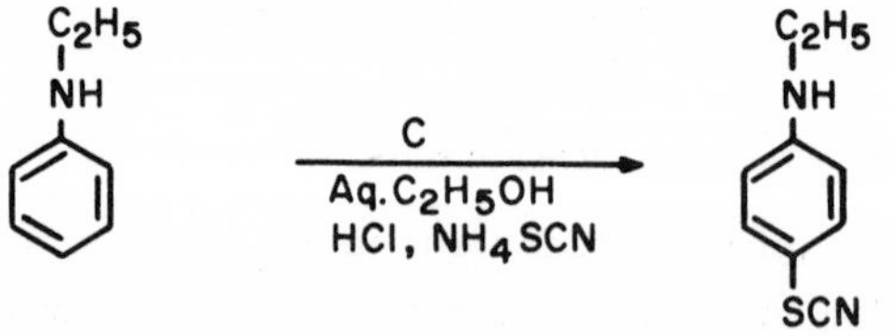

Some of the oxidation products of aniline may also be transformed into useful organic intermediates. For example, it was mentioned in the previous section that the oxidation of phenol under a variety of conditions failed to give the de-

sired 4,4'-dihydroxydiphenyl (42). However, this compound
was finally obtained from the oxidation of benzidine, Ref.
52, Eqs. (5.18) and (5.19).

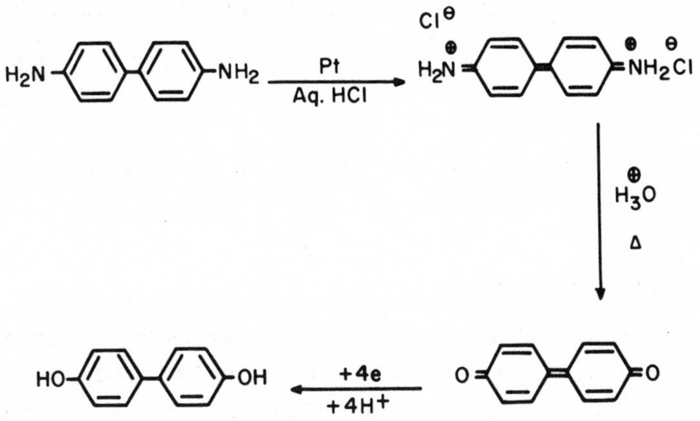

$$H_2N-\underset{}{\bigcirc}\underset{}{\bigcirc}-NH_2 \xrightarrow[\text{Aq. HCl}]{\text{Pt}} \overset{Cl^{\ominus}}{H_2N}\overset{\oplus}{=}\underset{}{\bigcirc}\underset{}{\bigcirc}\overset{\oplus}{=}NH_2Cl \quad (5.18)$$

$$\underset{}{\overset{\oplus}{H_3O}} \\ \Delta$$

$$HO-\underset{}{\bigcirc}\underset{}{\bigcirc}-OH \xleftarrow[+4H^+]{+4e} O=\underset{}{\bigcirc}\underset{}{\bigcirc}=O \quad (5.19)$$

4,4'-Dihydroxydiphenyl is useful in several polymerization
reactions.

The mechanism of the electrochemical oxidation of anilines was studied by Adams and co-workers (53-55), who postulated the following primary step:

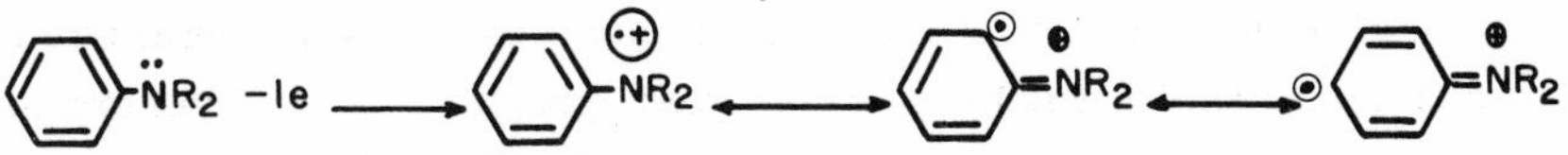

$$\underset{}{\bigcirc}\overset{..}{N}R_2 \xrightarrow{-1e} \underset{}{\bigcirc}\overset{\overset{\oplus+}{}}{N}R_2 \longleftrightarrow \underset{}{\bigcirc}\overset{\odot}{=}\overset{\ominus}{N}R_2 \longleftrightarrow \overset{\odot}{\underset{}{\bigcirc}}\overset{\ominus}{=}NR_2$$

$$[131]$$

The cation-radical intermediate [131], which has been detected by ESR spectra, may be the common intermediate for
the observed products from the oxidation of anilines, Eqs.
(5.20) and (5.21):

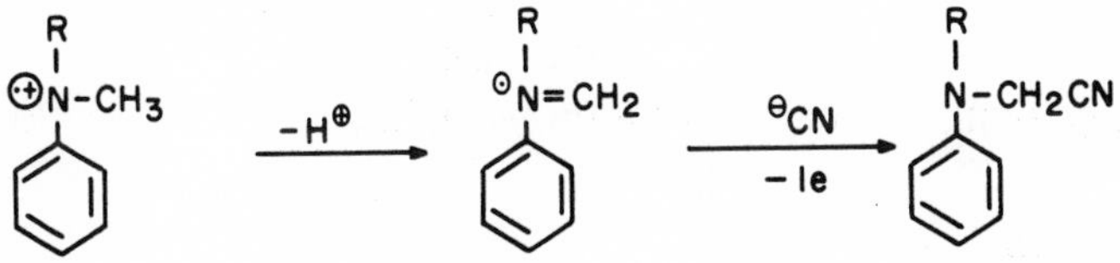

$$\overset{R}{\underset{}{\bigcirc}}\overset{|}{\overset{\oplus}{N}}-CH_3 \xrightarrow{-H^{\oplus}} \overset{R}{\underset{}{\bigcirc}}\overset{|}{\overset{\odot}{N}}=CH_2 \xrightarrow[-1e]{\ominus CN} \overset{R}{\underset{}{\bigcirc}}\overset{|}{N}-CH_2CN \quad (5.20)$$

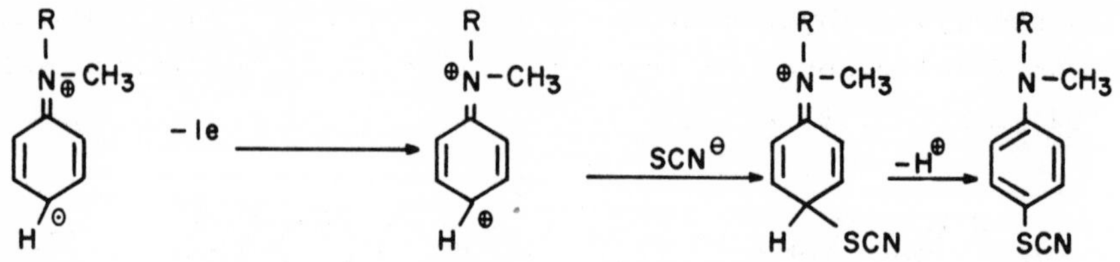

$$(5.21)$$

For further information on the oxidation of aromatic amines Refs. 56-62 are recommended.

5.2.3. OXIDATION OF OLEFINS

The electrochemical oxidation of olefins is of great synthetic utility since a slight variation in reaction conditions can lead to a variety of products. Table 5.5 describes some examples. These examples clearly demonstrate the versatility of this reaction and its sensitivity to reaction conditions. Furthermore, the feasibility of some of the reactions listed in Table 5.5 has been demonstrated on a large scale. Consider, for example, the synthesis of propylene oxide. This monomer, whose industrial consumption is well over 150 million pounds per year, is polymerized by conventional methods to afford polyols of different molecular weights, which find a large application in the urethanes as well as in the coating areas. Up to date, propylene oxide has been, for the most part, prepared conventionally via the halohydrin process, which involves the following reactions, Eqs. (5.22)-(5.24):

$$2Cl_2 + Ca(OH)_2 \longrightarrow 2HOCl + CaCl_2 \qquad (5.22)$$

$$CH_2{=}CH{-}CH_3 + HOCl \longrightarrow \overset{Cl}{\underset{}{C}}H_2{-}\overset{OH}{\underset{}{C}}H{-}CH_3 \qquad (5.23)$$

$$2\overset{Cl}{\underset{}{C}}H_2{-}\overset{OH}{\underset{}{C}}H{-}CH_3 + Ca(OH)_2 \longrightarrow 2CH_2{-}CH{-}CH_3$$

$$+ \; CaCl_2 + 2H_2O \qquad (5.24)$$

TABLE 5.5

Electrochemical Oxidation of Olefins

Olefin	Reaction Conditions	Products	References
$CH_2=CH_2$	Pt, Aq. H_2SO_4	$\overset{OH}{\underset{\vert}{CH_2}}-\overset{OH}{\underset{\vert}{CH_2}}$	63, 64
	Pt, Aq. H_2SO_4, Hg_2SO_4	$CH_3-\overset{OH}{\overset{\Vert}{CH}}$	65
	Pd Aq. H_2SO_4, $80°$	cis-butene-2 trans-butene-2	65
	C, LiOAc/ HOAc Pd	$CH_2=CH-OCH_3$	68
	AgO, Zn, Sodium benzoate, H_2O	Ethylene/Oxide	66
$CH_2=CH\overset{CH_3}{\underset{\vert}{C}}=CH_2$	Pt, NaSCN/ Aq. C_2H_5OH	$\overset{SCN}{\underset{\vert}{CH_2}}-CH=\overset{CH_3}{\underset{\vert}{C}}-CH_2-SCN$	70
	C, CH_3OH, NH_4NO_3	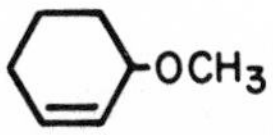	67
	Pt, CH_3CN	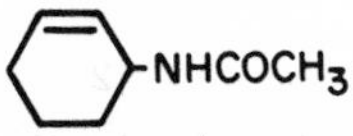	71
	Pt, CH_3CN, I_2, $LiClO_4$	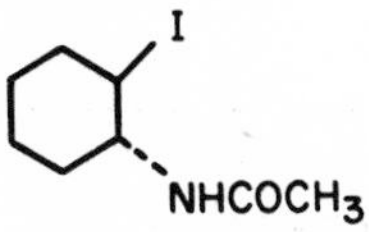	69

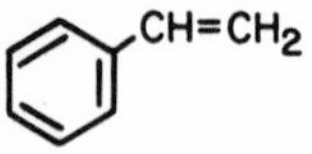

$$\text{(structure: phenyl)}-CH=CH_2 \quad \begin{array}{l} \text{Pt, } CH_3ONa/ \\ CH_3OH \end{array} \quad \begin{array}{l} \text{Styrene glycoldimethyl} \\ \text{ether} \end{array}$$

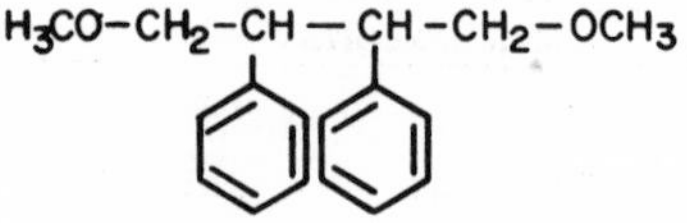

α-Methoxyethylbenzene   67

$$H_3CO-CH_2-CH-CH-CH_2-OCH_3$$

This process allows the formation of large amounts of $CaCl_2$ which has to be disposed of. The Kellog Company (72) has developed an electrochemical route for the synthesis of propylene oxide which does not form any salt by-product. A divided electrolysis cell is used and is charged, as is showm in Fig. 5.2.

The over-all reaction may be described by Eqs. (5.25)-(5.30):

Cathode

$$2H_2O + 2e \longrightarrow 2OH^{\ominus} + H_2 \qquad (5.25)$$

Anode

$$2Cl^{\ominus} - 2e \longrightarrow 2Cl^{\odot} \longrightarrow Cl_2 \qquad (5.26)$$

$$Cl_2 + H_2O \longrightarrow HOCl + H^{\oplus} + Cl^{\ominus} \qquad (5.27)$$

$$HOCl + CH_2=CHCH_3 \longrightarrow \overset{Cl}{\underset{}{CH_2}}-\overset{OH}{\underset{}{CH}}-CH_3 \qquad (5.28)$$

$$\overset{Cl}{\underset{}{CH_2}}-\overset{OH}{\underset{}{CH}}-CH_3 + NaOH \longrightarrow CH_2-CH-CH_3 + NaCl + H_2O \qquad (5.29)$$

$$CH_2=CH-CH_3 + H_2O \longrightarrow CH_2-CH-CH_3 + H_2 \qquad (5.30)$$

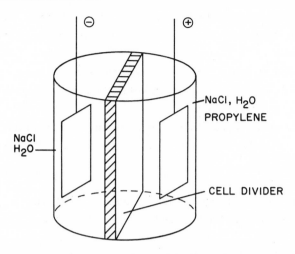

FIG. 5.2. Electrolysis cell for the preparation of
propylene oxide. [This picture of the cell is used strictly
for illustration and is not intended to represent the cell
used in Ref. 72.]

Thus, the over-all reaction [Eq. (5.30)] in the electrochemical preparation of propylene oxide involves the consumption
of propylene, water, and electricity. The preparation of
propylene oxide by the above route would be another step
forward in introducing electrolysis as a tool for the industrial preparation of organic molecules. (The first major
step was the announcement by the Monsanto Co. to prepare
adiponitrile from the electrochemical dimerization of acrylonitrile on a large scale.)

Mechanism of Reaction. Simple olefins, such as ethylene
and propylene, are difficult to oxidize on anodes such as
platinum and carbon. However, with an anode such as palladium, which may form a complex with the olefin, direct oxidation of the double bond is made easier. In general, the
products observed from the oxidation of olefins may be explained as follows:

1. $\ce{C = C}$ $-1e \longrightarrow$ $\overset{\ominus}{C} - \overset{\oplus}{C}$ $\longleftrightarrow$ $\overset{\oplus}{C} - \overset{\ominus}{C}$

[132]

The radical cation [132] may dimerize and/or react with a
variety of reagents to give the observed products listed
in Table 5.5.

2.   Abstraction of hydrogen from an allylic position
by a radical generated electrochemically from solvent or
supporting electrolyte.

3.   Chemical reaction between the olefin and a reagent
generated electrochemically (see preparation of propylene
oxide above).

For further discussion on the anodic oxidation of olefins,
Refs. 73-77 may be consulted.

5.2.4.   EXPERIMENTAL PROCEDURES

5.2.4.1.   Anodic Substitution

$$Ar-H + S^- - 2e \longrightarrow ArS + H^{\oplus}$$
$$Ar-H + SH - 2e \quad\quad ArS + 2H^{\oplus}$$

where S^- or SH are nucleophiles such as acetates, cyanides,
etc.

The electrolysis cell for anodic substitution reactions
is quite simple, and for most purposes, is a jacketed beaker
with the appropriate attachments, e.g., thermometer, stirrer,
condenser, etc.   A diagram of a general-purpose cell is shown
in Fig. 5.3.

It should be noted that the cell in Fig. 5.3 does not
include a divider.   This, of course, is possible where the
product is not reduced at the cathode.   The reader should
exercise his judgment for the use of a cell divider when
necessary.

The anode is usually made from platinum and can be in
the form of wires, gauze, or sheets.

The solvent and electrolyte will vary according to the
nature of the reaction in question.   Thus, for hydroxylation,
aqueous solutions are used, for acetoxylation, sodium ace-

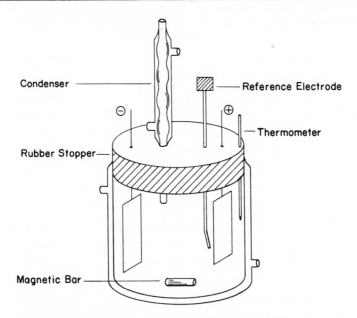

FIG. 5.3. General purpose electrolysis cell for anodic substitution.

tate in glacial acetic acid, and for methyoxylation, sodium methoxide in methanol.

   At the conclusion of the electrolysis reaction, the solvent is distilled under vacuum to prevent any decomposition of products, and the residue is extracted into ether or other suitable solvent. The ether is concentrated and the product identified by conventional techniques.

5.2.4.2. Alkoxylation Reaction (20)

$$H_3C \underset{O}{\bigcirc} CH_3 \xrightarrow[CH_3ONa/CH_3OH]{Pt} H_3CO \underset{O}{\overset{H_3C \quad CH_3}{\bigcirc}} OCH_3$$

(cis-trans-isomers)

   A jacketed beaker is used as the electrolysis cell. The beaker is equipped with a magnetic bar for stirring,

a thermometer, and a Teflon cover, which also serves as an
attachment for two platinum electrodes.

The beaker was charged with 125 ml methanol and sodium
metal, 23 g (1 g atom). To this solution there was added
22.2 g (2.31 mole) of 2,5-dimethylfuran. Electrolysis
proceeded at 2 A until a total of 43,200 C was passed
through the solution. At this time, the solution was made
strongly basic by adding a solution of sodium methoxide in
methanol. The methanol was then distilled under vacuum and
the residue extracted into 500 ml ether. The ether solution
was dried over magnesium sulfate and concentrated to 100 ml.
Analysis by vpc on a Golay column showed that the product
was a mixture of 20.5% cis- and 20.3% trans-isomers of 2,5-
dimethyl-2,5-dimethoxydihydrofuran [124].

5.2.4.3. Acetoxylation Reaction (27)

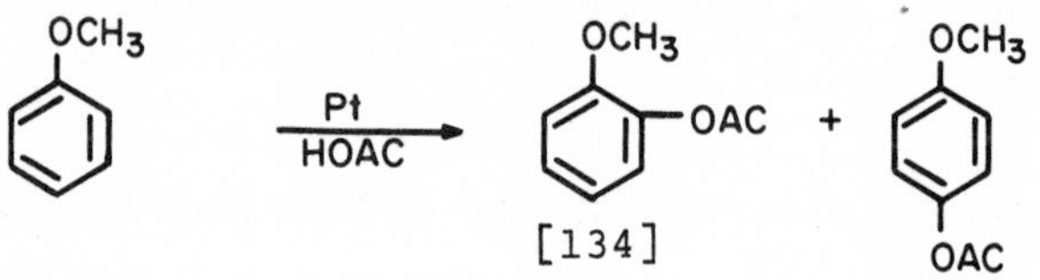

A jacketed beaker was used as the electrolysis cell
and was charged with a solution of 2.7 g (0.025 mole) anisole, 4.1 g (0.05 mole) anhydrous sodium acetate, and 150
ml glacial acetic acid. Platinum wire was used as the anode
(2 cm^2 surface area) and a platinum foil as the cathode.
A calomel electrode was used as the reference. The solution
was stirred by means of a magnetic bar.

Electrolysis proceeded at an anode potential of 1.5 V
(s.c.e.) and 0.05-0.1 A at room temperature until 50% of
the theoretical amount of coulombs was passed through the
solution. At this time, the acetic acid was removed by
means of a rotating evaporator at 40°. The residue was
treated with water and the organic material extracted into

ether, which was then washed thoroughly with sodium bicarbo-
nate and dried over magnesium sulfate. The ether was evaporated and the product distilled. The fraction bp 70° (0.5
mm) was collected. Analysis by conventional methods showed
that the product (40% yield) consisted of compounds [134]
and [135] in a 6:1 ratio.

5.2.4.4. Cyanation Reactions (35)

       Andreades and Zahnow (35) have described in detail the
construction of an electrolysis cell for cyanation reactions.
This cell is basically the same as the one described at the
beginning of this section, but has some good modifications.
The reader may well want to consider such a design.

[136]   [137]

       A solution of 0.5 g (3 mmole) anthrathene [136] and
6.0 g (35.5 mmole) of tetraethylammonium cyanide (see Ref.
35 for preparation] in 300 ml acetonitrile and 100 ml ether
was introduced into the cell. The anode and cathode were
made of platinum. A calomel electrode was used as the reference.

       Electrolysis proceeded at an anode potential of 2.0 V
(s.c.e.) for 36 min. This allowed the passage of 0.027
Faraday. At this time, the reaction product was concentrated by evaporating the ether and half the amount of acetonitrile. The mixture was diluted with water and filtered
to give 0.37 g of crude 9,10-dicyanoanthracene [137], which
can be purified by conventional techniques.

5.2.4.5. Anodic Acetamidation (32)

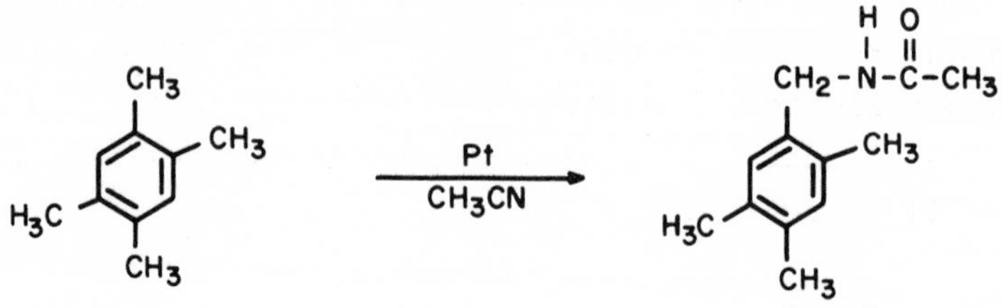

[139]

The electrolysis cell used for this reaction is similar to those described for the anodic substitutions reactions above. Platinum was used for both electrodes.

The anode compartment was charged with 150 ml acetonitrile, 7.1 g (0.05 mole) anhydrous sodium perchlorate, and 3.35 g (0.025 mole) 1,2,4,5-tetramethylbenzene. The cathode compartment was charged with the same solution, but without the tetramethylbenzene. Electrolysis was carried at an anode potential of 1 V (s.c.e.), which allowed the passage of 0.5 A through the solution. After 9 hr the acetonitrile was removed under vacuum and the residue was washed with water and extracted into ether. The ether was dried over magnesium sulfate and evaporated. Elution with a mixture of chloroform and ether gave 2.1 g solid mp 141.5-142.5° which was identified as 2,4,5-trimethylbenzyl-acetamide [139].

5.2.4.6. Oxidation of Aromatic Alcohols and Amines

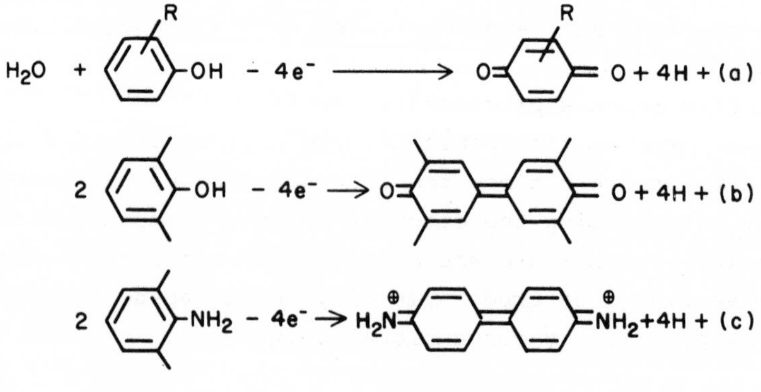

In general, the same procedures and considerations can be used for anodic substitution reactions, except for the following:

1. In reaction (a), to generate the quinone of the phenol, low concentrations of starting material and high concentration of water are desirable, to minimize coupling products. Also the use of PbO_2 anodes (in the presence of sulfuric acid) is recommended.

2. For reactions (b) and (c), high concentrations of starting material in acidic media are recommended. Also, note that in general the quinones and quinone imines are relatively unstable and should be handled with appropriate care.

3. The quinone-like products in all cases above are very susceptible to reduction to the corresponding alcohols or amines. These products can be expected to form in an undivided cell.

5.2.4.7. Oxidation of Phenol (38)

The electrolysis cell may be simply a beaker using lead electrodes (nondivided). An aqueous solution containing 3% sulfuric acid and 3% phenol is used. During electrolysis PbO_2 is formed on the anode. (More reproducible results will be obtained if the anode is preanodized in dilute sulfuric acid.) The electrolysis is conducted at 20 A/dm^2 until 4 Faradays/mole phenol have been consumed. This results in about 50% conversion to hydroquinone. The reaction temperature is kept between 40 and 60° to prevent precipitation of quinhydrone complexes and to avoid side reactions, respectively. At the end of the electrolysis, sufficient sodium sulfite is added to reduce residual benzoquinone.

A convenient work-up in the laboratory involves extraction with ether for ∿3 hr to remove the organics; the ether is evaporated to near dryness, and cold carbon tetrachloride is added. The hydroquinone precipitates as a tan, amorphous solid; yield, after drying, is ∿90% based on reacted phenol.

The hydroquinone is further purified by recrystalization from water.

5.2.4.8. Oxidation of Aromatic Amines (57)

The product(s) from the oxidation of aromatic amines may be susceptible to cathodic reduction. Consequently, it is recommended that a divided cell be used for the oxidation of these compounds. For most reactions, an H-type cell with a sintered glass disk as the divider, or a beaker into which a Coors porous cup is suspended (cathode compartment), can be used.

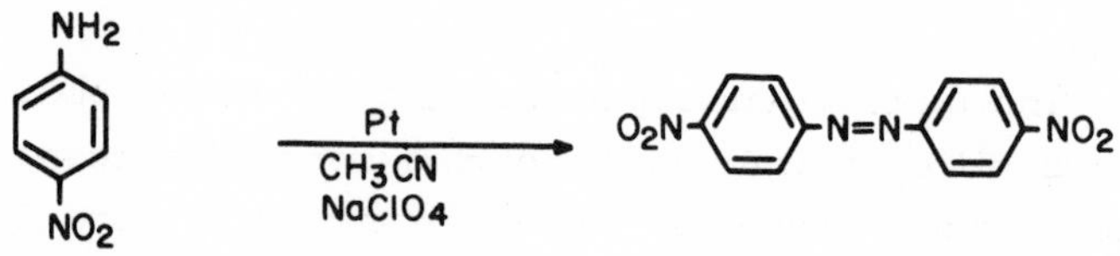

An H-cell with a sintered glass divider was equipped with a rotating cylindrical platinum gauze anode and a mercury pool cathode. A solution of 1 g (7 mmole) of p-nitro-aniline in 90 ml acetonitrile (0.5 M pyridine, 0.5 M sodium perchlorate) was added to the anode compartment. The cathode compartment was charged with the same solution but without the aniline. [The nature of the cathode solution was not mentioned in Ref. 57; however, the recommended solution should be suitable.]

The anode solution was degassed prior to and during electrolysis, which proceeded at 10 V and 0.3 A (decreased to 0.05 A) for 4 hr. The anolyte was concentrated to 20 ml and was hydrolyzed with 30 ml water. The precipitated product was collected and extracted into 100 ml acetone. Evaporation of the acetone gave a residue which was chromatographed on alumina using benzene as the solvent. The first fraction gave reddish crystals (0.26 g, 39.2%) mp 220-222°. The infrared spectrum of this compound was identical to that of an authentic sample of 4,4'-dinitroazobenzene.

5.2.4.9. Oxidation of Olefins

The general procedure for the oxidation of olefins is similar to that described for aromatic hydrocarbons. Olefins that boil below room temperature may need special handling. The reader may consult Refs. 63, 66, and 68 on the oxidation of ethylene.

REFERENCES ON OXIDATION OF UNSATURATED COMPOUNDS

1. H. Inoue and M. Shikata, Japan Ind. Chem. Soc., 24, 567 (1921); C. A., 16, 1046 (1922).

2. L. Miller, Tetrahedron Letters, 1968 1831.

3. J. D. Ross, M. Finkelstein, and R. Peterson, J. Am. Chem. Soc., 86, 4139 (1964).

4. A. Atansiu and C. Belcot, C. A., 32, 7830 (1938).

5. K. Koyama, T. Susuki, and S. Tsutsumi, Tetrahedron, 23, 2675 (1967).

6. K. Koyama, T. Susuki, and S. Tsutsumi, Tetrahedron Letters, 1965 627.

7. C. K. Mann and K. K. Barnes, Electrochemical Reactions in Nonaqueous Systems, Dekker, New York, 1969

8. R. Sasaki and W. J. Newby, J. Electroanal. Chem., 20, 137 (1969).

9. M. E. Peover, Electroanalytical Chemistry (A. J. Bard, ed.), Vol. 2, Dekker, New York, 1967.

10. N. L. Weinberg and H. R. Weinberg, Chem. Rev., 68, 449 (1968).

11. F. Fichter and R. Stocker, Chem. Ber., 47, 2003 (1914).

12. M. D. Law and F. M. Perkin, Trans. Faraday Soc., 1, 251 (1905).

13. B. E. Conway, N. Marincic, D. Gilroy, and E. Rudd, J. Electrochem. Soc., 113, 1144 (1966).

14. M. Fleishmann, J. Manfield, and W. Wynne-Jones, J. Electroanal. Chem., 10, 511 (1965).

15. L. Eberson, J. Am. Chem. Soc., 89, 4669 (1967).

16. V. D. Parker, Chem. Comm., 1164 (1968).

17. N. Clauson-Kass, F. Limborg, and K. Glens, Acta Chem. Scand., 6, 521 (1952).

18. N. Clauson-Kass, F. Limborg, and J. Fakstrop, Acta Chem. Scand., 2, 109 (1948).

19. N. Elming, Advances in Organic Chemistry, (R. A. Raphael, E. C. Taylor, and H. Wynberg, eds.), Interscience, New York, 1960, p. 67.

20. S. D. Ross, M. Finkelstein, and J. J. Uebel, J. Org. Chem., 34, 1018 (1969).

21. B. Belleau and N. L. Weinberg, J. Am. Chem. Soc., 85, 2525 (1963).

22. T. Inoue, K. Koyama, and S. Tsutsuni, Bull. Chem. Soc. Japan, 37, 1597 (1964).

23. L. Herk, M. Feld, and M. Szwarc, J. Am. Chem. Soc., 83, 2998 (1961).

24. H. H. Williams, Homolytic Aromatic Substitution, Pergamon, London, 1960, pp. 116-118.

25. L. Eberson, Chemistry of the Carboxyl Group, (S. Patai, ed.), Interscience, London, 1970.

26. L. Eberson, J. Am. Chem. Soc., 89, 4669 (1967).

27. L. Eberson and K. Nyberg, J. Am. Chem. Soc., 88, 1686 (1966).

28. J. F. O'Donnell and C. K. Mann, J. Electroanal. Chem., 13, 157 (1967).

29. L. Rand and A. F. Mohar, J. Org. Chem., 30, 3156 (1965).

30. S. D. Ross, M. Finkelstein, and R. C. Peterson, J. Am. Chem. Soc., 89, 4088 (1967).

31. L. Eberson and K. Nyberg, Acta Chem. Scand., 18, 1567 (1964).

32. L. Eberson and K. Nyberg, Tetrahedron Letters, 1966 2389.

33. L. Eberson and S. Nilsson, Discussions Faraday Soc., 242 (1968).

34. V. D. Parker and B. E. Burgert, Tetrahedron Letters, 1965 4065.

35. S. Andreades and E. Zahnow, J. Am. Chem. Soc., 91, 4181 (1969).

36. V. D. Parker and B. E. Burgert, Tetrahedron Letters, 1968 2411.

37. F. Fichter and E. Bruner, Bull. Soc. Chim. France, 19, 281 (1916).

38. F. H. Covitz, French Pat. 1,544,350 (1968).

39. A. Scott, P. A. Dodson, F. McCapra, and M. B. Meyers, J. Am. Chem. Soc., 85, 3702 (1963).

40. D. S. McKinney and J. P. Fugassi, U. S. Pat. 2,961,384 (1960).

41. R. S. Gregor and R. W. Bush, U. S. Pat. 3,477,924 (1969).

42. M. R. Rifi, unpublished results.

43. D. Hawley and R. N. Adams, J. Electroanal. Chem., 8, 163 (1964).

44. L. Papouchado, J. Bacon, and R. Adams, J. Electroanal. Chem., 24, Appendix 1 (1970).

45. K. Sasaki, K. Takehira, and H. Shiba, Electrochim. Acta, 13, 1623 (1968).

46. F. Fichter and W. Dietrich, Helv. Chim. Acta, 7, 131 (1924).

47. A. G. Perkin and F. M. Perkin, J. Chem. Soc., 93, 1186 (1908).

48. K. M. Johnston, Tetrahedron Letters, 1967 837.

49. S. Wawzonek and T. W. McIntyre, J. Electrochem. Soc., 114, 1025 (1967).

50. S. Andreas and E. Zahnow, J. Am. Chem. Soc., 91, 4181 (1969).

51. N. N. Mel'nikov, S. I. Sklyarenko, and E. Cherkasova, J. Gen. Chem. USSR, 9, 1819 (1939).

52. M. R. Rifi, Tetrahedron Letters, 1969 5089.

53. D. M. Mohilner, R. N. Adams, and W. J. Argersinger, Jr., J. Am. Chem. Soc., 84, 3618 (1962).

54. Z. Galus and R. Adams, J. Am. Chem. Soc., 84, 2061 (1962).

55. Z. Galus, R. M. White, F. S. Rowland, and R. N. Adams, J. Am. Chem. Soc., 84, 2065 (1962).

56. S. Goldschmidt and F. Nagel, Chem. Ber., 64, 1744 (1931).

57. S. Wawzonek, T. Plaisance, and T. McIntyre, J. Electro-Chem. Soc., 114, 588 (1967).

58. Z. Galus and R. Adams, J. Phys. Chem., 67, 862 (1963).

59. G. Cauquis and J. Billon, Compt. Rend., 255, 2128 (1962).

60. J. E. Dubois, P. C. Lacaze, and A. Aranada, Compt. Rend., 260, 3383 (1965).

61. G. Cauquis, J. Badoz-Lambling, and J. P. Billon, Bull. Soc. Chim. France, 1433 (1965).

62. E. T. Seo, R. Nelson, J. Fritsch, L. Marcoux, D. Leedy, and R. Adams, J. Am. Chem. Soc., 88, 3498 (1966).

63. E. W. Hultman, U. S. Pat. 1,992,309 (1935).

64. M. A. Kalinin and V. V. Stender, J. Applied Chem. USSR, 19, 1045 (1946).

65. G. O. Curme, Jr., U. S. Pat. 1,315,543 (1919).

66. J. A. LeDuc, U. S. Pat. 3,427,235 (1969).

67. T. Shono and T. Kosaka, Tetrahedron Letters, 1968 6205.

68. T. Inoue and S. Tsutsumi, *Bull. Chem. Soc. Japan*, **38**, 661 (1965).

69. N. Weinberg and K. Hoffman, *Can. J. Chem.*, **49**(5) 740 (1971).

70. G. Smith, U. S. Pat. 3,472,747 (1969).

71. G. Faita, M. Fleischmann, and D. Pletcher, *J. Electro-anal. Chem.*, **25**, 455 (1970).

72. M. W. Kellog Co., Belg. Pat. 637,691 (1962).

73. D. H. Geske, *J. Am. Chem. Soc.*, **81**, 4145 (1959).

74. V. D. Parker, K. Nyberg, and L. Eberson, *J. Electroanl. Chem.*, **22**, 150 (1969.

75. A. R. Blake, J. G. Sunderland, and A. T. Kuhn, *J. Chem. Soc. (A)*, 3015 (1969).

76. H. Schafer, *Chem. Ing. Tech.*, **42**, 164 (1970).

77. A. J. Baggaley and R. Brettle, *J. Chem. Soc. (C)*, 2066 (1968).

5.3. ANODIC HALOGENATION

5.3.1. GENERAL CONSIDERATIONS

When a solution containing a halide salt, an organic compound, and a solvent is oxidized at the appropriate anode, the organic substrate may be halogenated. This reaction, known as anodic halogenation, is, in general, similar to chemical halogenation reactions which involve the addition of X_2 (Cl_2, Br_2, I_2, F_2) to organic substrates under the appropriate conditions, i.e., at elevated temperatures or under the influence of light. Thus, anodic halogenation could, in principle, be simpler, easier, and perhaps more economical to use than chemical halogenation reactions which require the handling of dangerous chemicals (Cl_2 vs NaCl). Anodic fluorination, on the other hand, is run in liquid hydrogen fluoride (HF), which cannot be handled in simple glass laboratory equipment. It can, however, be used in plastic containers such as polyethylene, teflon polypropylene, Kel-F, etc. For a review of handling HF,

the reader should consult Refs. 1, 46. Hydrogen fluoride is
an excellent solvent for electrolysis. It has a high dielectric constant, bp ∿20°C, and dissolves a good many
organic compounds. However, its special handling has
limited its use as a solvent in electroorganic synthesis.

5.3.2. EFFECT OF REACTION CONDITIONS ON
COURSE OF REACTION

Nature of Anode. The choice of anode materials may
be quite critical in anodic halogenation, since elemental
halogens, e.g., bromine, react with such materials as platinum. Thus, while the most inert anode is carbon, platinum
containing traces of irridium or rhodium can be used. In
fluorination reactions the common anode used is nickel.
Platinum has been used; however, carbon anodes are not
suitable in molten inorganic fluoride, since they react
to form fluorinated products.

The anodic oxidation of toluene in a hydrochloric acid
suspension on a platinum anode at room temperature yields a
mixture of 70% o-chlorotoluene and 30% p-chlorotoluene (9).
However, when the reaction is carried out at reflux temperature using a carbon anode, the major product is the p-isomer.
The reader is well aware that the chemical chlorination of
toluene leads to benzylchloride.

Effect of Solvents. As mentioned above, hydrogen
fluoride is an excellent solvent in electrolysis; however,
extreme care should be exercised in its handling (37).
Other solvents, such as acetonitrile, can be used. Miller
(2) has described the use of this solvent and reported that
in iodination reaction it participates in the reaction to
form $CH_3-\overset{\oplus}{C}=NI$, which serves as an iodination reagent for
aromatic compounds.

pH of Medium. A control of the pH of the medium is
quite important in anodic halogenation. Thus, in basic

solutions, halogen molecules react with the hydroxyl anions
to form hypohalogen acid:

$$X_2 + {}^{\ominus}OH \longrightarrow HOX + X^{\ominus}$$

The consequence of this reaction was observed in the
chlorination of acetone. In dilute hydrochloric acid medium
monochloroacetone is formed, while under basic conditions
the "haloform" reaction takes place to give chloroform and
an acetate ion:

$$CH_3-\overset{\overset{\displaystyle O}{\|}}{C}-CH_3 + 3NaOCl \longrightarrow CHCl_3 + CH_3CO_2Na$$

<u>Temperatures</u>. In certain cases elevated temperatures
assist the halogenation of compounds that do not undergo
such a reaction at room temperature. For a discussion on
the effect of temperature on anodic substitution, the reader
may want to see Ref. <u>3</u>.

5.3.3. MECHANISM OF ANODIC HALOGENATION

The mechanism of anodic halogenation may involve one
or more reaction paths. Consider, for example, the halogenation of substrate A in a solvent S and a sodium halide as
the supporting electrolyte, Eqs. (5.31)-(5.35):

$$A -1e \longrightarrow A^{\oplus} \tag{5.31}$$

$$A^{\oplus} + X^{\ominus} \longrightarrow A-X \tag{5.32}$$

$$X^{\ominus} -1e \longrightarrow X^{\odot} \tag{5.33}$$

$$X^{\odot} + A \longrightarrow AX \tag{5.34}$$

$$2X^{\odot} \longrightarrow X_2 \xrightarrow{AH} AX + HX \tag{5.35}$$

Since the oxidation potentials of organic compounds are, in general, higher than the potential necessary to convert $X^{\ominus} \longrightarrow X^{\cdot} \longrightarrow \frac{1}{2} X_2$, Eq. (5.31) appears to be unlikely. However, Millington has recently found (4) that in the anodic bromination of anthracene in acetonitrile, the reaction was very slow when carried out at the oxidation potential of the bromide anion, but was substantially increased when carried out at a potential more anodic than the oxidation potential of anthracene.

5.3.4. ANODIC CHLORINATION

This reaction involves, in essence, the reaction of a chlorine molecule, or the product from a chlorine molecule and solvent, with an organic compound. Consider, for example, the preparation of chloroform from the oxidation of ethanol, Eq. (5.39), or acetone, Eq. (5.40), in an aqueous solution of chloride salt (5) and the preparation of

$$Cl^{\ominus} \ -1e \ \longrightarrow \ Cl^{\cdot} \qquad\qquad (5.36)$$

$$2Cl^{\cdot} \ \longrightarrow \ Cl_2 \qquad\qquad (5.37)$$

$$Cl_2 + H_2O \ \longrightarrow \ HOCl + HCl \qquad\qquad (5.38$$

$$CH_3CH_2OH + 5HOCl \ \longrightarrow \ CHCl_3 + 4H_2O$$

$$+ \ CO_2 + 2HCl \qquad\qquad (5.39)$$

$$CH_3 COCH_3 + 3 \ HOCl \ \longrightarrow \ CHCl_3$$

$$+ \ CH_3CO_2H + 2H_2O \qquad\qquad (5.40)$$

chlorohydrin compound from the oxidation of a solution containing aqueous chloride salt and an olefin, Ref. 6, Eq. (5.41):

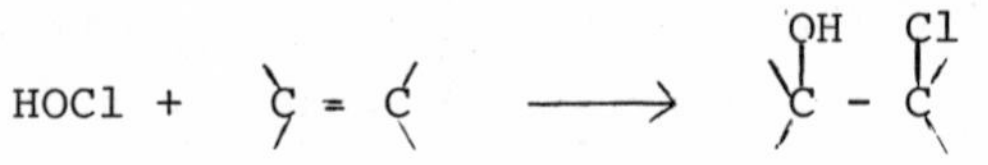

$$HOCl + \overset{/}{\underset{/}{C}} = \overset{\backslash}{\underset{\backslash}{C}} \quad \longrightarrow \quad \overset{OH}{\underset{/}{C}} - \overset{Cl}{\underset{\backslash}{C}} \tag{5.41}$$

The oxidation of benzene as a suspension in aqueous hydrochloric acid afforded chlorobenzene in 75% current efficiency (7). The efficiency was later improved (89%) by modifying the reaction conditions, such as the use of porous carbon anode and a medium consisting of benzene in hydrochloric acid-monochloroacetic acid solution, or using a temperature of $38°C$, and a current density of 4.3 A/dm^2 (8). For further information on the chlorination of aromatic compounds the reader should consult Refs. 10-12.

The anodic chlorination of ketones was studied by Szper (13), who reported the following reactions:

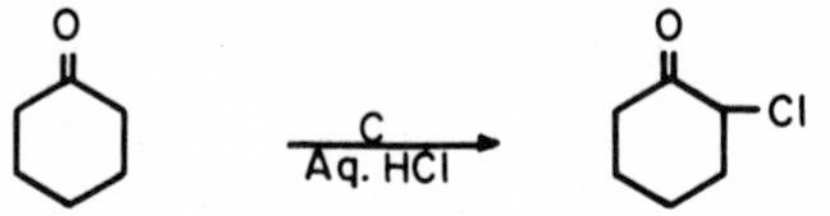

$$\underset{\text{Aq. HCl}}{\xrightarrow{C}} \tag{5.42}$$

$$\underset{\text{Aq. HCl, HOAc}}{\xrightarrow{C}} \tag{5.43}$$

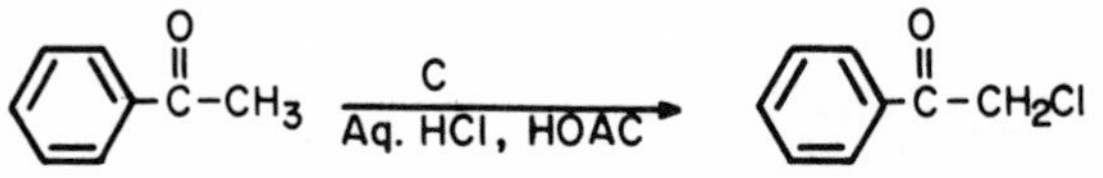

Methylamine can be converted to N-chloroamine in about 61% from the anodic oxidation of an aqueous potassium chloride (14). Aniline, on the other hand, can give a variety of products when oxidized in aqueous acid, Ref. 15, Eqs. (5.44)-(5.46):

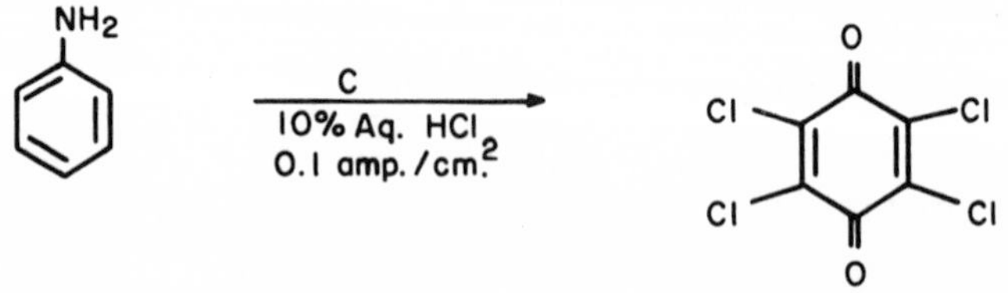

$$\underset{\substack{\text{10% Aq. HCl} \\ \text{0.1 amp./cm.}^2}}{\xrightarrow{C}} \tag{5.44}$$

$$\underset{\substack{\text{20% Aq. HCl} \\ \text{0.1 amp./cm.}^2}}{\xrightarrow{C}} \tag{5.45}$$

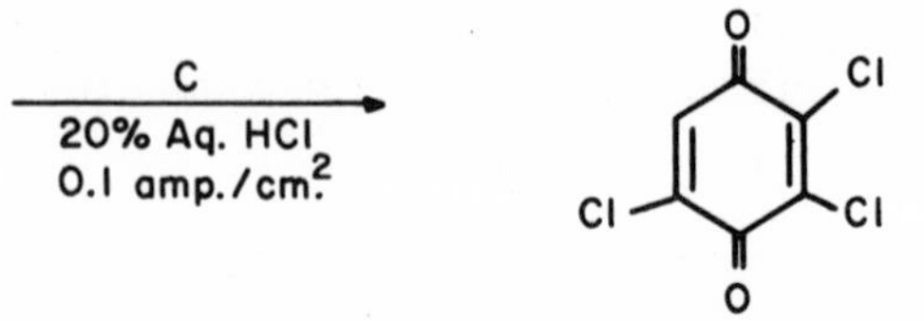

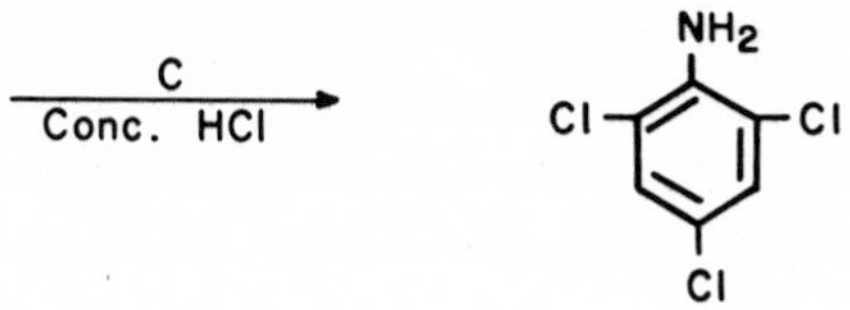

$$\xrightarrow[\text{Conc. HCl}]{\text{C}}$$

(5.46)

5.3.5. ANODIC BROMINATION

Like anodic chlorination, anodic bromination was studied with aromatic as well as aliphatic compounds. However, because of the ease of oxidation of bromide ion to bromine, the bromination of some aliphatic and aromatic compounds was not possible. Thus, while chloroform can be prepared from the oxidation of aqueous chloride salt in ethanol, it was not possible to prepare bromoform in a similar reaction (16). On the other hand, when a solution of acetone, potassium bromide, and water is oxidized on a platinum anode, a quantitative yield of bromoform was obtained (17).

In the anodic bromination reaction of aromatic molecules, it is questionable whether the bromination part is really electrochemical or chemical in nature. Thus, when toluene is electrolyzed in a hydrobromic acid medium on platinum or graphite, bromotoluene is obtained, but only if the reaction is carried out in the dark. If the reaction is exposed to light, the final product is benzyl bromide (18). For further discussion on the anodic bromination of aromatic compounds, e.g., benzene, phenols, and anilines, the reader should consult Refs. 19-26.

5.3.6. ANODIC IODINATION

The electrochemical preparation of iodoform is an excellent example which illustrates the advantage of this process over the chemical methods. Consider, for example, the preparation of iodoform from the reaction of iodine with ethanol, Eq. (5.47):

$$CH_3CH_2OH + 5I_2 + H_2O \longrightarrow CHI_3 + 7HI + CO_2 \quad (5.47)$$

It can be seen that a considerable amount of iodine (in the iodide form) is not consumed and would have to be reoxidized to be used in the above chemical reaction. Electrochemically, the regeneration of iodine is done in situ at the anode, and the over-all reaction, which involves the electrolysis of ethanol in aqueous potassium iodide, may be described as follows, Ref. 26, Eq. (5.48):

$$CH_3CH_2OH + 10I^- + H_2O \longrightarrow CHI_3 + 7HI$$

$$+ CO_2 + 10e \quad (5.48)$$

In Eq. (5.48). the iodide ion is converted to iodine, which reacts with the alcohol to afford iodoform, Eq. (5.49):

$$CH_3CH_2OH + 5I_2 + H_2O \longrightarrow CHI_3 + CO_2 + 7HI \quad (5.49)$$

The hydrogen iodide from the chemical reaction reacts with potassium hydroxide (formed at the cathode) to give potassium iodide. The iodide is reoxidized at the anode to give free iodine, which reacts with more alcohol to afford iodoform, thus making the reaction somewhat continuous until all the iodide is consumed.

The anodic iodination of aromatic compounds in acetonitrile solutions has been recently studied by Miller (2). Thus, ring substitution occurred with benzene, toluene, anisole, and triphenylmethane. No iodination, however, was obtained with nitrobenzene or anthracene. The lack of iodination of the latter compound was explained as due to its lower oxidation potential than that of iodine. (This seems a bit surprising.) The mechanism for the iodination of aromatic compounds was explained by Eqs. (5.50) and (5.51):

$$I_2 - 2e \longrightarrow 2I^{\oplus} \quad (5.50)$$

$$I^{\oplus} + \left[\bigcirc\right]\!\!-R \longrightarrow R\!\!-\!\!\left[\bigcirc\right]^{\oplus} \overset{-H^{\oplus}}{\longrightarrow} R\!\!-\!\!\left[\bigcirc\right] \qquad (5.51)$$

5.3.7. ANODIC FLUORINATION

The simplicity, as well as the diversity, of this reaction have made it one of the most widely studied and applied in organic, as well as inorganic, electrochemistry. For example, anodic fluorination is probably one of the mildest methods for the introduction of fluorine atoms into organic compounds. Chemical fluorination with F_2, on the other hand, can be quite hazardous, since it is susceptible to explosions, unless the reaction is diluted with appropriate solvents. The suitable use of liquid hydrogen fluoride as a solvent has greatly simplified anodic fluorination reactions. In practice, the electrochemical process consists of passing an electric current through a one-compartment cell containing liquid hydrogen fluoride together with the substrate to be fluorinated. Since HF attacks glass, the cell is usually constructed from iron or copper, which are inert to this solvent. The over-all voltage applied to the cell is generally in the order of 5-6 V and the current density about 2 A/dm^2. [While this process may sound too simple, the experimentalist at heart should be well acquainted with the hazardous properties of hydrogen fluoride (37).] Common anode materials used are nickel or some alloy forms of nickel such as Monel (33-35). Table 5.6 describes several types of anodic fluorination reactions.

The mechanism of anodic fluorination has been the subject of much controversy. The main reason for this controversy may be attributed to the lack of experiments carried out under controlled potential, as well as the lack of kinetic data on the reaction. For a comprehensive study of the mechanism of anodic fluorination the reader should consult Refs. 38, 45, and 46.

TABLE 5.6

Examples of Anodic Fluorination

Substrate	Reaction Conditions	Products	Yield %	References
CH_4	Ni, HF, NaF	CF_4	7.6	
		CF_3H	3.5	
		CF_2H_2	18.4	35
CH_3-CH_3	Ni, HF, NaF	F_3C-CF_3	28.6	
		F_3C-CHF_2	7.0	
		F_3C-CH_2F	6.6	
		F_3C-CH_3	2.2	
		$F_2CH-CHF_2$	9.7	36, 37
$CH_2=CH_2$	Ni, HF, NaF	F_3C-CF_3	36.0	
		$F_3-C-CHF_2$	13.3	
		F_3C-CH_2F	17.3	
		F_3C-CH_3	4.8	
		$F_2CH-CHF_2$	11.2	
Cyclo-propane	Ni HF, KF	1,2,3-Trifluoro-propane	--	38
		1,2-Difluorocyclo-propane	--	
		1,3-Difluoropropane	--	
		Monofluorocyclo-propane	--	
	Pt, AgF, CH_3CN	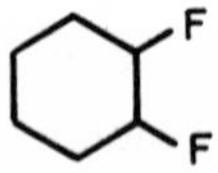	--	38
	Ni Aq. HF	Polymer	--	39

CH_3OH	Ni HF	CF_4	9.6	
		CF_3H	6.4	42
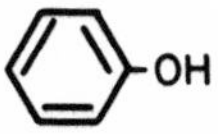 —OH	Ni HF	C_5F_{12} – C_6F_{12}	--	42, 43
R–S–H	Ni HF	RF	--	42
H_3C–O–CH_3	Ni HF	F_3C–O–CF_3	--	40
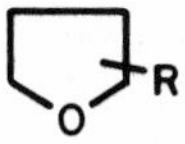	Ni	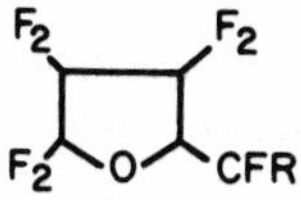	--	41
R_3N	Ni HF	$(FR)_3N$	--	42
R–$\overset{\text{O}}{\overset{\|}{C}}$R'	Ni HF	F_3C–$CO_2^{\ominus}$	--	43
R–$\overset{\text{O}}{\overset{\|}{C}}$OH(Cl)	Ni HF	F_3C–$CO_2^{\ominus}$, F_3C–$\overset{\text{O}}{\overset{\|}{C}}F$, CF_4	--	42
R–$\overset{\text{O}}{\overset{\|}{C}}$–OR'		F_3C–$\overset{\text{O}}{\overset{\|}{C}}O^{\ominus}$	--	34
R–$\overset{\text{O}}{\overset{\|}{C}}$–$N(CH_3)_2$	Ni HF	F–$\overset{\text{O}}{\overset{\|}{C}}$–$CF_3$	--	44

5.3.8. EXPERIMENTAL PROCEDURES

5.3.8.1. Anodic Halogenation

$$2X^- - 2e^- \longrightarrow X_2$$

X_2 + substrate $\longrightarrow$ halogenated product

In these reactions, the electrochemical aspects deal almost exclusively with the generation of the free halogen. Consequently, the apparatus and techniques should be designed accordingly. For example, chlorine and bromine are quite corrosive and easily reduced. Divided cells must be utilized. The most preferred anode materials are carbon or graphite; in general, platinum is not recommended, since it may itself be attacked by the free halogen. Alkali halides or other highly soluble halide salts are used both as the supporting electrolytes and the source of halogen.

Special considerations are required for anodic fluorinations. Glass apparatus cannot be used, and the vessel itself is preferably made of nickel or nickel alloys. In that case, the vessel serves as the anode. In no case should the experimenter work at high concentrations of free fluorine, since this may result in explosive reaction. Liquid HF is normally used as both solvent and source of fluorine. Sodium or potassium fluoride are generally employed as electrolytes.

5.3.8.2. Anodic Bromination (4)

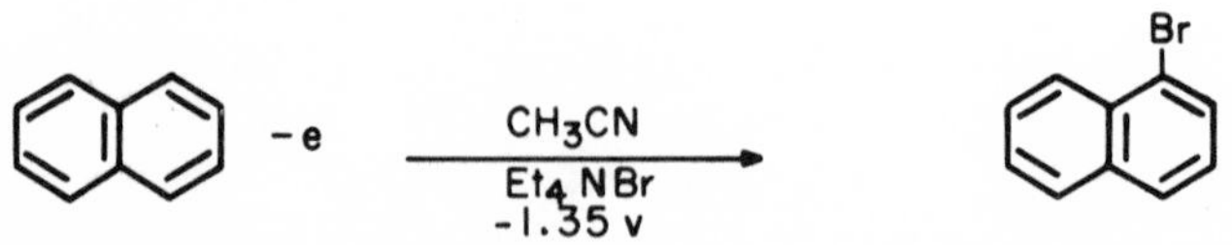

The electrolysis cell may be an H-type cell with the two compartments separated by means of a fritted disk. The electrodes are usually made from platinum and can be in the form of wire or foil.   Ag/AgClO$_4$ is used as the reference electrode.

The anode compartment is charged with 1 g (0.0078 mole) naphthalene in about 50 ml dry acetonitrile, which is 0.5 M in tetraethylammonium bromide.   [In order to dry the aceto- nitrile, the reader should consult Ref. 47.]   The cathode compartment was charged with the same solution, but without naphthalene.   Electrolysis was conducted at an anode poten- tial of +1.35 V (Ag/AgClO$_4$) and was allowed to run until 54.5% of the theoretical amount of coulombs was passed through the cell.   At this time, the anode solution was poured onto a large volume of water and extracted exhaus- tively with benzene.   The combined benzene extracts were washed successively with dilute sodium thiosulfate (in or- der to remove bromine) and dilute sodium hydrogen carbonate (to remove acid).   The final solution was then dried over sodium sulfate, and benzene was evaporated under vacuum at room temperature.   The residue was identified by mass spec- tral analysis and vpc (silicone gum rubber 1.5% on Chromo- sorb W↓)

With the above procedure 1-bromonaphthalene was formed in 38% yield.

Experimental procedures for anodic chlorination and iodination are quite analogous to those mentioned in the above example.   For specific reaction conditions the reader can refer to the appropriate references cited in this sec- tion.

REFERENCES ON ANODIC HALOGENATION

1. M. Kilpatrick and J. G. Jones, Chemistry of Nonaqueous Solvents (J. J. Lagowski, ed.), Vol. 2, Academic, New York, 1947.

2. L. L. Miller, E. P. Kujawa, and C. B. Campbell, J. Am. Chem. Soc., 92, 2821 (1970).

3.   H. M. Fox and F. N. Ruehlen, U. S. Pat. 3,393,249 (1968).

4.   J. D. Millington, J. Chem. Soc. (B), 1969 982.

5.   J. Feyer, Z. Elektrochem. Soc., 8, 283 (1905).

6.   C. K. Bhattacharyyar, M. S. Muthanna, and R. B. Patankar, J. Sci. Ind. Res. (India), (B), 365 (1952).

7.   A. Lowy and H. S. Frank, Trans. Am. Electrochem. Soc., 43, 107 (1923); C. W. Croco and A. Lowy, ibid., 50, 315 (1926).

8.   J. C. Ghosh, C. K. Bhattacharyya, R. A. Rao, M. S. Muthana, and R. B. Patnaik, J. Sci. Ind. Res. (India), 113, 361 (1952).

9.   J. B. Cohen, H. M. Dawson, and P. F. Crosland, J. Chem. Soc., 87, 1034 (1905).

10. F. Fichter and L. Glanzstein, Chem. Ber., 49, 2473 (1916).

11. F. Cutz and Z. Kucera, Chem. Listy, 47, 1166 (1953); C. A., 48, 3850 (1954).

12. I. M. Kolthoff and J. Jordan, J. Am. Chem. Soc., 75, 1571 (1953).

13. J. Szper, Bull. Soc. Chim. France, 51, 653 (1932).

14. E. J. Matzner, U. S. Pat. 3,449,225 (1969).

15. J. Erdelyi, Chem. Ber., 63, 1200 (1930).

16. J. Elbs and A. Herz, Z. Elektrochem., 4, 113 (1897).

17. P. Coughlin, J. Am. Chem. Soc., 27, 63-68 (1902).

18. L. Bruner and S. Czarnecki, Bull. Acad. Sci. Cracow, 322 (1909); B.C.A., A-1, 900 (1909).

19. C. W. Croco and A. Lowy, Trans. Am. Elect. Soc., 50, 315 (1926).

20. N. A. Isgarysev and W. S. Palikarpov, Chem. Zentr., 885 (1941).

21. G. Bionda and M. Civera, Ann. Chim. (Rome), 41, 814 (1951); C. A., 46, 7908 (1952).

22. L. Gilchrist, J. Phys. Chem., 8, 539 (1904).

23. R. Landsberg, H. Lohse, and N. Lohse, J. Prakt. Chem., 12, 253 (1961).

24. G. S. Kozak, Q. Fernando, and H. Freiser, Anal. Chem., 36, 296 (1964).

25. M. Lamchen, J. Chem. Soc., 1950 747.

26. M. J. Allen, Organic Electro de Processes, Reinhold, New York, 1958, p. 155.

27. L. Miller, Tetrahedron Letters, 1968 1831.

28.   E. Muller and R. Loebe, Z. Elektrochem., 10, 409 (1904).

29.   O. W. Brown and B. Berkowitz, Trans. Electrochem. Soc.,
      75, 385 (1939).

30.   A. Classen, U. S. Pat. 618,168 (1908).

31.   J. W. Shipley and M. T. Rogers, Can. J. Res., 17B, 147
      (1939).

32.   N. Nagase and H. Baba, Bull. Chem. Soc. Japan, 13, 1,
      20 (1964).

33.   S. Nagase and R. Kojima, Bull. Chem. Soc. Japan, 34,
      1468 (1961).

34.   S. Nagase, K. Tanaka, and H. Baba, Bull. Chem. Soc.
      Japan, 38, 834 (1965).

35.   P. Sartari, Angew. Chem. (Intern. Ed.), 2, 261 (1963).

36.   S. Nagase, K. Tanaka, and H. Baba, Bull. Chem. Soc.
      Japan, 39, 219 (1966).

37.   I. L. Knunyants, I. N. Rozhkov, A. V. Bukhtiarov, M. M.
      Goldin, and R. V. Kudryavtsev, Izv. Akad. Nauk SSSR,
      Ser. Khim., 1207 (1970); C. A., 73, 65752 (1970).

38.   A. F. Shepard and B. F. Dannels, U. S. Pat. 3,386,899
      (1968).

39.   J. H. Simons, U. S. Pat. 2,500,388 (1950).

40.   I. P. Kolenko, N. A. Ryabinin, and B. N. Lundin, Akad.
      Nauk SSSR, Ural'skii Filial Trudy Inst. Khim., 15, 59
      (1968); C. A., 70, 87422 (1969).

41.   J. H. Simons, U. S. Pat. 2,519,983 (1950).

42.   J. H. Simons, W. H. Pearlson, T. J. Brice, W. A. Wilson,
      and R. D. Dresdner, J. Electrochem. Soc., 95, 59 (1949).

43.   S. Nagase, H. Baba, and R. Kojima, J. Chem. Soc. Japan,
      64, 2126 (1961).

44.   J. A. Young, T. C. Simons, and F. W. Hoffman, J. Am.
      Chem. Soc., 78, 5637 (1956).

45.   H. Schmidt and H. Meinert, Angew. Chem., 72, 109 (1960).

46.   J. Burdon and J. C. Tatlow, Advances in Fluorine
      Chemistry (M. Staicy, J. C. Tatlow and A. G. Sharpe,
      eds.), Vol. 1, Academic, New York, 1960, p. 129.

47.   F. F. Coetzee, G. P. Cunningham, D. K. McGuire, and
      G. R. Padmanabham, Anal. Chem., 34, 1139 (1962).
      (Method C).

5.4.   MISCELLANEOUS OXIDATION

In order to acquaint the chemist (the organic chemist)
with anodic oxidation, and at the same time limit the length
of this book, several types of oxidation reactions are summarized in Table 5.7.

TABLE 5.7

Miscellaneous Oxidation Reactions

Type of Compound	Reaction Conditions	Products	References
I. Organometallic			
RMgBr	Pt, Hg, Ether	$\mathrm{C=C}$, R-R, R-H, ROH	1-5
 Li	Hg, Ether	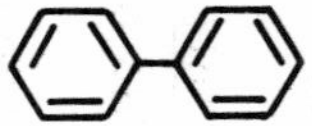 60%	3
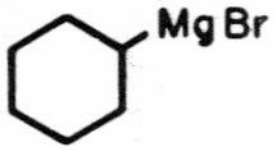 MgBr	Hg, Ether	 53%	3
$(H_5C_6)_3$GeNa	Pt, Liq. NH$_3$	$[(H_5C_6)_3Ge]_2$ 65-90%	6
Sodium tetraethylborate	Pb, H$_2$O	Tetraethyl lead 91%	7
	Bi, H$_2$O	Triethylbismuth	8
	Hg, H$_2$O	Diethylmercury	9

Table 5.7 (continued)

I. Organosulfur

RSR	Pt, Aq. CH₃OH HCL	$\overset{O}{\underset{\shortmid}{R\text{-}S\text{-}R}}$	9
2,2'-Dihydroxyethylsulfide	Pt, Aq. NaCl	2,2'-Dihydroxyethylsulfone 90%	9
Phenylsulfide	Pt, Aq. HOAc	Phenylsulfone 93%	
Benzylsulfide	Pt, H₂SO₄, HOAC	Benzylsulfoxide 90%	10-12
Phenyldisulfide	Pt, Aq. HOAc, HCl	Benzenesulfonic acid	12
Benzene sulfonic acids	Pt, PbO₂, H₂O	p-Hydroxybenzenesulfonic acid Benzoquinone	13-15

II. Organic Halides

RX RX=Cl, Br, I)	Pt, CH₃CN CH₃OH	$\overset{O}{\overset{\|}{R\text{-}C}}\text{NCH}_3$, $\overset{O}{\overset{\|}{R\text{-}C}}\text{NH}$, $\overset{}{>}\!C = C\!\overset{}{<}$, RH	16-18
CH₃CH₂CH₂Br	Pt, CH₃OH	Propylene 16% Cyclopropane 78% Propane 6%	17

(continued)

Table 5.7 (continued)

Substrate	Reagent	Product	Ref.
 (phenyl)–X	Pb, PbO$_2$	Quinone  (HO-C$_6$H$_4$-X); X(trace) biphenyl	19–22
o-iodotoluene (CH$_3$, I)	Pt, H$_2$SO$_4$ HOAc	o-Iodosobenzoic acid 83%	16, 21
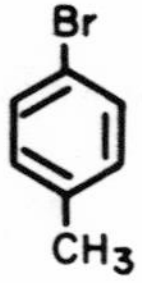 p-bromotoluene (Br–C$_6$H$_4$–CH$_3$)	Pt, HOAc, HNO$_3$	p-Bromobenzoic acid	23
IV Alcohols			
ROH	Pt, PbO$_2$	RCHO, CO$_2$, RCOH	24–31

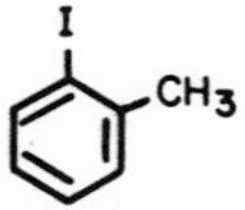

Table 5.7 (continued)

Isopropyl alcohol	Aq. Acids Pt, Aq. H_2SO_4	$\overset{}{C}=\overset{}{C}\ \ R-\overset{O}{\overset{\|}{C}}R$ Acetone, acetic acid, formic acid, CO_2	28
Glycerol	Rainy Ni Aq. KOH (8.14 V)	Dihydroxyacetone	37
	Rainy Ni Aq. KOH (0.22 V)	Hydroxypyruvic acid	
	Pd-C Aq. KOH (0.359 V)	Mesoxalic acid	
V. Ethers			
Tetrahydrofuran	C, CH_3ONa/CH_3OH	2-Methoxytetrahydrofuran 16.3%	29
	C, $EtOH/NH_4NO_3$	2-Ethoxytetrahydrofuran	29
	Pt, R_4NClO_4	Polytetrahydrofuran	30

(continued)

Table 5.7 (continued)

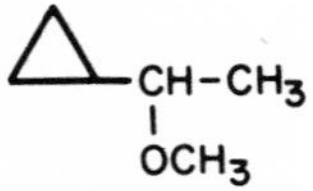

△ CH–CH$_3$, OCH$_3$ C, CH$_3$ONa/CH$_3$OH

△ CH–CH$_3$ 7.8% , OCH$_2$–OCH$_3$

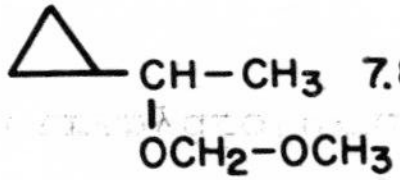

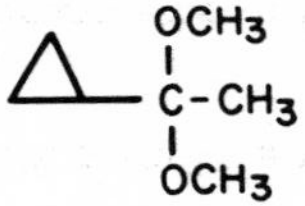

△ OCH$_3$ | C–CH$_3$ 24.3% | OCH$_3$

29

VI. Amides and Lactams, Lactones: 31–36

For a detailed discussion of each topic, the reader
should consult the suggested references (at the end of this
chapter.)

SUGGESTED REFERENCES

1. W. Evans and F. Lee, J. Am. Chem. Soc., 56, 654 (1934).

2. W. Evans and E. Field, J. Am. Chem. Soc., 58, 720 (1936).

3. J. L. Morgat and R. Pallaud, Compt. Rend., 260, 574,
    5579 (1965).

4. W. Evans and D. Braitwaite, J. Am. Chem. Soc., 61, 898
    (1939).

5. W. Evans, D. Braitwaite, and E. Field, J. Am. Chem.
    Soc., 62, 534 (1940).

6. L. S. Foster and G. S. Hooper, J. Am. Chem. Soc., 57,
    76 (1935).

7. K. Ziegler and H. Lehmkuhl, Ger. Pat. 1,212,085 (1966).

8. K. Ziegler and O. Stendel, Ann., 652, 1 (1962).

9. M. M. Nicholson, J. Am. Chem. Soc., 76, 2539 (1954).

10. F. Fichter and P. Sjostedt, Chem. Ber., 43, 3422 (1910).

11. F. Fichter and F. Braum, Chem. Ber., 47, 1526 (1914).

12. F. Fichter and W. Wenk, Chem. Ber., 45, 1373 (1912).

13. H. Lund, Collection Czech. Chem. Commun., 25, 3313
    (1960).

14. F. Fichter and E. Stocker, Helv. Chim. Acta, 7, 1064
    (1924).

15. J. Sebor, Z. Elektrochem., 9, 370 (1903).

16. L. Miller and A. K. Hoffmann, J. Am. Chem. Soc., 89,
    593 (1967).

17. J. T. Keating and P. S. Skell, J. Org. Chem., 34, 1479
    (1969).

18. J. R. Chadwick and E. Kinsella, J. Organometal. Chem.,
    4, 334 (1965).

19. L. Eberson, J. Am. Chem. Soc., 89, 4669 (1967).

20. H. Schmidt and H. Meinert, Angew Chem., 72, 109 (1960).

21. F. Fichter and P. Lotter, Helv. Chim. Acta, 8, 438
    (1925).

22. F. Fichter and M. Adler, Helv. Chim. Acta, 9, 279 (1926).

23. R. F. Dunbrook and A. Lowy, Trans. Electrochem. Soc., 45, 81 (1924).

24. V. D. Parker, Chem. Ind., 1363 (1968).

25. R. Kalvoda, J. Electroanal. Chem., 24, 53 (1970).

26. K. Elbs and D. Bruner, Z. Elektrochem., 6, 604 (1900).

27. K. I. Kryschenko, M. Y. Fioshin, N. G. Bakkchisaraits'yan, and G. A. Kokarev, Khim. Prom. (Moscow), 45, 496 (1969); C. A., 71, 101, 253 (1969).

28. T. Kurenniemi and E. Tommila, Suomen Kemistilehti, 913, 25 (1936); C. A., 31, 1299 (1937).

29. I. I. Radchenko, J. Applied Chem. USSR, 13, 1348 (1940).

30. S. Koizumi, J. Chem. Soc. Japan, 42, 928 (1921).

31. F. Fichter and F. Ackermann, Helv. Chim. Acta, 2, 583 (1919).

32. T. Shono and Y. Matsumura, J. Am. Chem. Soc., 91, 2803 (1969).

33. J. F. O'Donnell and C. K. Mann, J. Electroanal. Chem., 13, 157 (1967).

34. S. D. Ross, M. Finkelstein, and R. C. Petersen, J. Am. Chem. Soc., 88, 4657 (1966); ibid., 86, 2745 (1964).

35. S. Miguno, J. Electrochem. Soc. Japan, 29, 27 (1961); ibid, 29, 33, 112 (1961).

36. M. R. Rifi, unpublished results.

37. Siemens, to Schukertwerke Aktien Gesellschaft and Varta Pertrix Union Gmbh., Brit. Pat. 1,051,614 (1966); C. A., 66, 81920 (1967).

6

ELECTROINITIATED POLYMERIZATION

6.1.   GENERAL CONSIDERATIONS

The term "electroinitiated polymerization" simply means
the use of an electric current to "trigger" the polymerization of a monomer. The "triggering" of the reaction may be

done by the direct activation of the molecule, i.e., reduction or oxidation, or by generating an active species from the medium, which then starts the polymerization reaction. In such reactions, the activation energy is paid for by the potential of the operative electrode. Thus, compared to similar reactions which are triggered by light or heat, most electroinitiated polymerization reactions are carried out under mild conditions. Such conditions are of great importance in polymerization reactions where the preparation of initiator requires heat which could lead to branching, crosslinking, etc.

The polymerization of organic molecules by means of an electric current was recognized more than two decades ago (1). Since then, scattered papers and review articles have appeared which have described and surveyed the subject (2-6). However, these articles have, for the most part, dealt with the cathodic and anodic (via a Kolbe-type reaction) polymerization of activated olefins, with little emphasis on the utility of this technique in organic synthesis. Work in our laboratories has shown that a number of molecules such as cyclic-ethers (THF) esters, amines, and thietanes can be polymerized at the anode (10). Consequently, our discussion of electroinitiated polymerization will focus on what can be done with this technique, how to do it, and what advantages can be obtained from such work.

In order to be applicable on a large scale, electroinitiated reactions should be limited to the preparation of the active ingredient which triggers the polymerization reaction. This will obviate the necessity of carrying out bulk polymerization reactions on a multimillion-pound scale in electrolysis cells. This point must be considered and implemented if electroinitiated polymerization is going to penetrate the industrial world and encourage further research work in this area. However, this should not discourage any small-scale work in academic institutions or biochemical industries interested in small-scale reactions.

6.2. ADVANTAGES OF METHOD

One can visualize several advantages to electroinitiated polymerization.

1. Diversity of Reaction. If one oxidizes a monomer
to its corresponding cation, cationic polymerization may
take place. Conversely, if the monomer is reduced to its
anion or anion radical, then polymerization may occur via
anionic or radical mechanisms. Furthermore, electroinitiated polymerization is easily adapted for grafting (14, 15),
crosslinking (16, 17), and coating (discussed in detail in
a separate section).

2. Product Free from Catalyst. Certain catalysts
have to be removed at the conclusion of the polymerization
reaction for the product to be useful in certain application areas. For example, BF_3 must be removed from polyols
prior to their transformation to urethane, otherwise the
BF_3 will attack the isocyanate. The removal of such catalysts is costly and time consuming. We show later how
polyols can be easily prepared by electroinitiated polymerization and then transformed into urethanes without any purification. It should be emphasized that the removal of the
catalyst eventually may result in a pollution problem, since
these catalysts, in general, cannot be reused. It has been
the experience of the authors that the supporting electrolyte, which is a salt, does not interfere with the polymerization reaction; however, this should be ascertained by the
researcher in the particular reaction in question.

3. Control of Molecular Weight and Molecular Weight
Distribution of Product. In principle, it should be possible
to control the molecular weight and molecular weight distribution of the polymer (formed via ionic initiators) by simply
controlling the amount of coulombs passed through the reaction. This was indeed realized by Funt and co-workers (7-
9) in the anionic polymerization of styrene and in the
authors' work on the cationic polymerization of tetrahydrofuran and other cyclic compounds (10).

4. No Handling of Hazardous Catalysts. A number of
conventional polymerization catalysts, e.g., BF_3, R_3Al,
alkali metals, etc., require extreme caution in handling.
Electroinitiated polymerization eliminates such hazards.

5. Polymerization on Surface of Electrode. The role
of the electrode in electroinitiated polymerization and its
influence on the stereochemistry of the product has not been
investigated. Such a study may uncover important data pertaining to the formation of stereoregular polymers on the
surface of the electrode. A divided cell can be used to
carry out two types of polymerization reactions, i.e., anionic (cathode) and cationic (anode); see section on experimental procedure.

6. Ease of Formation of Block Copolymers. The formation of block copolymers from proper active homopolymers of
opposite charges (i.e., polymerization via living ions) has
rarely been used. This process should be particularly feasible in electroinitiated polymerization, where the positively charged homopolymer is produced at the anode and the
negatively charged homopolymer at the cathode. Thus, it
was shown in the authors' laboratories (10) that when styrene and tetrahydrofuran are homopolymerized in a divided
cell at the cathode and anode, respectively, and the two
homopolymers allowed to react, an A-B-A-type block copolymer
was formed, where A was tetrahydrofuran and B was styrene:

ANODE PRODUCT AND CATHODE PRODUCT

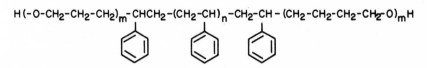

In principle, this type of product may be formed by using conventional techniques. Electrochemically, however, the operation is much simpler and would be quite economical.

6.3. DISADVANTAGES OF METHOD

Before the reader places his order to purchase equipment to carry out electroinitiated reactions and profit from all its advantages, he should be made aware that this method has some inherent problems:

1. <u>Reactions are Limited to Polar Media</u>. In order to allow the passage of current through a solution, this solution must have a minimum conductivity. This may rule out the polymerization of a number of monomers. Furthermore, the use of polar solvents may be detrimental to the polymerization reaction, since the propagating species may attack the solvent.

2. <u>Preparation of Active Ingredient is Slow</u>. This phenomenon is inherent in electrolysis in general, since the design of the electrolysis cell determines how fast an electric current, i.e., the number of coulombs, can be passed through a particular solution. Good agitation and an increase of the conductivity of the solution will help.

6.4. EFFECTS OF REACTION PARAMETERS ON COURSE OF REACTION

It was mentioned earlier that polymerization by electrolysis may proceeed via radical or cationic mechanisms. Con-

sequently, the researcher must take sufficient precautions
in order to favor one mechanism over another. The choice
of cationic vs anionic mechanism may be resolved by simply
allowing the reaction to occur at the anode or cathode.
However, anionic vs radical mechanisms may not be easily
distinguished. This point is discussed further under reaction mechanism.

Cell Dividers. The presence or absence of a cell divider plays an important role in the course of polymerization reactions, for example, work in the authors' laboratories (10) showed that styrene could not be polymerized
(to any extent) in a nondivided cell using mercury as the
cathode and platinum as the anode. However, polymerization
was quite feasible in a divided cell. On the other hand,
high molecular weight (in over 95% yield) was obtained from
the reduction α,α'-dihaloxylenes (11, 12) in an undivided
cell.

Choice of Solvent. One of the most important requirements that a solvent should have is that it be inert towards
the propagating species. For anionic and radical reactions,
the reader should consult the chapter on solvents in this
book.

It has been found in the authors' laboratories (10)
that, in many cases, e.g., polymerization of caprolactone,
THF, etc., the use of a solvent is not necessary, since
the monomer can act as the solvent. The product can be
easily separated from the monomer by the addition of the
appropriate nonsolvent for the polymer.

Supporting Electrolytes. In any ionic electroinitiated
polymerization, the supporting electrolyte acts as the
counter ion. Thus, it is the species that follows the propagating chain and, as such, may have a great influence on
its reactivity, such as branching and termination. Consequently, electroinitiated polymerization using different
supporting electrolytes should be considered (27).

6.5.   MECHANISM OF ELECTROINITIATED
POLYMERIZATION REACTION

The elucidation of reaction mechanisms in electroinitiated polymerization is complicated by many factors: (1)
The nature of the electrode for the supporting electrolyte
may be such that it forms a complex with the monomer; (2)
In ionic reactions, the counter ion (supporting electrolyte)
may have an influence on the propagating species; and (3)
The use of chemical agents as inhibitors is complicated by
the fact that these inhibitors (e.g., quinones) are electrochemically active. Yet, in spite of these complications,
enough data has been accumulated to allow us to appreciate
how electroinitiated polymerization takes place. This is
discussed under cathodic and anodic polymerization separately.

6.6.   CATHODIC POLYMERIZATION

Polymerization at the cathode may proceed via anionic
or radical mechanisms. The use of the latter has not been
investigated in detail but is commonly used in aqueous media,
where hydrogen radicals are generated and initiate the polymerization reaction (22). The current efficiency in these
reactions is generally low. This is due to the lifetime of
the hydrogen radical at the surface of the cathode. High
efficiencies, however, have been reported (18-20).

6.6.1. POLYMERIZATION OF ACTIVATED OLEFINS

Electroinitiated polymerization of activated olefins
at the cathode may occur via one (or more) of the following mechanisms:

1. <u>Direct Transfer of Electron from Cathode to Olefin</u>.
The half-wave potential of some activated olefins is listed

in Table 6.1 (10). It can be seen from the data in Table
6.1 that the polymerization of activated olefins using
tetraalkylammonium salts would be likely to proceed via a
transfer of electrons from the cathode to the olefins.
Thus, in nonaqueous solutions, the probable mechanism is
as follows:

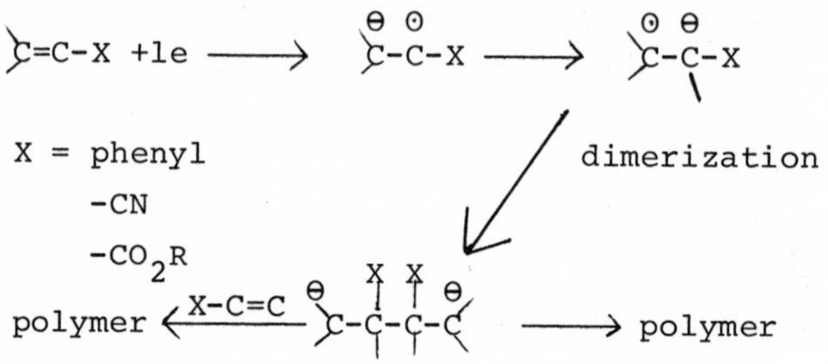

When the polymerization reaction is carried out in nonaqueous solutions, but in the presence of a proton source (methanol), polymerization is severely suppressed (10, 25, 28).
     In 1960, the cathodic polymerization of acrylonitrile,
using tetraethylammonium perchlorate under anhydrous conditions, was reported to proceed via an ethyl radical from
the supporting electrolyte (23). However, in the same year,

TABLE 6.1

Half-Wave Potentials of Some Activated
Olefins (DMF/nBu$_4$NClO$_4$)

Olefin	$-E_{1/2}$ V (s.c.e.)
Styrene	2.20
Acrylonitrile	1.74
Ethyl acrylate	1.80
Methyl methacrylate	1.91
DMF/nBu$_4$NClO$_4$	2.75

the same authors disproved this hypothesis (24) and proposed
an anionic mechanism (25). Similarly, Yamazaki (26) poly-
merized acrylonitrile in $DMF|Bu_4NClO_4$ and proposed a direct
electron transfer from the cathode to the monomer.

2. **Transfer of Electrons to the Supporting Electro-**
lyte. This reaction is quite common when sodium salts are
employed, since sodium ions reduce at a more positive potential than most olefins. Thus, the polymerization of acrylonitrile, using sodium nitrate as the supporting electrolyte, was postulated to proceed via a radical mechanism
(29). Similarly, styrene was polymerized using potassium
laurate as the supporting electrolyte (30). For other work
under similar conditions the reader may consult Refs. 20
and 31.

3. **Transfer of Electrons to Solvent.** The polymerization of methyl-methacrylate by hydrogen radicals generated
at the cathode from the reduction of water was among the
first electroinitiated reactions studied (32). Palit (33)
also described the polymerization of the same molecule in
aqueous sulfuric acid. For the polymerization of activated
olefins via hydrogen radical in aqueous solutions, the
reader may consult Refs. 22 and 35.

6.6.2. FORMATION OF LIVING ANIONS

In anionic polymerization, there are cases where the
propagating species has no termination step. Thus, the
propagating species possesses what is called a "living ion."
This phenomenon was first reported by Szwarc (36) and later
by Yamazaki (37) and Funt (7) in the polymerization of styrene. Funt and co-workers were able to control the molecular
weight distribution by controlling the concentration of the
living ion, which is directly proportional to the amount of
coulombs passed through the reaction medium (9). In our
laboratories, the use of the living anions (and cations
described later) was utilized (38) for the polymerization

of large quantities of monomers in conventional polymerization equipment by preparing the initiator (living ion) in a
small electrolysis cell. The apparatus used is described in
Fig. 6.1.

In Fig. 6.1, styrene was charged into the cathode compartment together with a tetraalkylammonium salt as the
supporting electrolyte. A lead sheet was used as the cathode.
The anode compartment was charged with acetonitrile and supporting electrolyte. Platinum gauze was used as the electrode. The three-necked flask contained only styrene, and
the temperature of the reaction could be controlled with

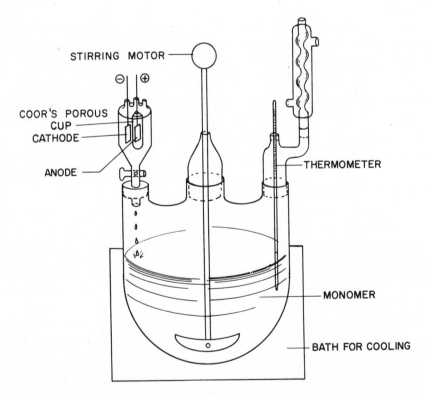

FIG. 6.1. Electroinitiated polymerization via the
living anions.

external cooling. The amount of styrene in the electrolysis
cell was about 1% or less of the total styrene used in the
reaction.

While only limited success was achieved with the above
setup, this technique opened the door to the investigation
of other systems which were considerably more successful
(e.g., living cations discussed later) and demonstrated
that electroinitiated polymerization can be adapted to
large-scale operations where only small electrolysis cells
are used for the preparation of the initiator.

The cathodic formation of other "living" ionic species
has been reported (41-43).

6.6.3. MISCELLANEOUS CATHODIC POLYMERIZATION

In order to limit the length of this chapter, the
electroinitiated polymerization of molecules that have not
been studied in as much detail as activated olefins is
listed in Table 6.2 with ample references.

TABLE 6.2

Miscellaneous Cathodic Polymerization

Monomer	Reaction Conditions	References
Acrylic acid	Hg, Aq. H_2SO_4	32
Acrylamide	Pt, Aq. Sol. $ZnCl_2$	49
Ethylene	C, Aq. Benzene, K_2TiF_6	20
	Dioxane, R_4NClO_4	39
Vinyl ethers	Nitrobenzene	25
Dienes (isoprene)	Pt, T.H.F., $NaB(C_6H_5)_4$	40, 26, 44
Caprolactam	$K^{\oplus}$	47
Isocyanates	DMF, $R_4N^{\oplus}$	52-54

6.7. ANODIC POLYMERIZATION

6.7.1. SCOPE OF TECHNIQUE

Compared to cathodic polymerization reactions, polymerization at the anode has received little attention. In
fact, most polymerization reactions at the anode have centered on the generation of radical species from acetate-
supporting electrolytes via the Kolbe reaction, which in
turn initiated the polymerization of activated olefins (21,
25, 56-59). Non-Kolbe-type anodic polymerization reactions
included the polymerization of tetrahydrofuran (55), vinyl
carbazol, and vinylethers (25). For the past four years,
Rifi has polymerized a number of different functional groups
at the anode, and has indications that the reaction mechanism involves a direct transfer of electrons from the monomer to the anode (10). Table 6.3 summarizes this work.

It can be seen from Table 6.3 that anodic polymerization is rather simple, since in most cases it requires no
solvent and affords products with good yields (50-95%).
Furthermore, in almost all cases reported in Table 6.3
Rifi found that polymerization proceeds via the living
cation. For example, in the polymerization of THF, a certain amount of electricity was passed through the cell and
then, with no current passing through, the change of molecular weight of the product with time was followed. The
results are shown in Fig. 6.2.

Further indication that the polymerization of THF (and
other monomers in Table 6.3) occurs via the living cation
was demonstrated as follows: A small amount of THF was
electrolyzed in a small cell, as shown in Fig. 6.1. The
solution from the anode compartment was then added to the
three-necked flask (see Fig. 6.1), which contained pure THF.
Stirring at room temperature caused the polymerization of
THF in the flask.

TABLE 6.3

Anodic Polymerization of Various Monomers

Monomer	Solvent	Temperature	F/mole %	Yield %	Reduced visc
 (oxetane structure)	None	25°	0.23	75	1.6
	None	25°	0.46	75	0.79
 (ethylene oxide structure)	None	-20°	1.7	30	--
	CH_2Cl_2	25°	0.3	<10	0.32
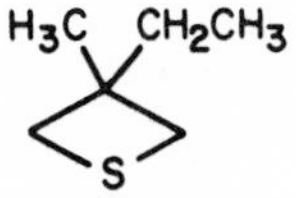 H_3C CH_2CH_3 (thietane structure)	None	25°	1.7	50	0.32
 (aziridine structure) N	None	25°	1.5	55	--
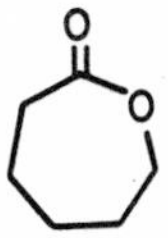 (caprolactone structure)	None	150°	0.2	>95	0.32
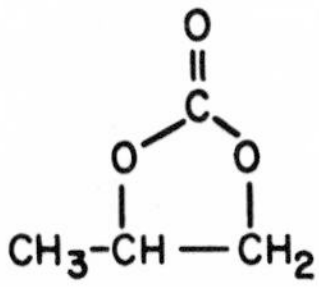 (cyclic carbonate) $CH_3-CH-CH_2$	None	25°	0.3	0	--
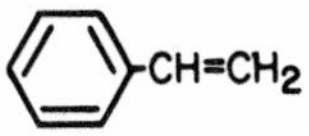 (phenyl)$-CH=CH_2$	None	25°	0.64	37	0.094

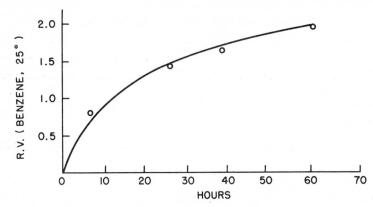

FIG. 6.2.   Effect of time on molecular weight of poly-THF.

6.7.2.   EFFECT OF REACTION PARAMETERS ON COURSE OF REACTION

The choice of reaction parameters for anodic polymerization are essentially the same as those described for cathodic polymerization.   One parameter that was found to be
important in anodic polymerization [this was also observed
by others for cathodic polymerization (9)] is the amount
of coulombs passed through the solution, which was inversely
proportional to the molecular weight of the polymer.   Fig.
6.3 illustrates this point in the polymerization of THF
(10).

In the absence of a termination reaction, the number
average molecular weight may be calculated from the following relationship: mol. wt = Weight of monomers/n(number
of Faradays) where n is the number of electrons involved in
the initiation step.

6.7.3.   MECHANISM OF REACTION

A detailed investigation of the kinetics of cationic
polymerization was recently advanced by Schulz and Strobel
(61).   Consequently, this topic is not discussed here.

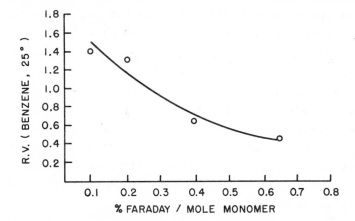

FIG. 6.3. Effect of coulombs on molecular weight of
poly-THF at a fixed time.

6.8.   EXPERIMENTAL PROCEDURE

6.8.1.   GENERAL CONSIDERATIONS

Polymerization reactions, whether initiated electrochemically or by conventional initiators, are very susceptible to traces of impurities. Consequently, monomers, solvents, and supporting electrolytes must be quite pure and
should be used shortly after purification. In most nonaqueous polymerization reactions, oxygen is excluded from
the setup. It is sometimes helpful to pre-electrolyze the
system, without the monomer, in order to destroy certain
electroactive impurities.
The electrolysis cell used for electroinitiated polymerization is, in general, similar to those used for oxidation and reduction reactions. Polymerization reactions
carried out in the authors' laboratories are described in
Fig. 6.1. Other authors have included some excellent dia-
grams of the electrolysis cells used in their investigations,
together with apparatus used for the purification of solvents and monomers (13, 25, 42, 60).

A simple laboratory apparatus for the polymerization of activated olefins is described in Fig. 6.4.

Purification of Reagents. For the purification of solvents and supporting electrolytes see Sec. 3.2.

Styrene. When obtained from commercial sources, this monomer should be first fractionally distilled under reduced pressure in an atmosphere of nitrogen. It is then stirred over calcium hydride (sodium metal will initiate polymerization) for 24 hr at room temperature under nitrogen. Finally, styrene is distilled under vacuum and nitrogen and is used shortly thereafter.

6.8.2. PROCEDURE

Anodic and Cathodic Polymerization of Styrene (10). The apparatus in Fig. 6.4, a 250 ml resin kettle, was used as the cell. It was flamed and cooled under nitrogen before use. Platinum gauze was used for both electrodes. The

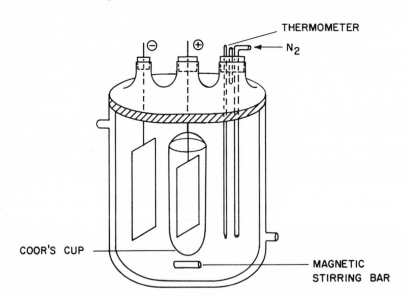

FIG. 6.4. Resin kettle for electroinitiated polymerization.

anode compartment was charged with 50 ml of pure styrene,
which was made 0.1 M with Bu_4NClO_4 (polarographic grade,
Matheson, Coleman, and Bell). The cathode compartment was
filled with 100 ml of the same solution. The cell was cooled
externally with an ice bath. Electrolysis proceeded for 6
hr at 100 V and 5-10 mA. The solution, bluish in the cathode and brownish in the anode, were left standing under
nitrogen for two days. At this time, the individual products (anode and cathode compartments) were purified separately by dissolving in benzene and precipitating from
methanol. This procedure was repeated once more. The precipitates were then dried in a vacuum oven (30 mm) at $65^\circ C$
overnight. Based on several runs, the average yield of
polystyrene at the anode was 37%, with a reduced viscosity
(benzene 25°) of 0.1. Polymers obtained at the cathode
varied in yields from 30-55% with reduced viscosities of
0.2 to 0.5 (benzene 25%).

REFERENCES

1. C. L. Wilson, Record Chem. Prog., 25, (1949).

2. M. Y. Fioshin and A. P. Tomilov, Plasticheskie Massy,
    10, 2 (1964).

3. C. L. Wilson, Encycl. Electrochem., 963, (1964).

4. B. L. Funt, Macromol. Revs., 1, 35 (1966).

5. H. Friedlander, Encycl. Polymer Sci. Technol., 5, 629
    (1966).

6. P. Arnaud, Ind. Chim. Belge, 31, 896 (1966).

7. B. L. Funt, S. N. Bhadani, and D. Richardson, J.
    Polymer Sci (A-1), 4, 2871 (1966).

8. B. L. Funt, U. S. Pat. 3,448,020 (1970).

9. B. L. Funt and D. Richardson, J. Polymer Sci. (A-1),
    8, 1055 (1970).

10. M. R. Rifi, U. S. Patent 3, 645, 986 (1972).

11. H. Gilch, J. Polymer Sci., 4, 1351 (1966).

12. F. Covitz, J. Am. Chem. Soc., 89, 5403 (1967).

13. B. L. Funt and S. M. Bhadani, J. Polymer Sci., (c), 1 (1968).

14. G. Smets, A. Poot, M. Mullier, and J. P. Bex, J. Polymer Sci., 34, 298 (1959).

15. J. A. Epstein and A. Bar.-Nun, Polymer Letters, 2, 27 (1964).

16. W. F. Borman, Can. Pat. 800,592 (1968).

17. W. R. Grace and Co., Brit. Pat. 1,156,309 (1968).

18. D. Georring, H. Jonas, and W. Moschel, Ger. Pat. 935,867 (1955).

19. D. Georring and H. Jonas, Ger. Pat. 937,919 (1956).

20. H. Gehrke and M. Fechenheim, Ger. Pat. 1,014,774 (1957).

21. W. B. Smith and H. G. Gilde, J. Am. Chem. Soc., 82, 659 (1960).

22. W. Kern and H. Quast, Makromol. Chem., 10, 202 (1953).

23. J. W. Breitenbach and H. Gabler, Monatsh. Chem., 91, 202 (190).

24. J. W. Breitenbach, C. Srna, and O. F. Olaj, Makromol. Chem., 42, 171 (1960).

25. J. W. Breitenbach and C. Srna, Pure Appl. Chem., 4, 245 (1962).

26. N. Yamazaki, I. Tanaka, and S. Nakahama, J. Macromol. Sci.(Chem.), 6, 1121 (1968).

27. T. Asahara, M. Seno, and M. Tsuchiya, Bull. Chem. Soc. Japan, 42, 2416 (1969).

28. B. L. Funt and S. N. Bhadani, Can. J. Chem., 42, 2733 (1964).

29. H. Chapiro and E. Henrychowski, J. Chim. Phys., 65, 616 (1968).

30. J. Y. Yang, W. E. McEwan, and J. Kleinberg, J. Am. Chem. Soc., 79, 5833 (1957).

31. J. D. Anderson, J. Polymer Sci. (A-1), 6, 3185 (1968).

32. E. Dineen, T. C. Schwan, and C. L. Wilson, Trans. Electrochem. Soc., 96, 226 (1949).

33. N. N. Das and S. R. Palit, Sci. Cult. (Calcutta), 16, 34 (1950).

34. N. S. Tsvetkov and Z. F. Glotova, Vysokomolekul Soedin., 5, 997 (1963); C. A., 59, 7657 (1963).

35. G. Parravano, J. Am. Chem. Soc., 73, 628 (1951).

36. M. Szwarc, Nature, 178, 1168 (1956).

37. N. Yamazaki, S. Nakahama, and S. Kambara, Polymer Letters, 3, 57 (1965).

38. M. R. Rifi, unpublished results.

39. J. W. Loveland, Can. Pat. 566,274 (1958).

40. B. L. Funt and S. N. Bhadani, Polymer Sci. (c), 23, 1 (1968).

41. G. M. Guzman and A. Bello, Makromol. Chem., 46 (1967).

42. M. Morton, A. Rembaum, and J. Hall, J. Polymer Sci., (A), 1, 461 (1963).

43. A. Gosnell, J. Gervasi, Mrs. D. K. Woods, and V. Stannett, J. Polymer Sci. (c), 611 (1969).

44. N. Yamazaki and S. Murai, Chem. Commun., 147 (1966).

45. W. A. Kornicker, Can. Pat. 829,881 (1968).

46. W. A. Kornicker, U. S. Pat. 3,474,012 (1969).

47. H. Gilch and D. Michael, Makromol. Chem., 99, 103 (1966).

48. H. Gilch, U. S. Pat. 3,419,482 (1969).

49. J. Sobieski and M. Zerner, U. S. Pat. 3,464,960 (1969).

50. D. Laurin and G. Parravano, Polymer Letters, 4, 797 (1966).

51. D. Laurin and G. Parravano, J. Polymer Sci. (c), 22, 103 (1968).

52. G. Shapoval, M. Skobets, and N. Markova, Sintez Fiz. Khim. Polimerov, 5, 76 (1968); C. A., 70, 4677 (1969).

53. G. Shapoval, M. Skobets, and N. Markova, Vysokomolekul. Soedin., 8, 1313 (1966); C. A., 67, 11894 (1967).

54. G. S. Shapoval, E. Skobets, and N. P. Markova, Dokl. Akad, Nauk. SSSR, 173, 392 (1967).

55. C. F. Heins, J. Polymer Sci., (13), 7, 625 (1969).

56. R. V. Lindsey and M. R. Peterson, J. Am. Chem. Soc., 81, 2073 (1959).

57. W. B. Smith and H. Gilde, J. Am. Chem. Soc., 81, 5325 (1959).

58. G. Smets, X. VanDerBorght, and G. VanHaeren, J. Am. Chem. Soc. (A), 2, 5187 (1964).

59. G. Smets and J. P. Bex, U. S. Pat. 3,330,745 (1967).

60. B. L. Funt, D. Richardson, and S. Bhadani, Can. J. Chem., 44, 711 (1966).

61. W. Strobel and R. C. Schulz, Makromol. Chem., 133, 303 (1970).

ELECTROCOATING

7.1. GENERAL CONSIDERATIONS AND BACKGROUND

In discussing the reduction and oxidation of organic
compounds, the authors have attempted to include the importance of such electrochemical reactions in organic synthesis. Yet one of the most important electrochemical reactions that has received so much attention in the past ten
years is that of water. For electrocoating, or electrodeposition (in its commercial application), is simply the deposition of an organic substrate (a polymer) from a water
solution through the oxidation or reduction of water. Specifically speaking, a polymer with charged groups is dissolved

337

(or dispersed) in water and is deposited at the anode or
cathode as explained below.

7.1.1. ANODIC DEPOSITION

Consider an aqueous solution of a polymeric chain
which contains pendent carboxylic acid salts. This chain
will remain soluble in water as long as it contains the
pendent salt groups (see Fig. 7.1).
Suppose, now, that an electric current is impressed on the
solution. The anode is immediately coated because of the
following simple reactions, Eqs. (7.1) and (7.2):

$$H_2O - 2e \longrightarrow 1/2\ O_2 + 2H^{\oplus} \tag{7.1}$$

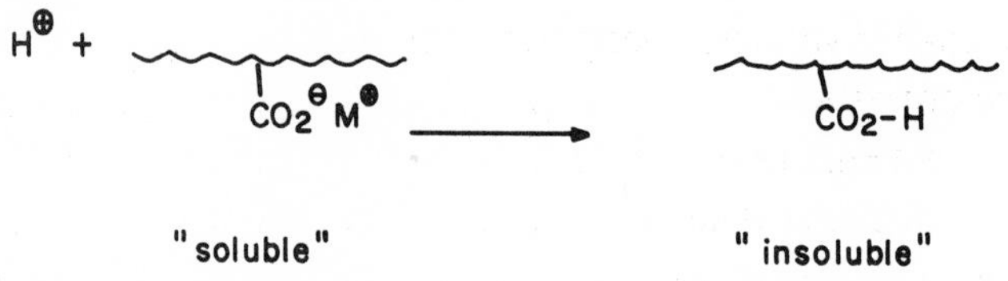

$$H^{\oplus} + \quad \sim\!\!\!\sim\!\!\!\sim\!\!\!\sim\!\!\!\sim \quad \longrightarrow \quad \sim\!\!\!\sim\!\!\!\sim\!\!\!\sim \tag{7.2}$$
$$\underset{CO_2^{\ominus}M^{\oplus}}{} \qquad\qquad \underset{CO_2\text{-}H}{}$$

"soluble"   "insoluble"

7.1.2. CATHODIC DEPOSITION

Cathodic deposition, which is not yet as widespread
as its sister counterpart, involves the deposition of a
polymeric chain as follows, Eqs. (7.3) and (7.4):

Cathodic reaction:

$$2H_2O + 2e \longrightarrow 2\overset{\ominus}{O}H + 2H_2 \tag{7.3}$$

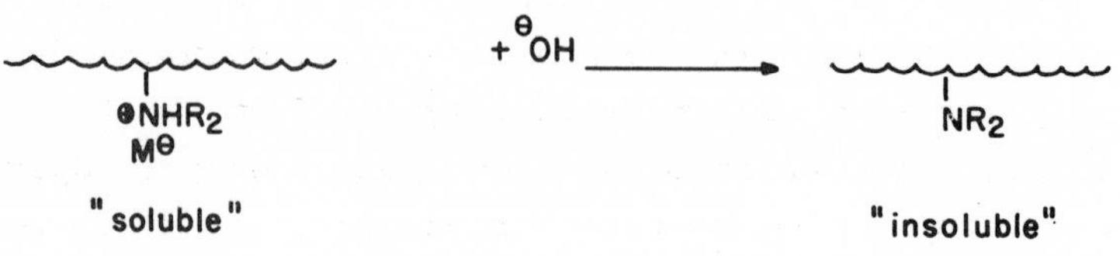

$$\sim\!\!\!\sim\!\!\!\sim\!\!\!\sim \quad + \overset{\ominus}{O}H \longrightarrow \quad \sim\!\!\!\sim\!\!\!\sim\!\!\!\sim \tag{7.4}$$
$$\underset{\underset{M^{\ominus}}{\oplus NHR_2}}{} \qquad\qquad\qquad \underset{NR_2}{}$$

"soluble"   "insoluble"

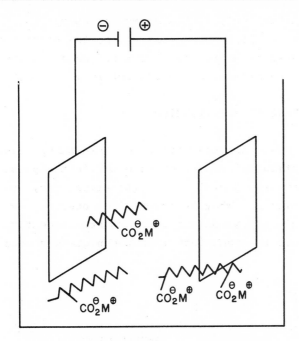

FIG. 7.1. Aqueous solution of polymeric chains with pendent carboxylic acid groups (⋀⋁⋀ represents the backbone of the polymer).

This type of deposition offers an advantage to anodic deposition in that the cathode (which is the object being coated) does not undergo any electrochemical reaction which might influence the coating. This may be illustrated with reactions (7.5) and (7.6):

Anodic reaction:

$$Fe - 3e \longrightarrow Fe^{+3} \tag{7.5}$$

$$Fe^{+3} + H_2O \longrightarrow Fe(OH)_3 \tag{7.6}$$
$$\text{Yellow rust}$$

Thus, when a steel object is being coated anodically, a yellow color may appear in the paint. This cannot be tole-

rated, particularly if the object is being coated with a
white paint. If, on the other hand, the steel object is
being coated cathodically, the final paint is not contami-
nated by any side reaction product from the cathode.

7.1.3. PROGRESS OF TECHNIQUE

What has made the utility of this simple electrochemical oxidation and reduction of water in the paint industry
so popular (see Fig. 7.2.) is its simplicity, as well as
its efficiency. What is surprising, however, is that this
utility did not become so popular until recently. This is
in spite of the fact that this technique was realized as

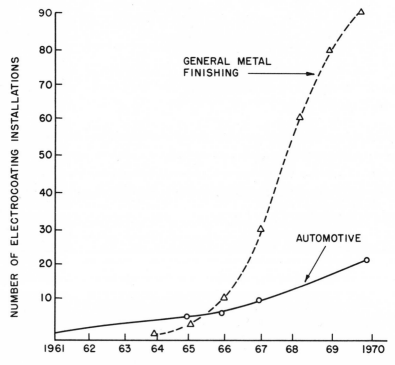

FIG. 7.2. Number of electrocoat installations in
U. S. (total number in Europe is about 250).

early as 1927 by Sheppard (1). Yet it was not until the
middle of 1960 that an upsurge in this activity was observed,
with most of the work carried out in industrial institutions.

A detailed discussion of electrocoating and its
application are beyond the scope of this book. Fortunately,
an excellent book on this subject has been written by Yeates
(2). Another good reference of interest to the reader is a
series of booklets entitled Electrophoretic Painting, which
appears periodically and is published by R. H. Chandler,
Ltd. Considering these available references, only a brief
description of electrocoating is presented in this section,
with some emphasis on laboratory experiments pertaining to
this subject. Thus, the following topics are discussed.

1. Formulation of electrocoating solution;
2. Deposition of paint on substrate;
3. Evaluation of coating on the substrate.

7.2. FORMULATION OF AN ELECTROCOATING SOLUTION OR DISPERSION

There are many available polymers that are suitable
for electrodeposition. When the choice of polymer has been
made, it is made water soluble by mixing with aqueous acid
or base* depending on the nature of deposition in question.
Prior to this step, it may be desired to pigment the polymer
with the desired color. This may be done by mixing the two
ingredients with the aid of a high-speed stirrer such as a
Cowles dissolver. Once the polymer is pigmented and neutralized, it is diluted with water to the appropriate concentration (usually 10%) and is now ready for deposition.
Because of the great insolubility of polymers in water, it
is often necessary to add a cosolvent such as cellosolves

---

*The amount, in moles, of acid or base in water is
about 80% of the number of moles of acid, or base, in the
polymer.

or alcohols. The amount of the cosolvent depends on the
nature of the polymer. In general, however, the amount is
less than 20% by weight of volatile portion of paint.

Examples of resins that can be used in electrodeposition
are as follows:

1. Ethylene-Acrylic Acid Copolymer

$$CH_2=CH_2 + CH_2=CH-\overset{\overset{O}{\|}}{C}OH \xrightarrow{\text{Catalyst}} (-CH_2-CH-)_n-(CH_2-\underset{CO_2H}{\underset{|}{CH}}-)_n$$

The random copolymer is totally miscible in water-amine
solution and can be deposited at the anode in the clear or
pigmented form.

2. Maleinized Oils

An oil may be defined as the ester resulting from the
reaction of a fatty acid with an alcohol. A naturally occurring oil, such as linseed oil, is the triester of linseed
acid* of glycerol. When the oil is treated with maleic anhydride, it becomes "maleinized:"

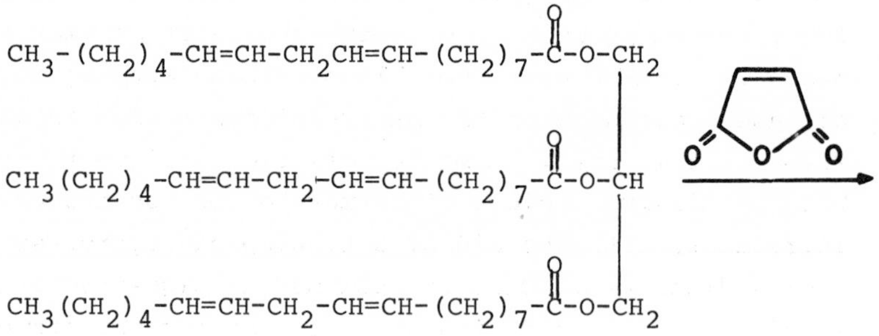

$$CH_3-(CH_2)_4-CH=CH-CH_2CH=CH-(CH_2)_7-\overset{\overset{O}{\|}}{C}-O-CH_2$$

$$CH_3(CH_2)_4-CH=CH-CH_2-CH=CH-(CH_2)_7-\overset{\overset{O}{\|}}{C}-O-CH$$

$$CH_3(CH_2)_4-CH=CH-CH_2-CH=CH-(CH_2)_7-\overset{\overset{O}{\|}}{C}-O-CH_2$$

----

*Linseed acid is a mixture of 50.8% linolenic acid,
19.5% linoleic acid, 20% oleic acid, and a 6% saturated
C18-acid.

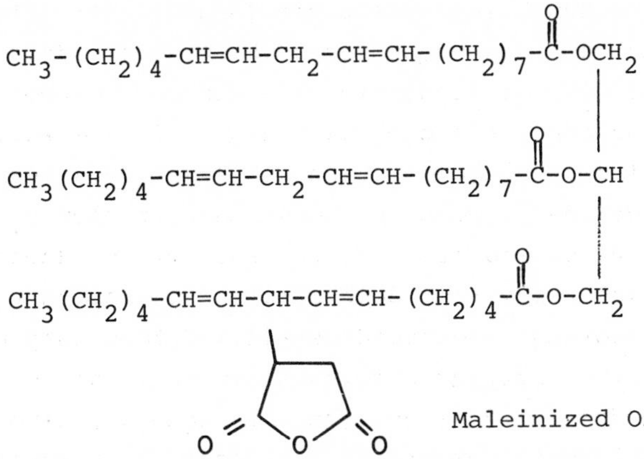

$$CH_3-(CH_2)_4-CH=CH-CH_2-CH=CH-(CH_2)_7-\overset{O}{\overset{\|}{C}}-OCH_2$$

$$CH_3(CH_2)_4-CH=CH-CH_2-CH=CH-(CH_2)_7-\overset{O}{\overset{\|}{C}}-O-CH$$

$$CH_3(CH_2)_4-CH=CH-CH-CH=CH-(CH_2)_4-\overset{O}{\overset{\|}{C}}-O-CH_2$$

Maleinized Oil

Hydrolysis of the maleinized oil gives a linseed oil with pendent carboxylic groups for water solubilization. Because of the availability and the low cost (8-10¢/lb.) of linseed oil, its use in electrocoating is quite popular.

7.3. DEPOSITION OF POLYMER

According to Faraday's law, the amount of polymer that is deposited is directly proportional to the amount of electricity passed through the system. However, some of the electricity passed through the system is used up in some side reactions. For example, in anodic deposition the Kolbe reaction may take place, Eq. (7.7):

$$R-CO_2^{\ominus} \ \overset{\oplus}{NR_3H} \longrightarrow R^{\odot} + CO_2 \tag{7.7}$$

Another side reaction is the oxidation of the anode. It should be noted, however, that both of the above reactions consume less than 10% of the electricity used for electrocoating.

In a typical electrocoating operation, the aqueous paint is placed in a metallic container, which normally

functions as the auxilliary electrode (Sec. 3.1.1). The
object to be coated is used as the working electrode. When
the voltage and time of deposition are chosen, the coating
operation may be done in one of two ways: (1) the working
electrode is placed in the solution at V = 0 (<u>Dead</u> <u>Entry</u>
of working electrode), and then the voltage is slowly raised
to the desired value and held for the rest of the deposition.
(2) Alternatively, prior to deposition, the desired voltage
is set and the working electrode is allowed to slowly enter
the solution (<u>Live</u> <u>Entry</u> of working electrode), and deposition continued for the desired time. In general, live en-
tries are practiced in commercial applications; however, it
may be more convenient for laboratory work to use the dead
entry method. In most electrodepositions no cell divider
is used.

During electrodeposition, the following observations
are made: when the power is on, there is a rapid upsurge
of current, indicating rapid deposition on the working
electrode. This deposition is, in general, essentially
complete after 10-15 sec. At this time, current gradually
drops (it should be kept in mind that the over-all voltage
during this operation remains constant). This is illustrated in Fig. 7.3. At the conclusion of deposition, the

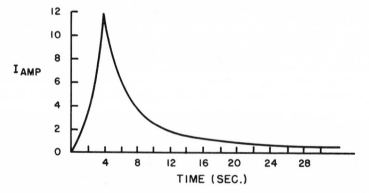

FIG. 7.3. Variation of current with time during
electrodeposition.

power is turned off, the coated object is rinsed with water,
and is evaluated.

7.3.1. FACTORS AFFECTING DEPOSITION

1. Effect of pH of Medium: For anodic deposition,
the pH is normally about 8.5. For cathodic deposition it
can vary from 4.5 to 6.5. In either type of deposition
(Secs. 7.1.1 and 7.1.2) H^+ and an ^-OH are generated at the
surfaces of the electrodes. Since one of them is used for
deposition, the other will remain in solution and changes
its pH. An increase in pH value during anodic deposition
will decrease the coulombic efficiency of deposition and
may affect the chemical structure (hydrolysis of ester
groups etc.) of the resin in solution.

The control of pH in electrodeposition may be accomplished in one of two ways: (a) use of a cell divider; or
(b) replenishment of the electrodeposition tank with a base-
deficient resin. The latter method is more common in the
U. S., while the first is widely used in Europe.

2. Conductivity of Medium: The conductivity of solutions used in electrodeposition vary between 500-3000 μmhos
(Sec. 3.3.3). These values depend to a large extent on the
degree of neutralization of the resin. Dispersions of low
conductivities tend to give rough coatings, and lowering of
the operating voltage and throwing power (see definition
below). High conductivity value will decrease the coulombic
efficiency and raise the operating voltage beyond a practical range.

3. Type of Metal Being Coated: In practice, most
metals used in electrodeposition are pretreated (for better
corrosion resistance) prior to coating.* This pretreatment

---

*A good reference to consult on pretreatment of metals
is: J. A. Murphy, Surface Preparation and Finishes for
Metals, McGraw-Hill, New York, 1971.

(mostly zinc phosphate) may include ingredients which may, during deposition, interfere with the paint (see Sec. 7.1.2). Consequently, the operator should familiarize himself with the history of the electrode being coated.

4. Current Density: In general, there is a minimum current density that has to be reached in order to coat a metallic object. Thus, in practice, a high current density is used. Low current densities tend to give soft (because of the presence of water), and hence unsatisfactory, coatings. It should be remembered, however, that if the current density is too high, excessive heat and gassing will occur, which will result in film rupture.

5. Temperature: Most electrodepositions are performed at ambient (or slightly below) temperatures. However, during deposition, heat is generated and should be controlled. This is particularly true in a continuous operation. In practice, cooling units are included in the electrodeposition tank. The effect of temperature on electrochemical reactions is discussed in Sec. 2.8.8.

6. Agitation: Most electrodeposition systems are made up of pigmented resins dispersed in water. To prevent this dispersion from settling, it is continuously agitated. The rate of agitation should be controlled, however, since it could affect the rate of deposition of paint. Thus, excessive agitation will (perhaps by sweeping the resin away from the surface of the electrode) decrease the rate of deposition. For the effect of agitation on electrochemical reactions, see Sec. 2.8.4.

In the electrodeposition of paint on a metal substrate, certain parameters have to be determined. A brief discussion of these parameters follows:

Operating Voltage: This is the voltage that is required to deposit a certain film thickness (usually 0.75 mil) in a certain time (usually 1 or 2 min). The thickness of the film is measured after the coated substrate is baked (see Sec. 7.4 on evaluation). Typical operating voltages

are 50-300 V. The factors that affect operating voltages
include: (a) number of pendent functional groups in the
backbone of the resin, e.g., carboxylic acid and their degree of neutralization [the higher the functionality, the
higher is the operating voltage]; (b) conductivity of medium [this in essence depends on point (a), the operating
voltage increases with an increase in conductivity of me-
dium]; and (c) molecular weight, the higher the molecular
weight, the higher the operating voltage.

Breakdown Voltage: If a thick film is desired to be
deposited on a substrate, this can be achieved by increasing the voltage of the cell. However, a voltage is eventually reached where the film ruptures during deposition.
This voltage is called the breakdown voltage. Typical
breakdown voltages are 200-400 V. Voltage rupture is attributed to the rapid evolution of gas (O_2 at anode, H_2
at cathode) which penetrates the deposited film. Consequently, the physical properties of the film will affect
the breakdown voltage. In general, the factors that influence breakdown voltage are: (a) molecular weight of resin
and its chemical nature; and (b) the cosolvent used in the
solution. Thus, if the cosolvent is in fact a good solvent
for the deposited film, this cosolvent will soften the film
and prevent its rupture by the evolved gas. Experimentally,
one knows that a rupture of the deposited film is taking
place from the following observation: A voltage is chosen
for deposition and the power is turned on. Immediately
(1 sec) there is a surge in current. This current, however,
drops down within the next 10-15 sec. This is due in the
high resistance developed at the working electrode, because
of film deposition. When this film ruptures, it exposes
part of the electrode. This causes an upward surge of current which can be observed on the ammeter (see Fig. 7.4).

Throwing Power: This is perhaps the most important
phenomenon in electrocoating and is defined as the ability
of the paint (under the influence of an electric field) to

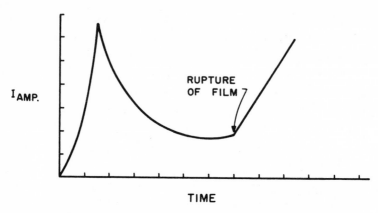

$I_{AMP.}$

RUPTURE
OF FILM

TIME

FIG. 7.4. Variation of current with time during
deposition. (Observation for breakdown voltage.)

coat hidden areas. There are several methods that can be
used to measure this phenomenon. These are described in
Ref. 2. In the laboratory, it can be measured as follows
(see Fig. 7.5).

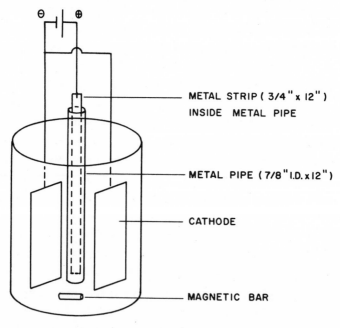

Θ ⊕

METAL STRIP (3/4" x 12")
INSIDE METAL PIPE

METAL PIPE (7/8" I.D. x 12")

CATHODE

MAGNETIC BAR

Fig. 7.5. Ford-Pipe Test for measuring throwing power.

In Fig. 7.5 two cathodes are used. A metal pipe is used
as the anode. A voltage and time of deposition are selected. For best results the voltage chosen should be close
to the operating voltage or slightly higher. Time of deposition is normally 1 or 2 minutes. Under these conditions,
the height of deposition inside the metal pipe is measured
in inches and recorded as the throwing power. To accurately
measure this height, the pipe has to be cut in half. This
means that the tube cannot be used again. To avoid cutting
the tube, a strip of metal which barely fits inside the
tube is used, as shown in Fig. 7.5. Since the strip of
metal is in contact with metal pipe, it will get coated.
The height of the coating on the metal strip is a measure
of the throwing power of the solution. Values of throwing
power will vary between 4 and 10 in; of course, the higher
the number the better. Some of the factors that affect
throwing power are: (a) concentration of resin (and pigment) in solution; (b) conductivity of solution; and (c)
molecular weight and type of resin. How good an insulator
a resin is will be reflected in its throwing power.

7.4. EVALUATION OF THE COATING

Once a polymer is deposited on a substrate, the substrate is rinsed with water and baked at a certain temperature for a length of time. At this time, the coating is
ready for evaluation, and since different substrates are
used in different application areas, the requirements of
the coating will vary over a wide range. Nevertheless, as
a standard, the following properties are determined:

Impact: This test may be carried out in a number of
ways. In the laboratory it is generally done by allowing
a steel ball of a certain weight to drop on the coated object from a certain height. The maximum height from which
the impact of the ball does not chip the coating is recorded.

Both faces of the substrate are examined and the forward (front) and reverse (back) impacts are recorded in inch-pounds.

Salt Spray: This test relates to the resistance of the coating to corrosion. Thus, a metallic panel is coated and placed in an environment of 5% sodium chloride for several time intervals and the condition of the paint examined.

Chemical Resistance: There is no standard chemical resistance test for substrates which are coated electrochemically, since these differ for different applications. In general, however, it is desirable to know the resistance of the coating to a certain caustic concentration for a period of time. Other tests, e.g., resistance to acid and certain chemical solvents, may be considered for certain applications.

7.5. ADVANTAGES AND DISADVANTAGES
OF ELECTROCOATING

It was mentioned earlier that the application of electrocoating has become quite popular in recent years, particularly in the automotive industry. Below are some of the points which have contributed to this popularity.

7.5.1. ADVANTAGES

1. Coating is performed from aqueous solutions. This eliminates pollution problems.

2. The coating formed by this process is pinhole free, i.e., it coats hidden areas and is quite uniform.

3. The process is very efficient in utilizing the paint in solution with minimum loss.

4. The process is easily automated.

In spite of its advantages, electrodeposition has some disadvantages. These, however, have been simplified and some will hopefully be eliminated in the future.

7.5.2. DISADVANTAGES

1. To perform electrocoating in the laboratory, one needs a dc power supply, an oven, and some other basic equipment. On a large-scale operation, however, the initial cost for purchasing electrical equipment is still high.

2. The process involved in the preparation of the electrocoating solution is often long and somewhat complicated.

3. For a resin to be suitable for electrocoating, it must have certain groups in its backbone. This point is not too serious, since most polymers can be made to react with other compounds which have such necessary groups.

7.6. CURRENT APPLICATIONS

Electrocoating is widely used in the automotive industry for the application of primers. Recently, its use in the appliance area (air conditioners, washing machines, etc.) have become quite popular. Other areas include general metal finishes such as metal toys, coat hangers, etc.

REFERENCES

1. S. E. Sheppard, Trans. Electrochem. Soc., 52, 47 (1927).
2. R. L. Yeates, Electropainting, Robert Draper, Teddington, England, 1966.
3. M. R. Rifi, unpublished results.

APPENDIX A
QUESTIONS AND ANSWERS

QUESTIONS (CHAPTER 2)

2.1 and 2.2. FARADAY'S AND OHM'S LAWS, UNITS

1. Define the following:
   a. A faraday
   b. A coulomb
   c. A mho

2. How many coulombs would be consumed in the reduction of 100 g of 1,3-dibromopropane to produce cyclopropane?

$$Br{-}CH_2{-}CH_2{-}CH_2{-}Br + 2e \longrightarrow \triangle \quad + 2Br^{\ominus}$$

3. How long would it theoretically take to complete the reaction in question 2, if the electrolysis is carried out at 5 A?

4. Assuming that the completion of the reaction in question 2 actually required 7 hr, calculate the electrical efficiency of the reaction.

5. If the iR drop of the medium in question 4 was found to be 50 V, calculate the amount of energy in kilowatt-hours dissipated in the solution.

6. How many electrons are there in 1 coulomb?

2.3-2.6. ELECTRON TRANSFER

1. Explain the difference between electrode potential and cell voltage.

2. a. What is the cell voltage of a system which
has an iR drop of 10 V, and a cathode and anode potential
of -2 V and + 1.5 V respectively, versus s.c.e.

b. What would the cell voltage of the above system be if the reference electrode is Ag/AgCl?

3. What is the effect of current density on the anode
and cathode potentials?

4. Using the Nernst equation, explain how a glass
electrode, quinhydrone, etc., electrodes can be used as pH
electrodes and all give the same response, i.e., 0.059 V/pH
unit.

5. How does the electrochemical transfer coefficient
α resemble the Bronsted coefficient in acid-base catalysis?

6. Define overvoltage.

7. For an irreversible one-electron reduction reaction,
how does a change of -1 V in the electrode potential affect
the activation energy of the reaction? Assume the transfer
coefficient equals 0.5.

8. Define the term "potential of zero charge" (p.z.c.).

2.7-2.9. PHYSICAL CONCEPTS AND CONTROLLED POTENTIAL
ELECTROLYSIS

1. What is the role of diffusion in an electrochemical
cell?

2. The oxidation of hydroquinone occurs at a less
positive potential than that for phenol; however, little
if any oxidation of hydroquinone occurs in the presence of
excess phenol; explain.

3. a. Assuming that the double layer extends about
7$\overset{o}{A}$ into the bulk of the solution, what would the electric
field (V/cm) be if a 2 V drop occurred in that region?

b. How might this field affect electrochemical
reactions?

4.   List the important variables that should be considered in carrying out an electrochemical reaction.

5.   Distinguish the following electrodes: (a) working;
(b) reference; and (c) auxiliary.

6.   List some of the important properties of a cell
divider which make it suitable for use in electrochemical
reactions.

7.   Why is a calomel electrode used as a reference
electrode?

8.   Draw a simple sketch illustrating the relationship
between the current passing through the cell, as a function
of time: (a) at constant cell voltage; and (b) at constant
electrode potential.

9.   What is the primary function of a supporting
electrolyte?   What other functions might it have?

10.   Under what circumstance is the use of controlled
potential electrolysis of value in organic synthesis?

ANSWERS (CHAPTER 2)

2.1.   FARADAY'S AND OHM'S LAWS, AND 2.2. UNITS

1.   a.   A Faraday is the amount of electricity in
coulombs required to oxidize or reduce one electron-
equivalent of a substance.

b.   A coulomb is the amount of electricity passed
when 1 A flows for 1 sec.

c.   The mho is the unit of conductivity and equals
the reciprocal of the resistance.

2.   95,500 C.

3.   19,100 sec or 318 min or 5.3 hr.

4.   75.6%.

5.   1.25 kwh.

6. 6.237×10^{18}.

2.3-2.6.   ELECTRON TRANSFER

1.   The electrode potential is the voltage difference across an electrode/solution interface, measured with respect to a reference electrode.   Cell voltage is simply the difference in voltage between anode and cathode.

2.   a.   13.5 V = ([1.5-(-2.0)]+ 10).

b.   13.5 V since ($E_{anode} - E_{cathode}$) is independent of the reference electrode, providing the same reference electrode is used to measure both potentials.

3.   Increasing the current density will always make the cathode potential more negative and the anode potential more positive.

4.   Most reversible electron transfer reactions involving protons can be expressed in the following form:

$$A+B+ne^- + nH^+ = C+D$$

The Nernst equation predicts the following:

$$(E-E_o) = -0.059/_n \ \log \left[\frac{(C)(D)}{(A)(B)(H^+)^n} \right]$$

$$= 0.059 (pH) + const.$$

Therefore, the measured potential is always a linear function of pH, and the slope (response) is always 0.059 V per pH unit.

5.   For general acid catalysis, a proton is involved in the transition state.   A common interpretation of the Bronsted coefficient α is that it represents the relative position of the proton, between reaction centers, in the transition state.   For example, a value $\alpha = \frac{1}{2}$ would imply that the proton lies midway between reaction centers, and

that for acids with ionization constant differing by a
factor of ten, the rate constant would differ by a factor
of $10^{\frac{1}{2}}$, or 3.16. The electrochemical transfer coefficient
α has been interpreted in an exactly analogous manner,
i.e., that it represents the position of the electron in
the transition state.

6. Overvoltage is the difference in potential between
the measured electrode potential and the theoretical value
at equilibrium (no current flow and no concentration gradients).

7. The activation energy barrier would be lowered by
11.5 kcal/mole (0.5 V).

8. The potential of zero charge is the measured
electrode potential at which there is no net charge at the
electrode surface. At potentials more positive than the
p.z.c., the double layer will consist of anions first, then
cations. Conversely, at potentials more negative than the
p.z.c., a cationic layer is formed as the first layer.

2.7-2.9. PHYSICAL CONCEPTS AND CONTROLLED POTENTIAL
ELECTROLYSIS

1. When current flows, a concentration gradient forms
at the electrode/solution interface. Material will diffuse
(either toward or away from the electrode) in the direction
of the concentration gradient.

2. Phenol is much more strongly <u>adsorbed</u> than hydroquinone, and therefore prevents oxidation of hydroquinone.
This is presumably due to the fact that hydroquinone is
more compatible than phenol toward water.

3.   a. 2 V/7×10^{-8} cm $= 2.86 \times 10^{7}$ V/cm.

b. In the presence of an electric field of that
magnitude, polarizable molecules, or molecules which have
a dipole moment, can be significantly distorted or oriented,
and hence, the field can have an effect on the nature of
the reaction product.

4. Cell design, electrode materials, agitation, electrode potential, current density, cell voltage, solvent,
electrolyte, temperature.

5. The working electrode is the one at which the reaction of interest is occurring; the auxiliary is the other
electrode which completes the cell. The reference electrode
is only used to measure electrode potential.

6.   (1) Permeability of materials through it.

   (2) Selectivity in discriminating in favor of
        or against certain materials.

   (3) Resistivity toward passage of ionic current.

   (4) Chemical resistance toward medium.

   (5) Thermal stability.

   (6) Rigidity.

7. The half-cell reaction in question is :

$$Hg_2Cl_2 + 2e = 2Hg^{O} + 2Cl^{-}$$

a. The half-cell reaction is highly reversible.

b. The activity of Hg_2Cl_2 and Hg^{O}, both being
insoluble, are constant.

c. The Cl^{-} concentration is maintained constant
by the use of saturated KCl solution.

8.

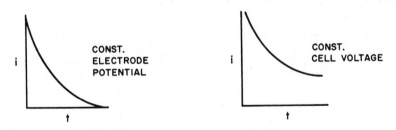

Decays to zero current;   Does not, in general,
100% electrical effi-   decay to zero current;
ciency is possible.   100% electrical effi-
                                  ciency impossible.

9. The primary function of the electrolyte is to
provide a source of ions to conduct current across the cell.
It may also take part in the electrochemical/chemical reactions; it may also preferentially adsorb on the electrode
surface.

10.   a. When the molecule contains more than one distinguishable electroactive site.

b. When the product formed during electrolysis
is itself susceptible to further electrolysis.

c. When the electrolysis of solvent and/or electrolyte can interfere with the desired reactions.

QUESTIONS (CHAPTER 3)

1.   a. List the general elements of an electrolysis
cell.

b. List the requirement of a cell divider.

2. Arrange the following metals in order of increasing hydrogen overvoltage:

a. lead   d. iron
b. mercury   e. platinum
c. nickel

3. What are the general requirements of solvents and
supporting electrolytes used in electrochemical reactions?

4. An organic compound is being electrolyzed under
the following conditions: dimethylformamide as the solvent,
tetraethylammonium bromide as the supporting electrolyte,
platinum electrodes 10 cm^2 each placed 10 cm apart. The
conductivity of the system, as measured by means of a conductivity cell (cell constant = 0.1 cm^{-1}) was 0.2 mhos.
At what voltage would the electrolysis be carried out in
order to obtain a current density of 10 A/dm^2?

5. An electromechanical coulometer (precision dc motor
type) was designed to have a response of 0.2 count/sec when
4 V are applied to its terminals. What shunt resistance

should be used across the motor terminals in order that the meter read out directly in coulombs?

6. What are the benefits of utilizing voltammetry in electroorganic synthesis?

7. What are the advantages and disadvantages of the use of mercury as dropping electrode in polarography?

8. What special considerations should the organic chemist be aware of in graduating from a laboratory-scale to a large-scale (industrial) electrochemical operation?

ANSWERS (CHAPTER 3)

1a. (1) A container, inert and nonconductive (unless it is to be used as one of the electrodes).

(2) Provision for inclusion of cell divider.

(3) Provision for electrode placement (preferably symmetrical).

(4) Suitable agitation.

(5) Reference electrode.

b. (1) Permeability to ions (preferably impermeable to other species).

(2) Stability toward electrolytic medium at the electrolysis temperature.

(3) Mechanical strength and integrity (pinhole-free) toward any pressure difference between compartments.

2. Pt, Fe, Ni, Pb, Hg.

3. (1) Stability toward electrolysis conditions (range of accessible potentials).

(2) Solubility of starting material.

(3) Conductivity.

(4) Reactivity toward products of intermediates (either nonreactive or reacts to give desired product).

(5) Ease of purification and separation.

(6) Degree of adsorption on electrodes.

(7) Toxicity and ease of handling.

4. Specific cond. = 0.1 x 0.2 = 0.02 mho/cm

conductance of electrolysis cell $\left(10 \ cm/10cm^2\right)$
= 0.02 mho

current = $(10 \ A/dm^2)$ x $(0.1 \ dm^2)$ = 1 A

∴ V = current/conductance = 1/0.02 = <u>50 V</u>

5. $R_s = \left(\dfrac{counts}{coulomb}\right)/K$

K = 0.2/5 = 0.05 counts/sec/V

∴R_s = 1/0.05 = 20 ohms

6.   (1) Determines whether compound is itself
electroactive.

   (2) Gives information concerning range of
electrode potentials to be used.

   (3) Provides data concerning the number of
distinct electrochemical steps; therefore provides the
basis for controlled electrolysis.

   (4) Useful in elucidating reaction mechanism.

   (5) Allows quantitative measure of disappearance
of starting material and/or appearance of product during
electrolysis.

7. Advantages:

   (1) Since a fresh surface is constantly being
formed, no surface contamination of the electrode occurs.

   (2) Since mercury is readily available in very
pure form, and since no pretreatments are required, it is
an extremely convenient material for use in voltammetry.

   (3) Its high hydrogen overvoltage allows it to
be used at highly cathodic potentials in the presence of
water or hydronium ions.

   Disadvantages:

   (1) Because of its ease of oxidation, mercury
has a limited range of accessible anodic potentials.

   (2) Has the possibility of forming organomercury
compounds (e.g., with sulfur-containing compounds).

(3). Mercury is hazardous because of the combination of high toxicity and high vapor pressure.

8.   (1) Must convert to continuous operation, which implies complete rethinking about the chemistry involved, and about the cell design.

(2) Minimum electrode spacing to maximize power efficiency may cause mass transfer problems; can be cured by fast recirculating loop; flow pattern within cell.

(3) High current density operation and its effect on chemical and electrical efficiency.

(4) Cell divider and support.

(5) Material of construction (plumbing, etc.).

(6) Electrode materials--cost, fabrication, long-term effects.

(7) Continuous recovery system, integrated within framework of continuous electrolysis.

(8) Possible need for computer control of entire plant.

(9) Maintenance, number of operators required, availability of power, cost of power.

QUESTIONS AND ANSWERS (CHAPTER 4)

4.1. REDUCTION OF CARBONYL COMPOUNDS

QUESTIONS

1.   In the reduction of ketones, what conditions tend to favor the formation of pinacols over alcohols?

2.   What product is formed from the following reduction?

a. 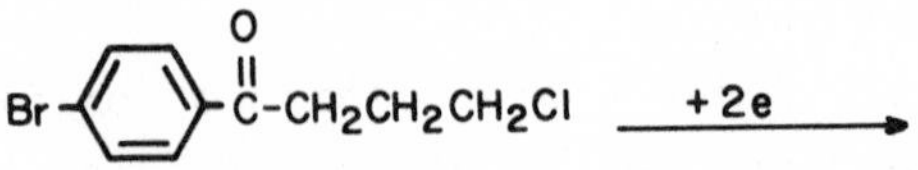 $Br\text{-}C_6H_4\text{-}\overset{\displaystyle O}{\overset{\|}{C}}\text{-}CH_2CH_2CH_2Cl \quad \xrightarrow{+2e}$

b. 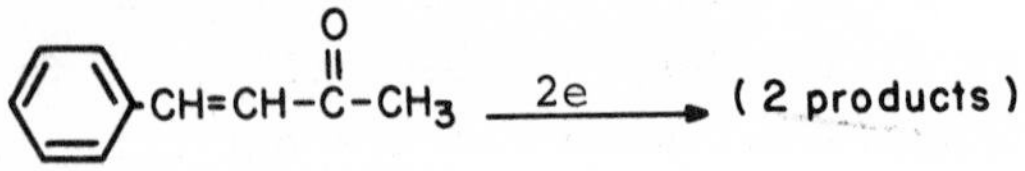 $C_6H_5\text{-}CH=CH-\overset{\displaystyle O}{\overset{\|}{C}}\text{-}CH_3 \quad \xrightarrow{2e} \quad (\text{2 products})$

3.   List in order of ease of reduction:

    a. $CH_3\overset{\overset{\text{O}}{\|}}{C}-OEt$

    b. $CH_3\overset{\overset{\text{O}}{\|}}{C}-CH_3$

    c. 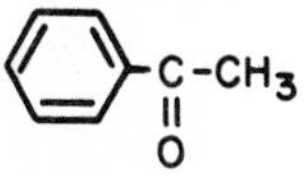   ⟨benzene ring⟩$-\overset{\overset{\text{}}{}}{\underset{\underset{\text{O}}{\|}}{C}}-CH_3$

    d. CH_3CHO

    e. $H-\overset{\overset{\text{O}}{\|}}{C}-N(CH_3)_2$

4.   Explain the following observations:

    a.   Reduction of 1,4-diacetylbenzene is much easier than acetophenone.

    b.   The half-reduced (keto-alcohol) product can be formed exclusively even though α,ω-diketones normally reduce to glycols.

5. Reconcile the following apparently contradictory observations concerning the reduction of p-aminoacetophenone:

    a.   The use of Hg (high hydrogen overvoltage) as the cathode favors alcohol formation, while the use of tin (lower hydrogen overvoltage) favors the pinacol.

    b.   At a given electrode, increasing electrode potentials favor pinacol formation.

ANSWERS

1.   a.   pH, under acidic conditions the formation of pinacols is favored.

    b.   High concentration of the ketone favors the formation of pinacol.

    c.   Operation at a low hydrogen overvoltage cathode favors the formation of pinacol.

2.   a. 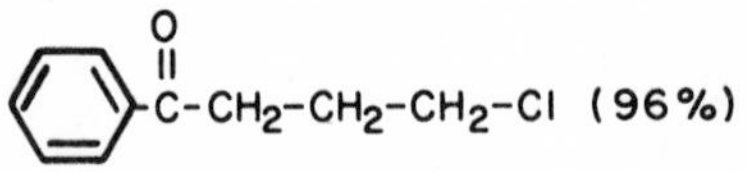 $\overset{\overset{\displaystyle O}{\parallel}}{C}$-CH$_2$-CH$_2$-CH$_2$-Cl (96%)

b. 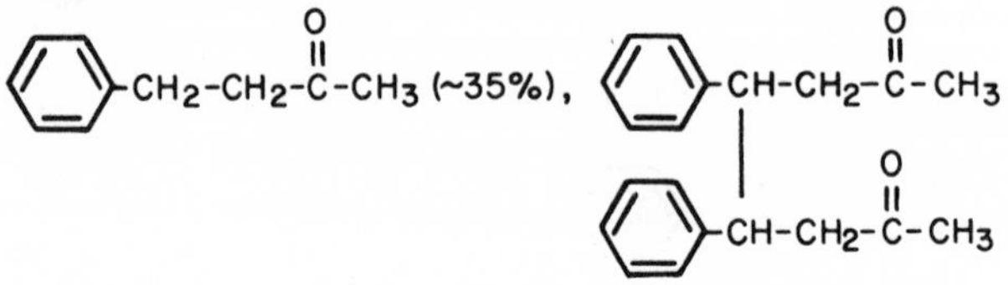   -CH$_2$-CH$_2$-$\overset{\overset{\displaystyle O}{\parallel}}{C}$-CH$_3$ (~35%),   -CH-CH$_2$-$\overset{\overset{\displaystyle O}{\parallel}}{C}$-CH$_3$

-CH-CH$_2$-$\overset{\overset{\displaystyle O}{\parallel}}{C}$-CH$_3$

3.   (e), (a), (b), (d), (c).

4.

(More stable than)

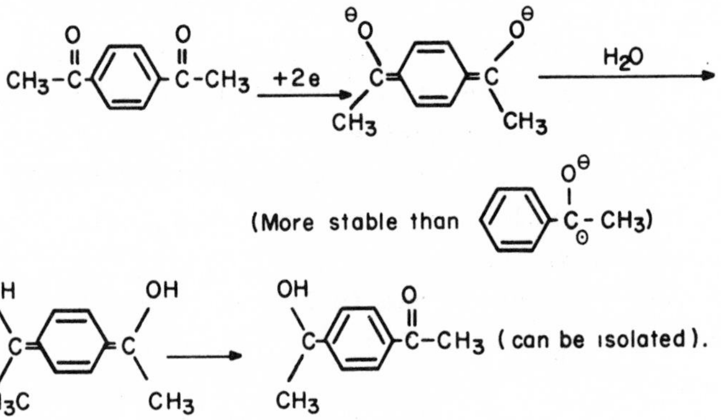

5.   Since Hg has a high hydrogen overvoltage, a high negative potential can be achieved to afford the dianion

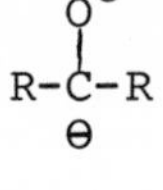

$$R-\overset{\overset{\displaystyle O^{\ominus}}{|}}{\underset{\ominus}{C}}-R$$

which does not dimerize, but affords the corresponding alcohol, whereas the use of tin allows the formation of the radical anion only, which can dimerize to the pinacol. The use of a higher electrode potential increases the current density, which in turn forms the radical anion in high concentrations, thus favoring dimerization.

4.2. NITRO GROUPS

QUESTIONS

1. What role does the pH of the medium play in determining the nature of the product(s) obtained from the electrochemical reduction of nitrobenzene?

2. Under what reaction conditions could you prepare
the following compounds from the reduction of nitrobenzene:

    a. p-Aminophenol   c. Hydrazobenzene
    b. Azobenzene   d. Aniline

3. Propose a procedure for the preparation of 4,4'-
dihydroxydiphenyl (diphenol) from the electrochemical reduction of nitrobenzene.

4. How can p-ethoxyaniline be prepared from the reduction of nitrobenzene?

ANSWERS

1. Reaction conditions with pH values under 7 favor
the formation of anilin or p-aminophenol. Reactions with
pH values above 7 favor the formation of coupling products,
e.g., azobenzene, etc.

2. a.

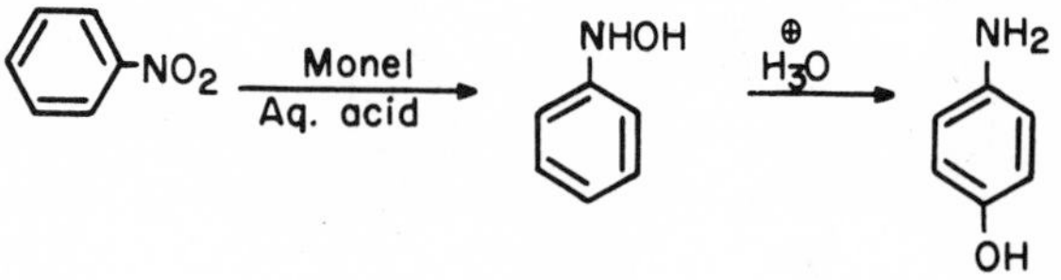

b.

c.

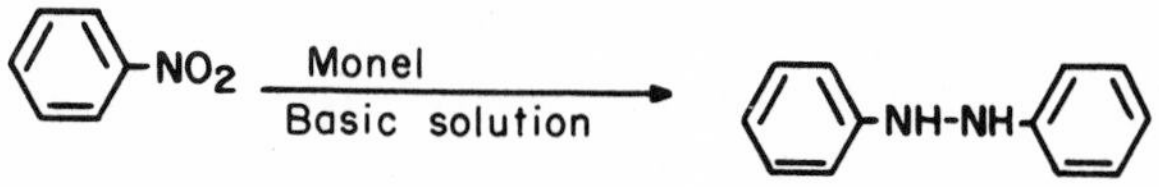

d.

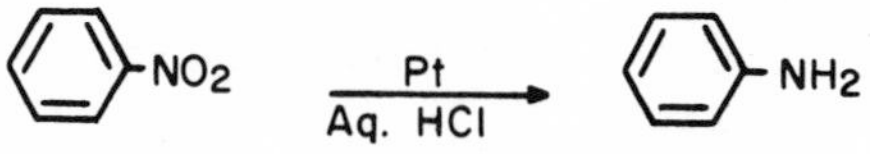

3.

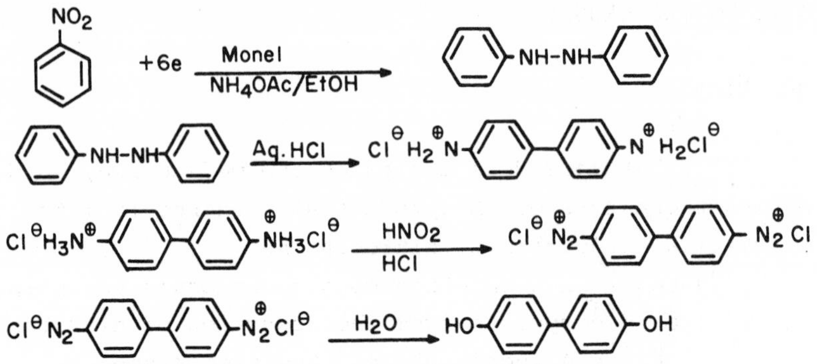

$$NO_2 \quad +6e \quad \xrightarrow[\text{NH}_4\text{OAc/EtOH}]{\text{Monel}} \quad \text{—NH-NH—}$$

$$\text{—NH-NH—} \quad \xrightarrow{\text{Aq. HCl}} \quad Cl^{\ominus} H_2\overset{\oplus}{N}\text{—}\text{—}\overset{\oplus}{N} H_2 Cl^{\ominus}$$

$$Cl^{\ominus}H_3\overset{\oplus}{N}\text{—}\text{—}\overset{\oplus}{N}H_3 Cl^{\ominus} \quad \xrightarrow[\text{HCl}]{\text{HNO}_2} \quad Cl^{\ominus}\overset{\oplus}{N}_2\text{—}\text{—}\overset{\oplus}{N}_2 Cl$$

$$Cl^{\ominus}\overset{\oplus}{N}_2\text{—}\text{—}\overset{\oplus}{N}_2 Cl^{\ominus} \quad \xrightarrow{\text{H}_2\text{O}} \quad HO\text{—}\text{—}OH$$

An alternate procedure which does not involve the preparation of diazonium salts (hazardous) was used by the authors:

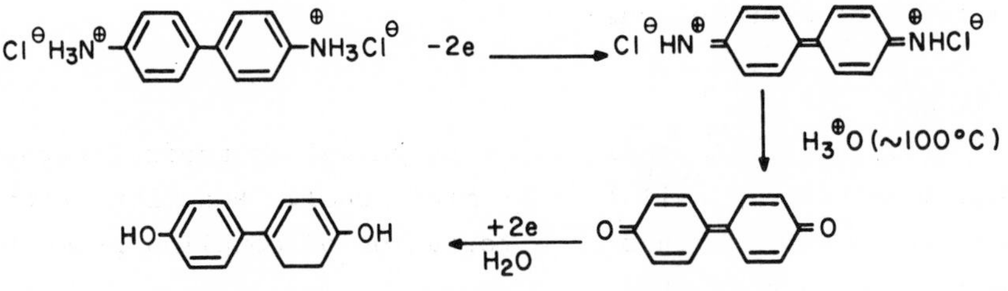

$$Cl^{\ominus}H_3\overset{\oplus}{N}\text{—}\text{—}\overset{\oplus}{N}H_3 Cl^{\ominus} \quad \xrightarrow{-2e} \quad Cl^{\ominus}\overset{\oplus}{H N}=\text{=}=\overset{\oplus}{N}H Cl^{\ominus}$$

$$\downarrow H_3\overset{\oplus}{O}\,(\sim100°C)$$

$$HO\text{—}\text{—}OH \quad \xleftarrow[\text{H}_2\text{O}]{+2e} \quad O=\text{=}=O$$

4.

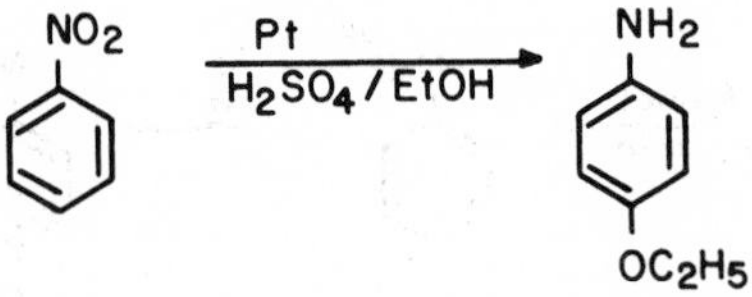

$$NO_2 \quad \xrightarrow[\text{H}_2\text{SO}_4/\text{EtOH}]{\text{Pt}} \quad NH_2 \quad \text{(see ref. 19)}$$

$$OC_2H_5$$

4.3. CARBON-HALOGEN BONDS

QUESTIONS

   1. List the following organic halides in order of ease of electrochemical reduction:

(a) CH_3Br, (b) 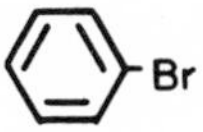 ⟨◯⟩–Br (c) 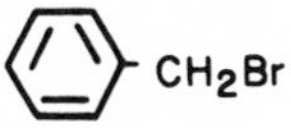 ⟨◯⟩–CH_2Br

(d) CH_2Br_2 (e) $Br-CH_2-CH_2Br$ (f) $CH_2=CH-CH_2I$

(g) $CH_2=CH-Br$ (h) 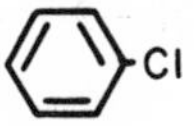 ⟨◯⟩–Cl

2. Propose a mechanism for the preparation of α-naphthol from 1,2-dibromobenzene and furan.

3. Complete the following reactions:

a. 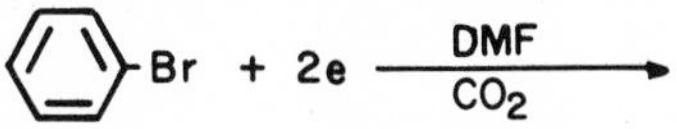   ⟨◯⟩–Br + 2e $\xrightarrow[CO_2]{DMF}$

b. 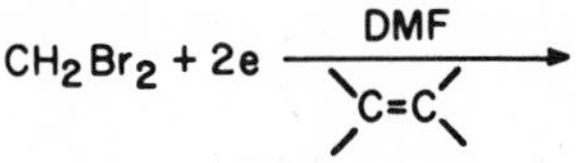 CH_2Br_2 + 2e $\xrightarrow[\diagup C=C \diagdown]{DMF}$

c.
$$\underset{BrCH_2}{\overset{BrCH_2}{\diagdown}}\underset{CH_2Br}{\overset{CH_2Br}{\diagup}}C \quad + \ 2e \xrightarrow{-1.8v\,(s.c.e.)}$$

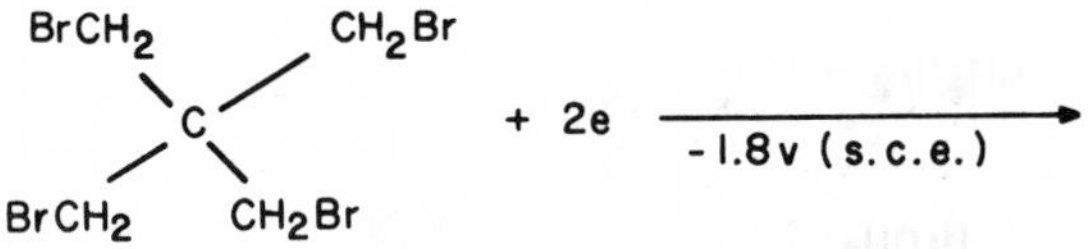

4. Which of the following compounds reduces easier?

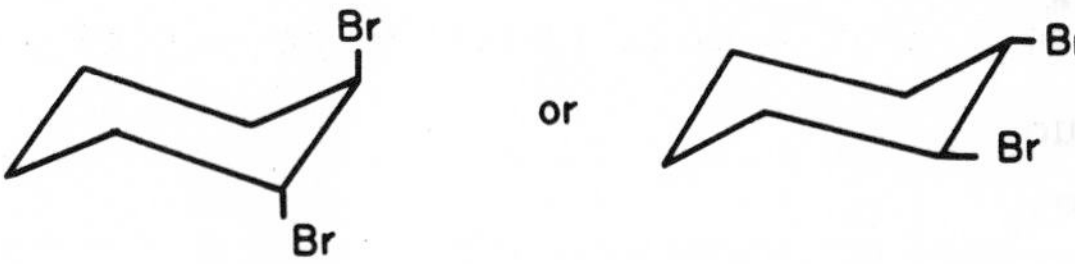

or

5. What would be the most likely product of the following reaction?

+ 2e $\longrightarrow$

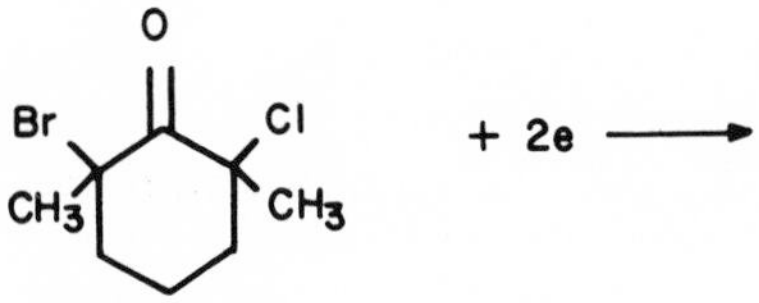

ANSWERS

1.   (h), (g), (b), (a), (d), (e), (c), (f).

2.

3.   a.

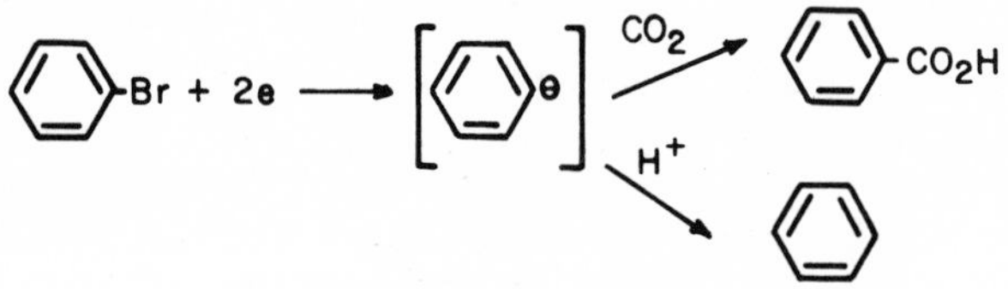

b.

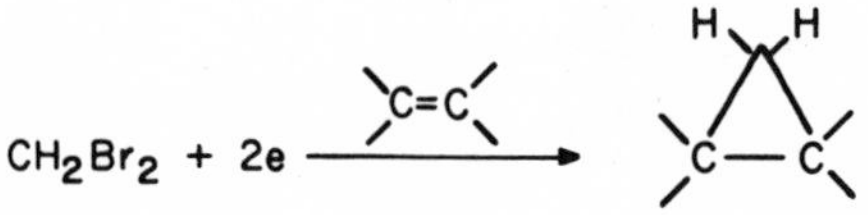

c.

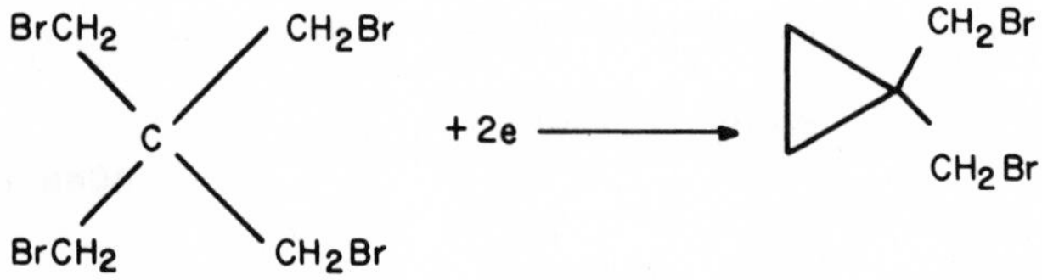

further reduction at a more negative potential leads to spiropentane.

4. 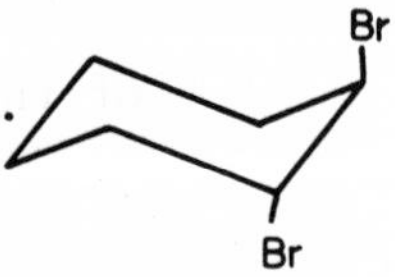   easier than the cis-form

5. 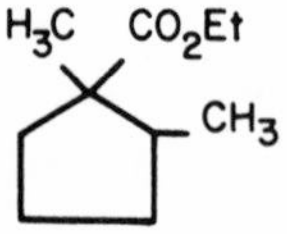   this reaction is similar to the Favorski rearrangement.

4.4. REDUCTION OF UNSATURATED COMPOUNDS

QUESTIONS

1.   a.   What is the significance of adiponitrile in
Nylon 6,6 production?

   b.   What are some possible advantages of producing
adiponitrile electrochemically?

2.   Compare the electrochemical reduction of benzene
with the Birch reduction of benzene.

3.   Propose a mechanism which explains the different
products obtained from electrochemical reduction of benzene
in divided vs undivided cells.

4.   In principle, four separate coupling products could
be formed from electrolytic reduction of R-CH=CH-R'.   What
types of substituents would you expect to favor?

   a.   Predominant formation of a single product.

   b.   Formation of the several possible products.

5.   a.   Discuss, in terms of molecular orbital theory,
the ease of reduction of aromatic compounds.

   b.   Arrange in order of ease of reduction:

   (1) anthracene, (2) napthalene, (3) tropenium
fluoroborate, (4) benzene.

ANSWERS

1.   a.   Adiponitrile is an intermediate to both adipic
acid and hexamethylene diamine, which are the comonomers
used to prepare Nylon 6,6.

   b.   The conventional preparation of adiponitrile
from adipic acid (as compared to its preparation from the
electrochemical dimerization acrylonitrile) has two short-
comings: (1) Adipic acid is almost twice as expensive as
acrylonitrile; and (2) On a weight basis, more adipic acid
needs to be used than acrylonitrile to prepare the same
amount of adiponitrile.

2.   Electrochemical:

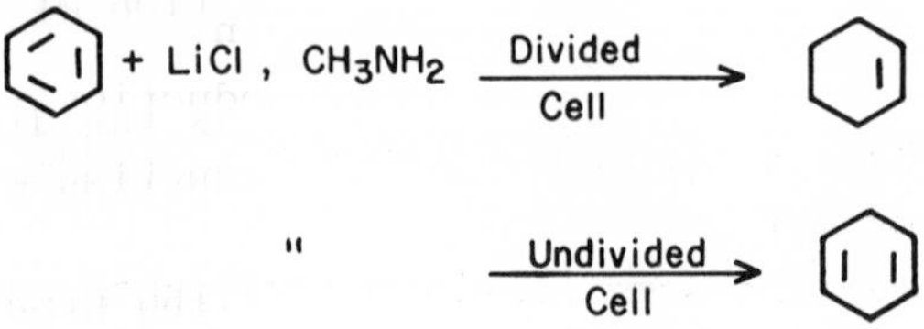

Chemical:

3.   Divided Cell

(a) $Li^{\oplus}$ + $1e$ $\longrightarrow$ Li^{0}

(b) 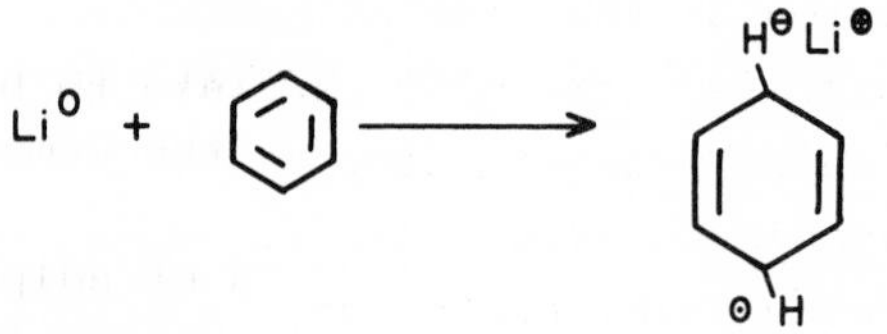 Li^{0} +

(c)

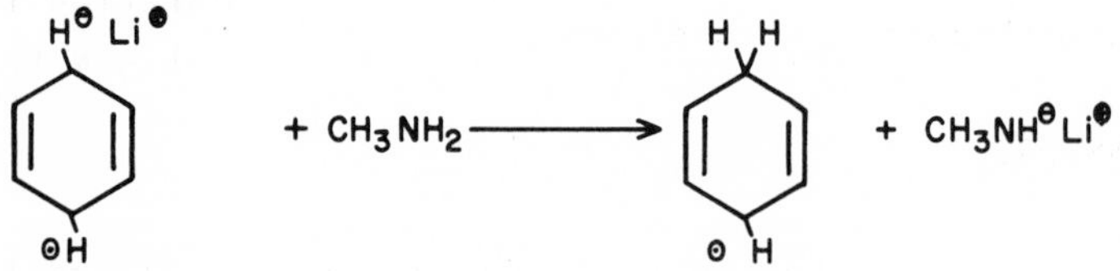

+ CH_3NH_2 $\longrightarrow$

+ $CH_3NH^{\ominus}Li^{\oplus}$

(d)

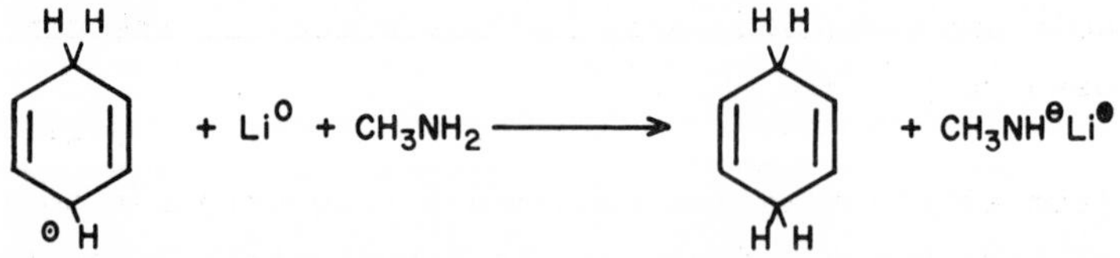

+ Li^{0} + CH_3NH_2 $\longrightarrow$

+ $CH_3NH^{\ominus}Li^{\oplus}$

(e)

+ $CH_3NH^{\ominus}Li^{\oplus}$ $\longrightarrow$

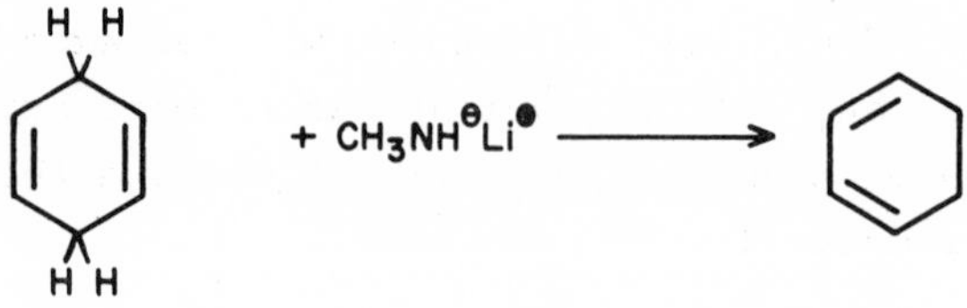

(f) 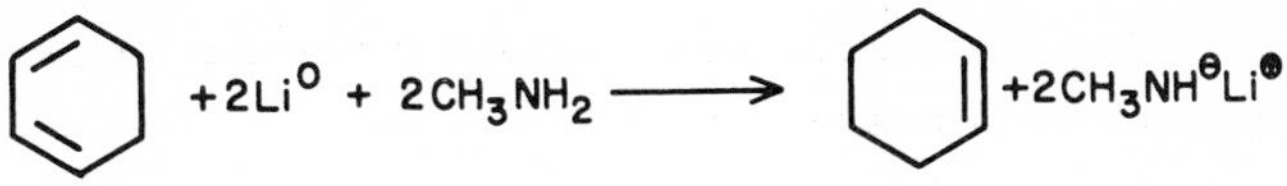 $+2Li^0 + 2CH_3NH_2 \longrightarrow$ $+2CH_3NH^\ominus Li^\oplus$

Undivided Cell:

Under these conditions methylamine is oxidized to methylamine hydrochloride, which neutralizes the methylamide anion, thus stopping the reaction from proceeding beyond 3d above.

4. a. For compounds where R and R' differ greatly in their ability to stabilize anions, e.g., acrylonitrile, one product (adiponitrile) will predominate.

b. For compounds where R and R' have comparable abilities to stabilize anions, a mixture of products is observed.

5. a. In the reduction of aromatic compounds, an electron is added to the lowest unoccupied molecular orbital. Thus it follows that the closer this orbital is to the zero energy level, the easier it is to reduce the aromatic compound.

b. (1) Tropenium fluoroborate; (2) anthracene; (3) naphthalene; (4) benzene.

4.5. REDUCTION OF CARBON-NITROGEN GROUPS

QUESTIONS

1. Complete the following reactions:

(a) 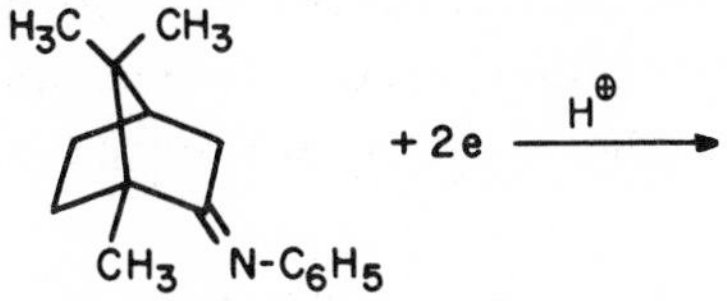 $+2e \xrightarrow{\ H^\oplus\ }$

(b) 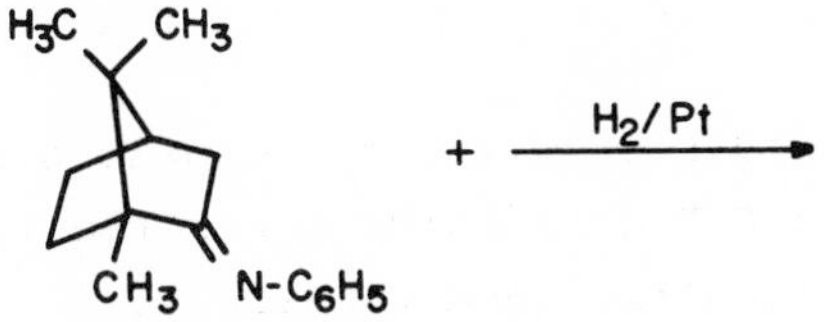 $+ \xrightarrow{\ H_2/Pt\ }$

2. How could one easily distinguish between the syn-
and anti- forms of oximes?

3. Explain the following observations:

a. $CH_3-CN \xrightarrow[\text{Aq.HCl}]{\text{Pt}} CH_3-CH_2-NH_2$
                                              100%

b. $CH_3-CN \xrightarrow[R_4N^+X^-]{\text{Hg}} (-\underset{\underset{CH_3}{|}}{C}=N-)_n$

4. Explain how the pH of a reaction medium affects
the mechanism of reduction of oximes.

ANSWERS

(1) (a)

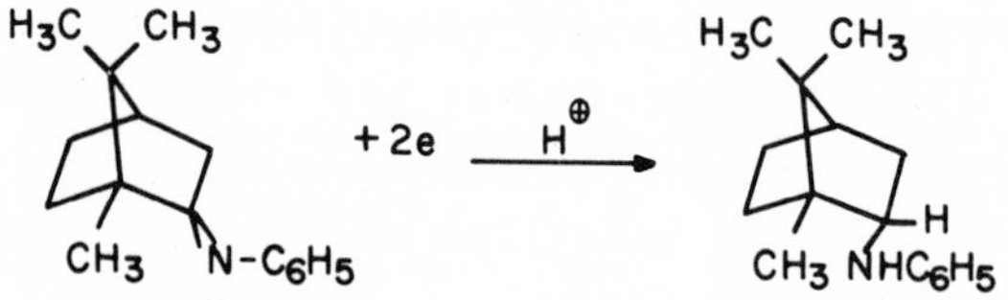

(b)   " H_2/Pt $\longrightarrow$

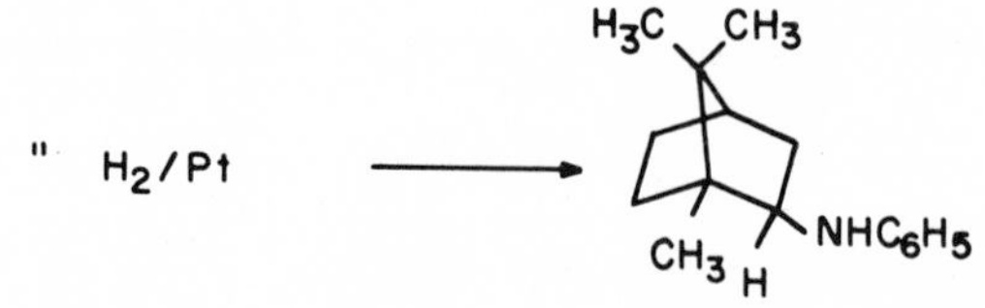

2. By polarography; under basic conditions, only the
syn-isomer exhibits a polarographic wave.

3.   a.   This reaction may involve the formation of
hydrogen molecules which affect the reduction of nitriles
to amines.

b. The reduction involves the formation of
anionic intermediate $CH_3-C=N^\ominus$ which in the absence of a
proton source induces the polymerization of acetonitrile:

$$CH_3-\overset{\ominus}{C}=N^\ominus \longrightarrow \underset{\underset{CH_3}{|}}{C}\equiv N \longrightarrow CH_3-\overset{\ominus}{C}=N-\underset{\underset{CH_3}{|}}{C}=N^\ominus \xrightarrow{CH_3CN} \text{etc.}$$

4. Under acidic conditions the oximes are first pro-
tonated to afford $C-N-\overset{\oplus}{OH}_2$, which makes the reduction much
easier than the reduction under basic conditions.

$$\rangle C = NOH + H_3 \overset{\oplus}{O} \rightleftharpoons \rangle C = \overset{\oplus}{N} - OH_2$$

$$\rangle \overset{\oplus}{C} = NOH_2 + 2e \xrightarrow{H_3 \overset{\oplus}{O}} \rangle C = NH + 2e \xrightarrow{H_3 \overset{\oplus}{O}} \rangle CH - NH_2$$

$$\rangle C = NOH \underset{Basic}{\rightleftharpoons} \rangle C = N\overset{\ominus}{O} \quad (\text{difficult to reduce})$$

4.6. MISCELLANEOUS REDUCTIONS

QUESTIONS

1. Complete the following electrochemical reduction reactions:

a. $R - \overset{\overset{O}{\|}}{S} - R + 2e \xrightarrow{H^{\oplus}}$

b. 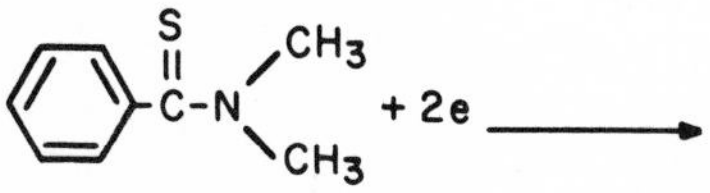   + 2e ⟶

c. $R - S - S - R + 2e \xrightarrow{H^{\oplus}}$

d. $R_3 SnCl + 2e \longrightarrow$

e. $R - O - O - R + 2e \xrightarrow{H^{\oplus}}$

f. $R_4 \overset{\oplus}{N} Cl^{\ominus} + 2e \xrightarrow{H^{\oplus}}$

g.    + 2e $\xrightarrow{H^{\oplus}}$

ANSWERS

1.   a. $R - \overset{\overset{O}{\|}}{S} - R + 2e \xrightarrow{H^{\oplus}} R - S - R + H_2O$

b. 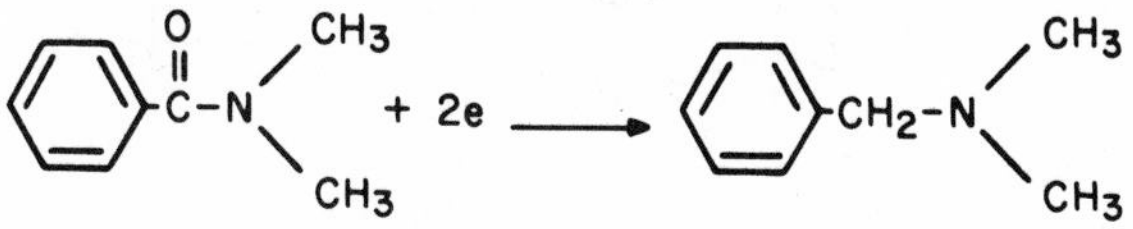   + 2e ⟶

c. $R-S-S-R + 2e \xrightarrow{H^{\oplus}}$ 2RSH

d. $R_3SnCl + 1e \longrightarrow \frac{1}{2}R_3Sn-SnR_3 + Cl^{\ominus}$

e. $R-O-O-R + 2e \xrightarrow{2H^{\oplus}}$ 2ROH

f. $R_4N^{\oplus}Cl^{\ominus} + 2e \xrightarrow{H^{\oplus}} R_3N + RH$

g.

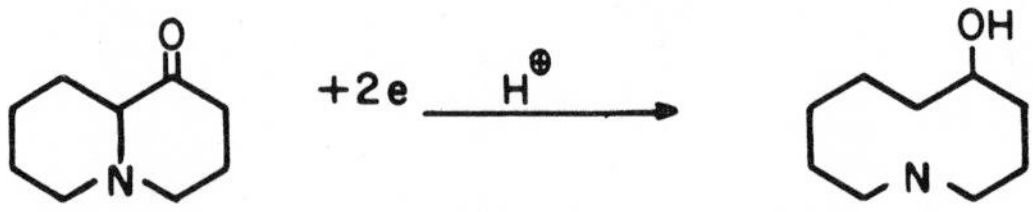

QUESTIONS AND ANSWERS (CHAPTER 5)

5.1.   THE KOLBE REACTION

QUESTIONS

1.   The electrochemical oxidation of carboxylic acids is affected by the nature of anode used. What is the predominant intermediate generated when carbon is used versus platinum?

2.   Why is the choice of solvent very critical in the Kolbe reaction?

3.   Propose a mechanism for the following reaction:

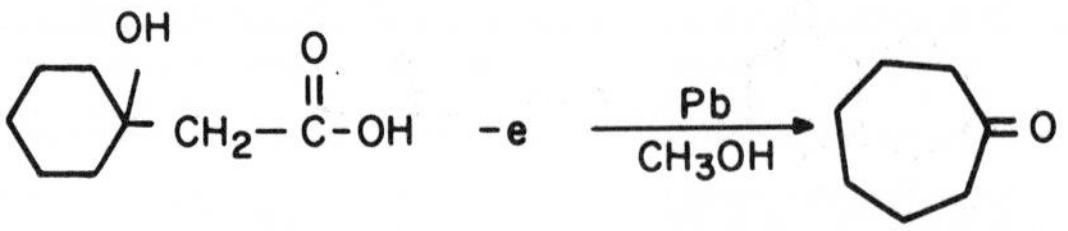

4.   Complete the following reactions:

a.

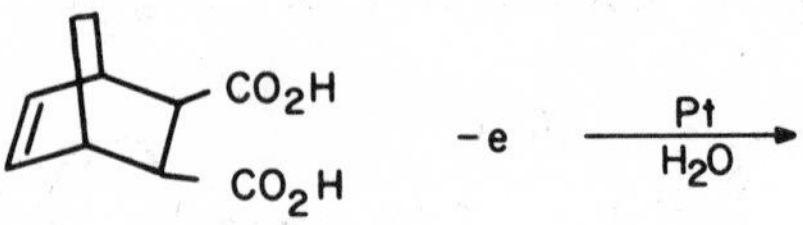

b. 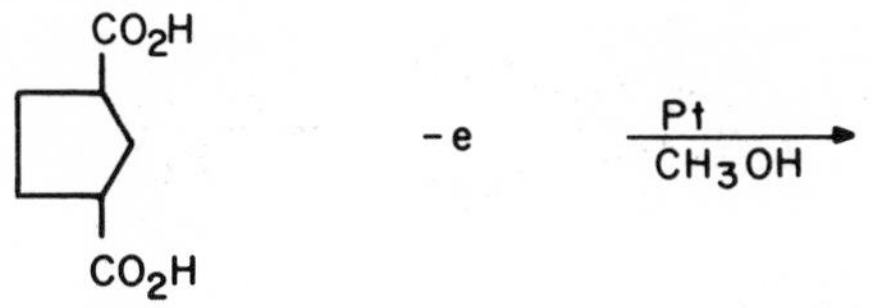 (cyclopentane with CO_2H groups) $-e$ $\xrightarrow[CH_3OH]{Pt}$

c. 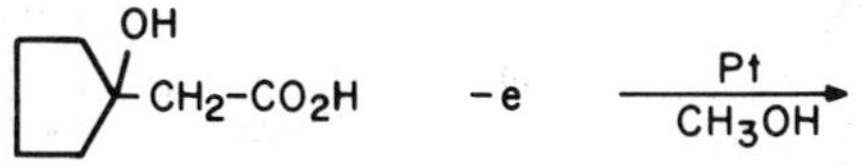 (cyclopentane with OH and CH_2-CO_2H) $-e$ $\xrightarrow[CH_3OH]{Pt}$

5. In the Kolbe reaction, how does the structure of a carboxylic acid affect the formation of coupling products?

ANSWERS

1. The use of carbon anodes favor the formation of carbonium ions, while the use of platinum favors the formation of radical intermediates.

2. Since carbonium ions can be formed in the Kolbe reaction, solvents such as methanol, ethanol, water, etc., can react with this intermediate to give a variety of products. In general, when radical intermediates are involved in the reaction, the choice of the solvent does not become critical.

3.

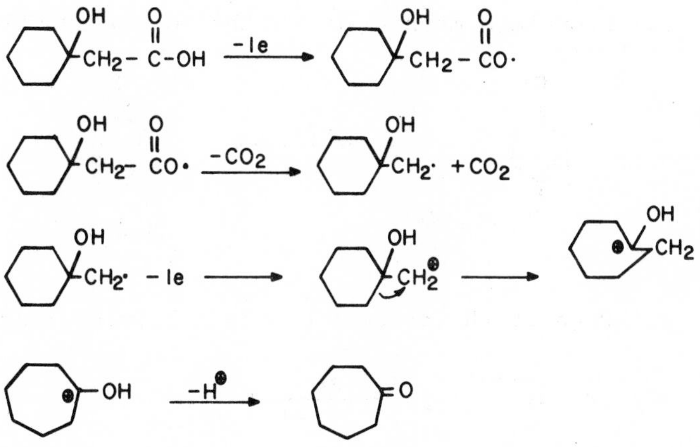

4. a.

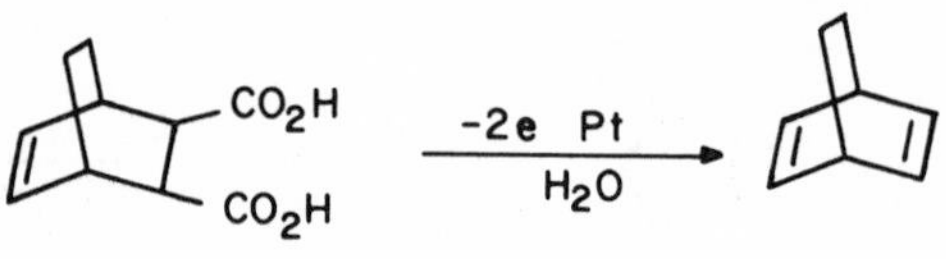
CO2H
CO2H
-2e Pt
H2O

b.

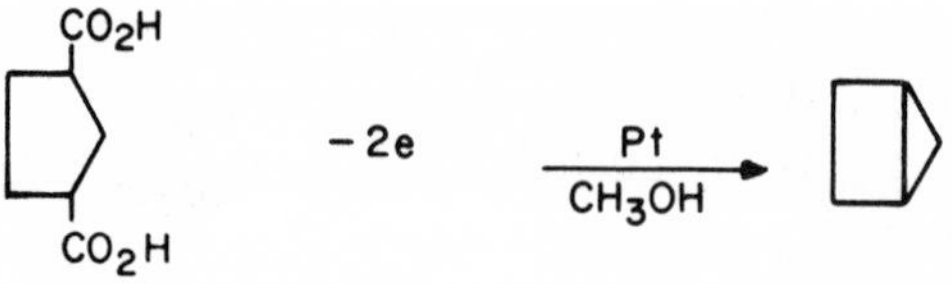
CO2H
CO2H
-2e
Pt
CH3OH

c.

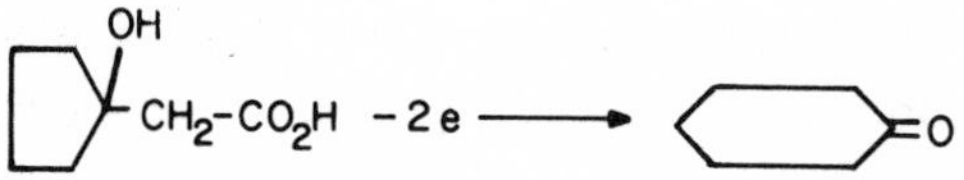
OH
CH2-CO2H -2e
O

5. The oxidation of a carboxylic acid proceeds via radical or carbonium intermediates, thus, substituents which destabilize these intermediates will impede the oxidation of the acid. Thus, α, β-unsaturated acids are very difficult to oxidize. Furthermore, no coupling products are formed from acids with the following α-substituents: cyano, methoxy, and amino.

5.2. OXIDATION OF UNSATURATED COMPOUNDS

QUESTIONS

1. Propose a mechanism for the following oxidation:

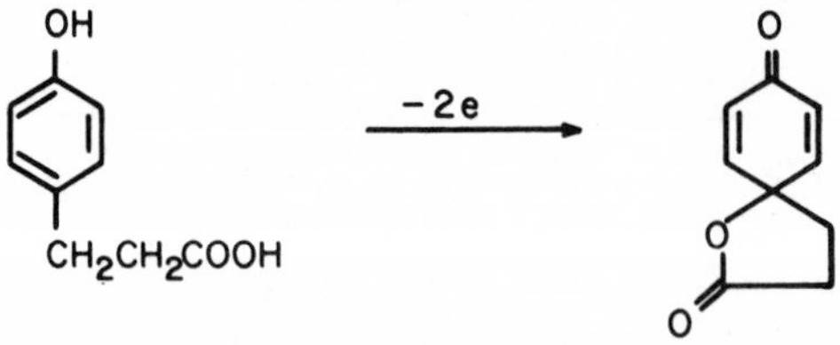
OH
CH2CH2COOH
-2e
O
O
O

2. Predict the product of the following reactions:

a.

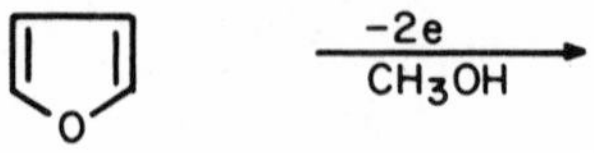
O
-2e
CH3OH

b.

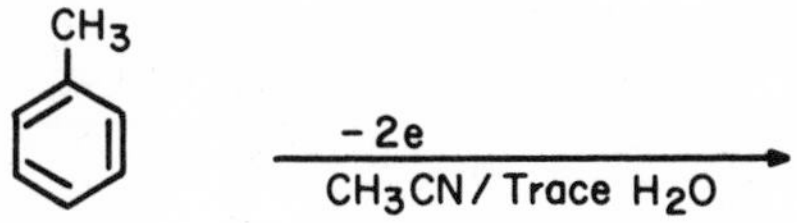

$$CH_3 \quad \xrightarrow[\text{CH}_3\text{CN/Trace H}_2\text{O}]{-2e}$$

c.

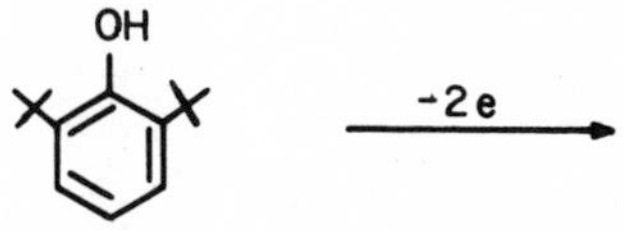

$$\xrightarrow{-2e}$$

d. $CH_2=CH-CH_3 \xrightarrow[\text{2) NaOH}]{\text{1) } -2e, \text{ NaCl}}$

e.

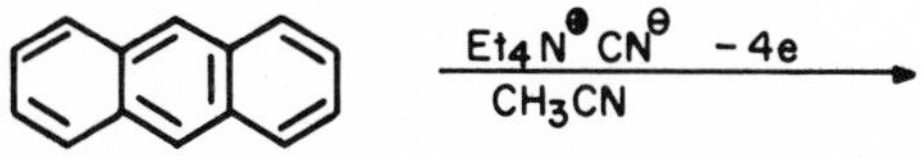

$$\xrightarrow[\text{CH}_3\text{CN}]{\text{Et}_4\text{N}^\oplus \text{ CN}^\ominus \ -4e}$$

3.   List in order of ease of oxidation:

a.    b. 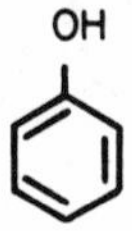   c.    d. $CH_2=CH_2$

4.   Propose a reasonable electrochemical reaction sequence to prepare diphenol from dimethyl aniline.

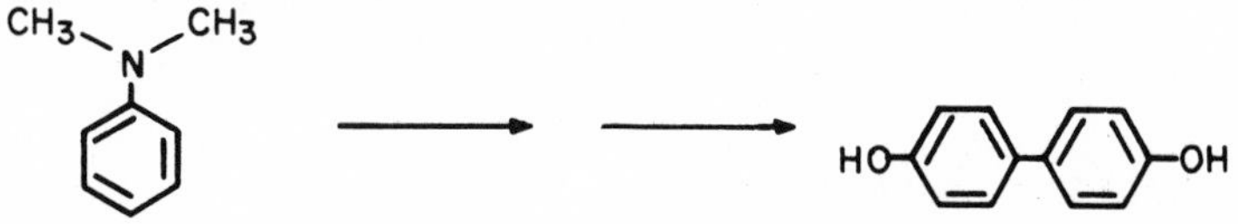

5.   What is the evidence that anodic cyanation does not proceed via the cyano radical?

ANSWERS

1.

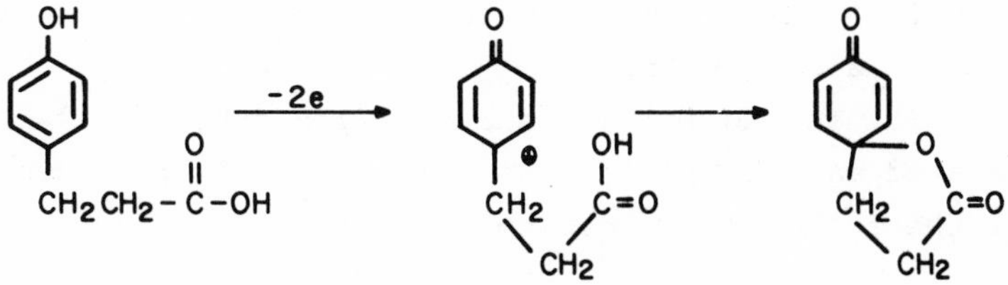

2. a.

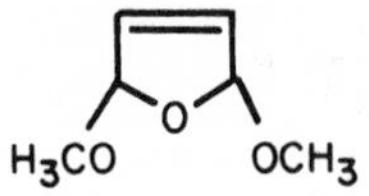

b. 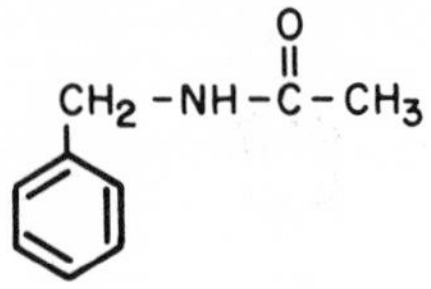
$$CH_2-NH-\overset{\overset{\displaystyle O}{\|}}{C}-CH_3$$

c. HO— ... —OH

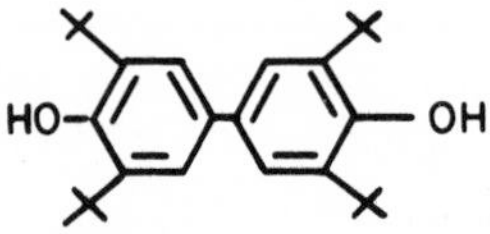

d. 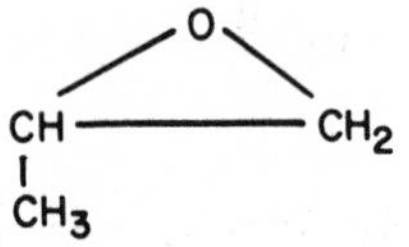 CH———CH_2 (e) 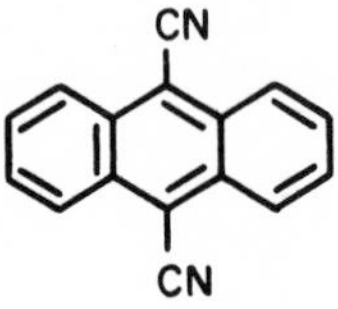
    |
    CH_3

3. d, b, c, a.

4. 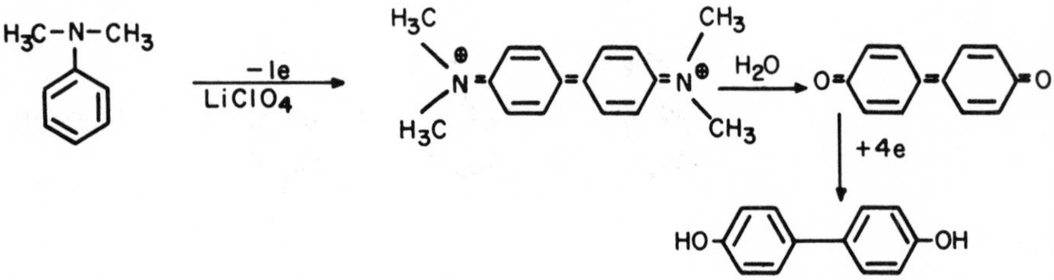 H_3C–N–CH_3 $\xrightarrow[\text{LiClO}_4]{-1e}$ ... $\xrightarrow{\text{H}_2\text{O}}$ O=...=O

    $\downarrow$ +4e

    HO— ... —OH

5. In spite of the fact that cyanide anions oxidize at a less positive potential than aromatic compounds, no cyanation occurs until the aromatic compounds are oxidized.

5.3.   ANODIC HALOGENATION

QUESTIONS

1. How does anodic halogenation compare with conventional chemical halogenation?

2. Outline the importance of the pH of the medium in anodic halogenation.

3. In anodic halogenation, the substrate being halogenated may itself undergo oxidation. Predict the product of the following oxidation:

4. Several metals are not stable in the presence of molecular halogen. Name some metals that have been used in anodic halogenation.

5. Outline the important precautions and choice of apparatus for anodic fluorination.

ANSWERS

1. Chemically speaking, the two reactions (i.e., addition of halogen to substrate) are essentially the same. However, anodic halogenation has the following advantages:

a. It allows the in situ preparation of halogens from salts such as sodium halides. This is quite important in anodic fluorination reactions where the handling of fluorine is quite difficult.

b. When a halide anion is a by-product from anodic halogenation, it can be electrochemically transformed to molecular halogen and reused.

2. Under acidic conditions anodic halogenation proceeds normally. However, under basic conditions, halogen molecules react with hydroxyl ions to form hypohalogen acid.

3.

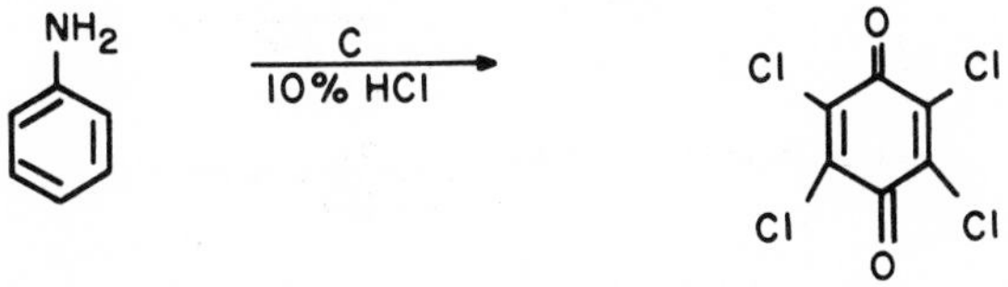

4. Carbon, nickel, platinum, monel.

5. In anodic fluorination the most important thing to remember is: (a) what material should be used to construct the electrolysis cell. For most purposes monel is

quite sufficient. The next thing to consider is (b) the
kind of metal to be used as electrodes. Nickel has been
the metal of choice. Next comes the (c) reference electrode,
and one that has been used successfully is the Hg/Hg_2F_2
electrode [see G. G. Koerber and T. DeVries, J. Am. Chem.
Soc., 74, 5008 (1952)]. In most electrochemical fluorination, a cell divider is not necessary.

The setup of the above apparatus will, naturally, depend on the type of reaction in question. However, if hydrogen fluoride is used as the solvent, the researcher should
keep the following precautions in mind: (1) The hazardous
nature of HF; and (2) because of its low boiling point, a
cooling device must be included in the apparatus. Some excellent references on experimental electrochemical fluorination are: S. Nagase and R. Kojima, Bull. Chem. Soc. Japan,
34, 1468 (1961); and S. Nagase, K. Tanaka, and H. Baba, Bull.
Chem. Soc. Japan, 38, 834 (1965).

5.4. MISCELLANEOUS OXIDATION

QUESTIONS

1. Complete the following electrochemical oxidation
reactions:

a.

b. Tetraethylborate $\xrightarrow[H_2O]{Pb}$

c. Phenyl sulfide $\xrightarrow[\text{Aq. HOAc}]{Pt}$

d.

e.

ANSWERS

1. a.

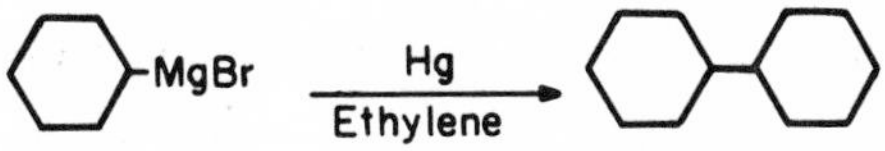

b. **Tetraethylborate** $\xrightarrow[H_2O]{Pb}$ **Tetraethyl lead**

c. **Phenylsulfide** $\xrightarrow[\text{Aq. HOAc}]{Pt}$ **Phenyl sulfone**

d. o-Iodotoluene $\xrightarrow[H_2SO_4]{Pt}$ Iodosobenzoic acid

e.

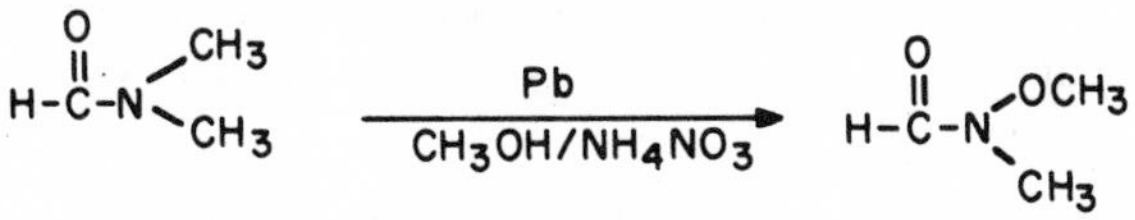

QUESTIONS (CHAPTER 6)

ELECTROINITIATED POLYMERIZATION

1. List some of the advantages of electroinitiated polymerization as compared to conventional polymerization techniques.

2. List the disadvantages of electroinitiated polymerization. What are some of the aspects of this technique that require further investigation?

3. How does cathodic polymerization differ from anodic polymerization? Is there an advantage in using one over the other?

4. How may both cathodic, as well as anodic, polymerization be effectively used at the same time to prepare specific types of polymers?

5. For a one-electron polymerization reaction which does not include a termination step, what would the molecular weight of a one-pound polymer be if the reaction consumes 0.05 Faraday?

ANSWERS (CHAPTER 6)

1.   a.   Diversity of reaction. Polymerization may
proceed via cationic (anodic), as well as anionic (cathodic)
intermediates.

b.   Product free from catalyst. (It should be
noted, however, that in electroinitiated polymerization, a
small amount of a supporting electrolyte is present in the
system.)

c.   Ease of control of molecular weight of polymer.

d.   No handling of potentially hazardous materials.

e.   Ease of formation of block copolymers.

f.   Possible polymerization on the surface of the
electrode.

2.   a.   Reaction is limited to polar media.

b.   Preparation of active ingredient is slow.

c.   Certain polymers may adhere on the electrode
and hinder further electrolysis.

3.   Cathodic polymerization proceeds via the generated
anion (or perhaps the radical) while anodic polymerization
proceeds via cations.

The advantage of one process over the other will
depend, to a large extent, on the nature of the monomer, and
thus, the researcher should consider both processes. It has
been the experience of the author that styrene, for example,
will afford a higher molecular weight polymer under cathodic
electrolysis.

4. Let us assume that it is desired to prepare a
block copolymer from monomers A and B. Monomer A is placed
in the cathode compartment of a divided cell, while monomer
B is placed in the anode compartment. Each monomer will
undergo a polymerization reaction but via different mechanisms, i.e., anionic and cationic:

At the end of the reaction the two polymers are allowed to
mix with each other, thus producing an A-B-type block copolymer (see the example of styrene and tetrahydrofuran
in Chap. 6).

5. Molecular weight = $\dfrac{\text{weight of monomer}}{n(\text{\# Faradays})}$

Molecular weight = $\dfrac{454g}{1(0.05)}$ = 9000

QUESTIONS (CHAPTER 7)

1. What are the advantages of electrocoating over
conventional coating techniques? What are the disadvantages?

2. List the electrochemical (and chemical, if any)
reactions that take place during anodic and cathodic electrocoating.

3. Define the following electrocoating parameters:

   a. Operating voltage

   b. Breakdown voltage

   c. Throwing power

4. Which of these polymeric products would require
the least amount of electricity to be deposited (i.e., deposit one mole polymer)? Explain.

   a. A polymer with molecular weight of 1000 and
containing ten carboxylate salts per mole.

   b. The same polymer with five carboxylate salts
per mole.

ANSWERS (CHAPTER 7)

1. Advantages:

   a. Pinhole-free uniform coatings are obtained.

   b. Coating is performed from an aqueous solution.
This eliminates pollution problems.

   c. This process allows the utilization of the
paint with minimum loss.

d.   The process is easily automated.

Disadvantages :

a.   The preparation of resins suitable for electro-coating is time consuming.

b.   On a large-scale operation, initial investment for equipment is high.

c.   For a resin to be suitable for electrocoating, certain functionality must be present in its backbone.   This limits the type of resins that can be used in electrocoating.

2.   Anodic:

a. $H_2O - 2e \longrightarrow \frac{1}{2}O_2 + 2H^{\oplus}$

b.

"Soluble resin"   "Insoluble resin"

c.

"Kolbe reaction

d.   Fe $-3e \longrightarrow Fe^{+3} \xrightarrow{3H_2O} Fe(OH)_3$
     (anode)

Reactions a and b are the most important in anodic electro-coating.

Cathodic:

a. $H_2O + e \longrightarrow \frac{1}{2}H_2 + {}^{\ominus}OH$

b.

"Soluble resin"   "Insoluble resin"

3.   a.   Operating Voltage: It is the voltage that will allow the deposition of a certain paint thickness in a certain time. For the automotive industry, the thickness is 0.75 mil in 1 or 2 minutes.

b. Breakdown Voltage: This is the minimum voltage at which the paint coating begins to rupture.

c. Throwing Power: The ability of the paint to
coat hidden areas.

4. As the operating voltage is, in general, proportional to the number of carboxylate groups in the backbone
of the polymer, polymer b will require the least amount of
electricity to be deposited. Although this amount would
neutralize the same number of carboxylate groups in polymer
a this polymer may still have enough carboxylate groups to
render it partially soluble.

APPENDIX B

GLOSSARY OF TERMS COMMONLY USED
IN ELECTROCHEMISTRY

Adsorption: The process whereby molecules in a fluid phase
 are collected onto the surface of a solid.

Adsorption isotherm: The relationship between the concentration of the material in the fluid phase and
 that at the solid surface (at a given temperature).

Ampere: The unit of current (rate of flow of electricity)
 defined as one coulomb of electricity per second.

Anode: The electrode at which oxidation takes place.

Anolyte: That part of the electrolyte surrounding the anode.

Auxiliary electrode: The electrode not directly involved in
 the reaction of interest. Also commonly referred
 to as the counter electrode.

Capacitance: The property of a system which is a measure
 of its ability to allow charge separation.

Cathode: The electrode at which reduction takes place.

Catholyte: That part of the electrolyte surrounding the
 cathode.

Cell divider: A permeable barrier that separates the
 catholyte from the anolyte.

Cascade cells: Electrolysis cells connected in series.

Cell voltage: Total voltage difference between the cathode
 and anode.

Concentration gradient: The difference in concentration
 of a substance between different regions in the
 solution.

Concentration polarization: The difference between the
 measured electrode potential and that calculated
 using bulk concentration (caused by a concentration gradient at the electrode surface).

387

<u>Conductivity</u>: The ability of a substance to conduct an
        electric current; measured in units of mhos.

<u>Controlled electrode potential</u>: The externally controlled
        working electrode potential.

<u>Coulomb</u>: Amount of electricity equal to one ampere flowing
        for one second (6.24×10^{18} electrons).

<u>Counter-electrode</u>: Same as auxiliary electrode.

<u>Current density</u>: Current per unit electrode area (commonly
        expressed in amps per square decimeter).

<u>Cyclic conversion</u>: The consumption of electricity by al-
        ternate oxidation and reduction of substrates with
        no net chemical change (most likely to occur in
        undivided electrolysis cells).

<u>Cyclic voltammetry</u>: A current-voltage relationship obtained
        by sweeping the working electrode potential first
        in one direction, then in the opposite direction.

<u>Depolarizer</u>: Any substance that tends to lower the elec-
        trode potential.

<u>Diffusion coefficient</u>: A characteristic of a molecule
        which depends on its size as well as the nature
        and temperature of the surrounding medium. It
        normally is defined as the proportionality between
        the flux of the substance and its concentration
        gradient (Fick's law).

<u>Diffusion-limiting current</u>: That current which depends only
        on the rate of diffusion of molecules towards and
        away from the electrode.

<u>Diffusion layer</u>: The region near the electrode over which
        a substantial concentration gradient occurs.

<u>Double layer</u>: The region adjacent to the electrode, con-
        sisting of oppositely charged ionic layers.

<u>Electric field</u>: Force vectors felt by a unit charge lo-
        cated at given points within a region of space.

<u>Electric efficiency</u>: The ratio between the calculated and
        actual amount of electricity used to form a pro-
        duct.

Electrocoating: The deposition of a coating onto an elec-
        trode in an electrolytic process.

Electrode potential: The measured potential difference be-
        tween a working electrode and a suitable reference
        electrode.

Electroinitiated polymerization: The initiation of a poly-
        merization reaction by an electrolytic process.

Electrolysis: The chemical alteration of molecules by the
        action of electron transfer at an electrode.

Electrolyte: The medium in which an electrolysis takes place.

Faraday: One equivalent of charge; 95,600 coulombs.

Faraday's law: One equivalent of electricity is required
        to affect the transformation of one equivalent
        of material.

Fick's law: The flux of a material is proportional to the
        concentration gradient of that substance.

Flux: The amount of material crossing a unit area per unit
        time.

Half-cell: That section of an electrolysis cell encompas-
        sing one of the two electrodes.

Half-wave potential: The potential (characteristic of the
        particular compound) on a polarographic curve
        where the current is equal to one-half the limit-
        ing current.

Hydrogen overvoltage: The difference between the measured
        electrode potential and that calculated from the
        Nernst equation, at a given current density, for
        the reaction: $H^+ + e = \frac{1}{2} H_2$

Ilkovic equation: In polarography, the relationship between
        the diffusion-limiting current and the number of
        electrons involved in a reaction, concentration,
        and diffusion coefficient of substrate and the
        characteristics of the capillary tube.

Joule: The basic unit of energy in the metric system.
        $1 J = 10^7$ ergs = 0.239 cal = 1 W-sec.

Kolbe reaction: An electrolytic oxidation of an organic
        carboxylate salt normally leading to dimers, and
        first studied extensively by Kolbe. Normally:
        $2RCO_2 \longrightarrow R-R+2CO_2+2e$

Linear sweep voltammetry: A type of voltammetry in which
        the electrode potential is constrained to vary
        linearly with time.

Liquid junction potential: A difference in electric poten-
        tial which occurs across the junction between two
        dissimilar liquid electrolytes.

Luggin capillary: A slender electrolyte-filled capillary
        normally used as a salt bridge to minimize the
        effect of a large reference electrode on the
        measurement of electrode potential.

Mho: The unit of electric conductivity. 1 mho = 1 A/V =
        $1 \ \Omega^{-1}$.

Nernst equation: The fundamental relationship, derived by
        Nernst, between the electrode potential and activi-
        ties of the various species involved in a reversible
        electron transfer.

$$E = E_o - \frac{RT}{nF} \ln Q$$

        where E_o is the standard potential of the system
        and Q is the equilibrium expression for the
        system.

Normal hydrogen electrode (n.h.e.): The universally ac-
        cepted reference electrode consisting of a plati-
        num electrode in contact with hydrogen gas at one
        atmosphere, and hydronium ion at unit activity.
        The n.h.e. is assigned a potential of zero volts
        at all temperatures.

Ohm: The unit of electric resistance. $1 \ \Omega$ = 1 V/A.

Ohm's law: The linear relationship, discovered by Ohm,
        between current and voltage in an electric con-
        ductor (both electronic and ionic): V = iR,
        where R is the resistance in ohms.

Operational amplifier: A type of dc amplifier, character-
        ized by a very high negative voltage gain, and
        having great general utility in measuring and
        transforming electric signals.   It is usually
        symbolized by

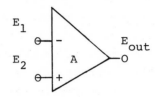

        where E_{out} is $-A$ (E_1-E_2) and A is usually $>10^3$.
Overvoltage: The difference between the actual measured
        electrode potential at specified electrolysis
        conditions and the electrode potential which
        would apply if the system were at equilibrium.
Oxygen overvoltage: The overvoltage at specified conditions
        of a given electrode for the reaction
        $$2H_2O \longrightarrow 4H^+ + O_2 + 4e$$
Passivation: Normally, the process of forming a thin,
        inert oxide layer on a metal surface by oxidation.
        Passivation has a large effect on the normal vol-
        tammetric behavior of the metal.
Platinization: The process of depositing a finely divided
        coating of platinum on the surface of an electrode.
        Usually effected by reduction of chloroplatinic
        acid onto the base electrode.
Polarography: The type of voltammetry restricted to the
        case where the working electrode is a dropping
        mercury capillary.
Potential of zero charge (p.z.c.):   The electrode potential
        where no net charge exists at the surface of a

specific electrode. At the p.z.c., the capacitance at the electrode/solution interface is normally at its minimum value. For the special case
where the electrode is mercury, the p.z.c. can be
measured at the point where the interfacial tension is at a maximum.

Reference electrode: A stable, reversible half-cell which
is convenient to use as a reference system for
the measurement of electrode potentials.

Resistance: The property of a substance to resist the passage of electric current. Resistance is measured
in ohms.

Rotating disk voltammetry: A type of voltammetry where the
working electrode is a planar disk rotating uni-
formly about an axis perpendicular to, and passing
through, its center.

Saturated calomel electrode (s.c.e.): A commonly used
reference electrode, where the half-cell comprises
mercury in contact with an aqueous solution saturated with respect to mercurous chloride (calomel)
and potassium chloride.

Shunt: A resistance placed in parallel with a two-terminal
device; usually used to change the range, or cali-
brate in convenient units, an electric measuring
device.

Standard potential: The potential at which the species
involved are at unit chemical activity and are
at chemical and electrical equilibrium.

Surface coverage: The fraction of a given surface covered
by an adsorbed species.

Tafel plot: A graph of the logarithm of the current density
vs electrode potential. Usually useful only when
mass transfer is not rate limiting.

Transfer coefficient (α): That fraction of the change in
applied electrode potential which affects the
forward reaction.

Triangular wave voltammetry: A type of voltammetry where the
        working electrode potential is constrained to vary
        linearly, first in one direction, and then in the
        opposite direction. Usually interchangeable with
        the term "cyclic voltammetry."

Volt: The unit of electron "pressure" or activity. 1 V =
        1 J/C. In electrolytic work, one volt in electrode
        potential is equivalent to 22.9 kcal/mole.

Vacuum tube voltmeter (VTVM): Characterized by a high
        (usually 10 MΩ) input impedance.

Voltammetry: Any technique which involves the simultaneous
        measurement of both electrode potential and cur-
        rent.

Working electrode: The electrode at which the electrode
        reaction of interest is occurring.

Zeta potential: The measurable potential difference between
        the bulk of an electrolyte and the region near the
        electrode where bulk motion is possible (usually
        just outside the compact or inner double layer).

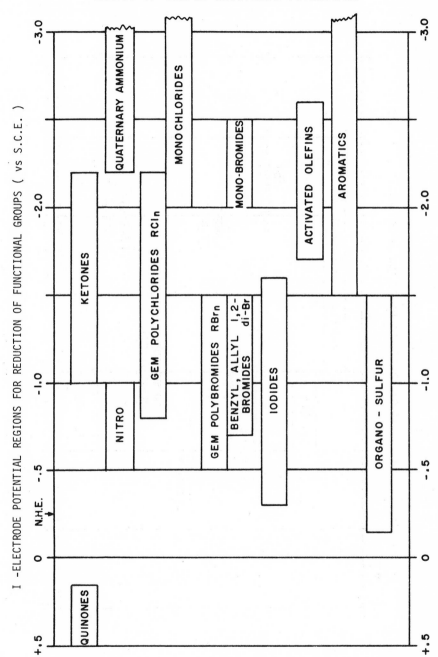

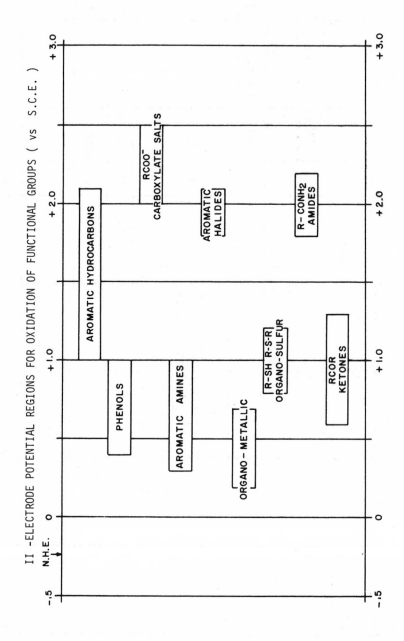

II - ELECTRODE POTENTIAL REGIONS FOR OXIDATION OF FUNCTIONAL GROUPS (vs S.C.E.)

III – RECOMMENDED POTENTIAL LIMITS OF COMMON SOLVENTS*

SOLVENT	-3	-2	-1	N.H.E.0 / S.C.E.	1	2	3
WATER (Hg)		Na^+	Zn	Pb Cu	Hg Ag	Au	
WATER (Pt)					I^-	Br^- Cl^- F^-	
DIMETHYL FORMAMIDE R_4N^+		Na^+		Cu	I^- Hg		
ACETONITRILE R_4N^+	Li^+	Na^+			Hg	ClO_4^-	
LIQUID AMMONIA	SOLVATED ELECTRON			Hg I^-	Cl^-		
HEXAMETHYL PHOSPHORANIDE $R N^+$		Na^+		Hg I^-			
ETHANOL / METHANOL							
METHYLENE CHLORIDE							

$\rbrack$ = REDUCTION   [⊠] = REDUCTION OF SOLVENT   OXIDATION OF SOLVENT = [⧄] OXIDATION = $\lbrack$

*Based on authors' experience and literature values when available. Data on other solvents may be found in section 3.2.

AUTHOR INDEX

Numbers in parentheses are reference numbers
and indicate that an author's work is referred
to although his name is not cited in the text.
Underlined numbers give the page on which the
complete reference is listed.